Progress in Mathematics

SADLIER-OXFORD

Rose Anita McDonnell   Catherine D. LeTourneau

Anne Veronica Burrows   Francis H. Murphy   M. Winifred Kelly

with
Dr. Elinor R. Ford

Series Consultants

Tim Mason
Math Specialist
Palm Beach County School District
West Palm Beach, FL

Margaret Mary Bell, S.H.C.J., Ph.D.
Director, Teacher Certification
Rosemont College
Rosemont, PA

Dennis W. Nelson, Ed.D.
Director of Basic Skills
Mesa Public Schools
Mesa, AZ

Sadlier-Oxford
A Division of William H. Sadlier, Inc.

The publisher wishes to thank the following teachers and administrators, who read portions of the program prior to publication, for their comments and suggestions.

Mrs. Maria Bono
Whitestone, NY

Mrs. Jennifer Fife
Yardley, PA

Ms. Donna Violi
Melbourne, FL

Sr. Lynn Roebert
Covina, CA

Ms. Anna Cano-Amato
Brooklyn, NY

Mr. Galen Chappelle
Los Angeles, CA

Mrs. Ana M. Rodriguez
Miami, FL

Sr. Ruthanne Gypalo
East Rockaway, NY

Sr. Anita O'Dwyer
North Arlington, NJ

Mrs. Madonna Atwood
Creve Coeur, MO

Mrs. Marlene Kitrosser
Bronx, NY

Acknowledgments

Every good faith effort has been made to locate the owners of copyrighted material to arrange permission to reprint selections. In several cases this has proved impossible. The publisher will be pleased to consider necessary adjustments in future printings.

Thanks to the following for permission to reprint the copyrighted materials listed below.

"A Lot of Kids" (text only) from THE BUTTERFLY JAR by Jeff Moss. Copyright © 1989 by Jeff Moss. Used by permission of Bantam Books, a division of Bantam Doubleday Dell Publishing Group, Inc.

Excerpt from A REMAINDER OF ONE (text only). Text copyright © 1995 by Elinor J. Pinczes. Reprinted by permission of Houghton Mifflin Company. All rights reserved.

Excerpt from "Arithmetic" (text only) from THE COMPLETE POEMS OF CARL SANDBURG, copyright © 1950 by Carl Sandburg and renewed 1978 by Margaret Sandburg, Helga Sandburg Crile, and Janet Sandburg, reprinted by permission of Harcourt Brace & Company.

"Dividing" (text only) by David McCord. From ONE AT A TIME by David McCord. Copyright © 1986 by David McCord. By permission of Little, Brown and Company.

"Little Bits" (text only) by John Ciardi. Used by permission of HarperCollins Publishers.

"Math Class" (text only) by Myra Cohn Livingston. From THE MALIBU AND OTHER POEMS by Myra Cohn Livingston. Copyright © 1972 by Myra Cohn Livingston. Reprinted by permission of Marian Reiner.

Excerpt from MATH CURSE (text only) by Jon Scieszka. Copyright © 1995 by Jon Scieszka. Used by permission of Viking Penguin, a division of Penguin Putnam Inc.

"Math Makes Me Feel Safe" (text only) is reprinted by permission of the author, Betsy Franco, who controls all rights.

"Popsicle Sticks and Glue" (text only) by Leslie Danford Perkins. Used by permission of the author, who controls all rights.

"Sheepshape" (text only) by X.J. Kennedy. Reprinted with the permission of Margaret K. McElderry Books, an imprint of Simon & Schuster Children's Publishing Division from GHASTLIES, GOOPS, & PINCUSHIONS by X.J. Kennedy. Copyright © 1989 X.J. Kennedy. Reprinted also by permission of Curtis Brown, Ltd.

"Is Six Times One a Lot of Fun?" (text only) copyright © 1975 by Karla Kuskin. Reprinted by permisison of Scott Treimel New York.

"Take a Number" (text only) from TAKE A NUMBER by Mary O'Neill. Copyright © 1968 by Mary O'Neill. © Renewed 1996 by Erin Baroni and Abigail Hagler. Reprinted by permission of Marian Reiner.

"Who Hasn't Played Gazintas?" (text only) from ONE AT A TIME by David McCord. Copyright 1952 by David McCord. By permission of Little, Brown and Company.

"Willis C. Sick" (text only) from THE HOPEFUL TROUT AND OTHER LIMERICKS by John Ciardi. Text copyright ©1989 by Myra J. Ciardi. Reprinted by permission of Houghton Mifflin Company. All rights reserved.

Anastasia Suen, Literature Consultant

All manipulative products generously provided by ETA, Vernon Hills, IL.

Photo Credits

Manipulatives supplied by ETA. Diane J. Ali: 25. Cate Photography: 81,133, 148, 218, 248. Corbis: 109. Courtesy of the Field Museum, Chicago: 84–85. Florida Department of Transportation: 347.

FPG/ Bill Cummins: 18; Tanaka Associates: 75; Jeffrey Meyers: 153; Dean Siracusa: 214 right. Neal Farris: 53. Richard & Amy Hutchings: 38 top, 38–39, 187, 220, 269. The Image Bank/ Steve Allen: 208. International Stock: 188. Greg Lord: 150.

Clay McBride: 23, 35, 67, 95, 125, 165, 205, 212, 217, 239, 265, 295, 325, 342, 357, 381, 411, 441. Nancy Sheehan: 326. The Stock Market/ David Wood: 179. Superstock: 239 (Babe Ruth). Tony Stone Images: Lori Adamski Peek: 33; Gary Vestal: 44 top; Ralph Wetmore: 44 bottom; Ken Biggs:

107; David Madison: 119; Philip Habib: 183; Peter Cade: 194; Gary Nolton: 209, 297; Tony Aruza: 390; Manoj Shah: 393; Gary Hush: 431; Johnny Johnson: 442; Chad Ehlers: 459. Visuals Unlimited/ Thomas Gula: 44 center. Westlight/ Jim Richardson: 210.

Illustrators

Diane Ali
Batelman Illustration
Bob Berry
Don Bishop
Robert Burger

Ken Coffelt
Adam Gordon
Dave Jonason
Robin Kachantones

Bea Leute
Blaine Martin
Kathy O'Connell
Wendy Pierson

John Quinn
Fernando Rangel
Sintora Vander Horst
Dirk Wunderlich

Contents

* Algebraic Reasoning

✱ **Algebraic Reasoning**

CHAPTER 3
Addition and Subtraction

* Algebraic Reasoning

CHAPTER 4
Multiplying by One and Two Digits

CHAPTER 5
Dividing by One Digit

       * Algebraic Reasoning

CHAPTER 6
Measurement

CHAPTER 7
Statistics and Probability

CHAPTER 8
Fraction Concepts

* Algebraic Reasoning

CHAPTER 9
Fractions: Addition and Subtraction

✱ **Algebraic Reasoning**

CHAPTER 10
Geometry

CHAPTER 11
Perimeter, Area, and Volume

∗ **Algebraic Reasoning**

CHAPTER 12
Dividing by Two Digits

End-of-Book Materials

✱ **Algebraic Reasoning**

Welcome to

Progress in Mathematics

Whether you realize it or not, you see and use mathematics every day!

The lessons and activities in this textbook will help you enjoy mathematics *and* become a better mathematician.

This year you will build on the mathematical skills you already know, as you explore *new* ideas. Working in groups, you will solve problems using many different strategies. You will also have opportunities to make up your own problems and to keep a log of what you discover about math in your own personal Journal. You will learn more about multiplication and division, fractions and decimals, geometry, measurements, probability, and statistics. There will also be many opportunities to use algebraic reasoning.

You will become a *Technowiz* by completing the computer and calculator lessons and activities. These not only will teach you valuable skills, but also expose you to how technology is used in different situations by many people.

And you can use the Skills Update section at the beginning of this book throughout the year to sharpen and review any skills you need to brush up on.

We hope that as you work through this program you will become aware of how mathematics really is a *big* part of your life.

An Introduction to Skills Update

Progress in Mathematics includes a "handbook" of essential skills, Skills Update, at the beginning of the text. These one-page lessons review skills you learned in previous years. It is important for you to know this content so that you can succeed in math this year.

If you need to review a concept in Skills Update, your teacher can work with you, using ideas from the Teacher's Edition. You can practice the skill using manipulatives, which will help you understand the concept better.

Your class may choose to do these one-page lessons at the beginning of the year so that you and your teacher can assess your understanding of these previously learned skills. Or you may choose to use Skills Update as a handbook throughout the year. Many lessons in your textbook refer to a particular page in the Skills Update. This means you can use that Skills Update lesson at the beginning of your math class as a warm-up activity. You may even want to practice those skills at home.

If you need more practice than what is provided on the Skills Update page, you can use exercises in the *Skills Update Practice Book*. It has an abundance of exercises for each lesson.

Hundreds

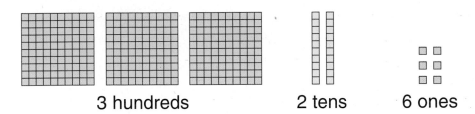

3 hundreds   2 tens   6 ones

Standard Form: 326
Word Name: three hundred twenty-six

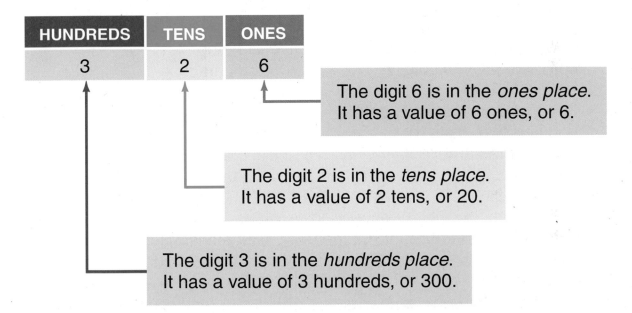

HUNDREDS	TENS	ONES
3	2	6

The digit 6 is in the *ones place*.
It has a value of 6 ones, or 6.

The digit 2 is in the *tens place*.
It has a value of 2 tens, or 20.

The digit 3 is in the *hundreds place*.
It has a value of 3 hundreds, or 300.

Write the number in standard form.

1.

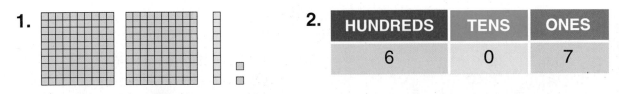

2.

HUNDREDS	TENS	ONES
6	0	7

3. 1 hundred 8 tens 3 ones   **4.** five hundred sixty-two

Write the place of the red digit. Then write its value.

5. 482   **6.** 369   **7.** 141   **8.** 965   **9.** 174   **10.** 218

11. 522   **12.** 697   **13.** 742   **14.** 831   **15.** 420   **16.** 505

Numeration II

Comparing Whole Numbers

> **> means "is greater than"**   **< means "is less than"**
> **= means "is equal to"**

To compare numbers:
- Align the digits   6453
  by place value.   6459 ↓

- Start at the left. Compare   6453   6 = 6
  the digits in the greatest place. 6459

- If these digits are the same,   6453   4 = 4
  compare the next digits.   6459

- Keep comparing digits until   6453   5 = 5
  you find two digits that   6459   9 > 3
  are *not* the same.

So 6459 > 6453. You could also say 6453 < 6459.

Study this example.

423 _?_ 2423   Think: There are no
                             thousands in 423.
423 |
2423 ↓

0 < 2
So 423 < 2423   **or**   2423 > 423.

Compare. Write <, =, or >.

1. 57 _=_ 57   **2.** 65 _?_ 62   **3.** 48 _?_ 56   **4.** 82 _?_ 28

5. 325 _?_ 523   **6.** 649 _?_ 841   **7.** 127 _?_ 134   **8.** 525 _?_ 522

9. 6241 _?_ 9246   **10.** 7983 _?_ 7983   **11.** 9015 _?_ 9012

12. 2704 _?_ 2714   **13.** 8619 _?_ 8617   **14.** 1844 _?_ 1846

Recognizing and Counting Money

ten-dollar bill	five-dollar bill	one-dollar bill

$10.00	$5.00	$1.00

half-dollar	quarter	dime	nickel	penny
50¢ or $.50	25¢ or $.25	10¢ or $.10	5¢ or $.05	1¢ or $.01

To count bills and coins, arrange in order from greatest to least value. Then count on.

$10.00 + $5.00 + $.25 + $.10 + $.01

$10.00 ⟶ $15.00 ⟶ $15.25 ⟶ $15.35 ⟶ $15.36

Write each amount. Use the dollar sign and decimal point.

1.

2.

3. 1 five-dollar bill, 3 quarters, 1 dime, 3 nickels, 2 pennies

4. 4 dollars, 1 quarter, 2 nickels

3

Whole Number Operations I

Addition and Subtraction Facts

▶ Add: 5 + 4 = __?__

$$\begin{array}{r} 5 \\ +\,4 \\ \hline 9 \end{array}$$ addends sum   or   5 + 4 = 9 addends sum

▶ Subtract: 11 − 5 = __?__

$$\begin{array}{r} 11 \\ -\,5 \\ \hline 6 \end{array}$$ ← difference   or   11 − 5 = 6 difference

Remember:
5 + 4 = 9 is a **number sentence** for addition.
11 − 5 = 6 is a **number sentence** for subtraction.

Add or subtract. Watch the signs.

1. $\begin{array}{r}8\\+8\\\hline\end{array}$	**2.** $\begin{array}{r}4\\+9\\\hline\end{array}$	**3.** $\begin{array}{r}16\\-\,9\\\hline\end{array}$	**4.** $\begin{array}{r}6\\+5\\\hline\end{array}$	**5.** $\begin{array}{r}14¢\\-\,7¢\\\hline\end{array}$	**6.** $\begin{array}{r}12¢\\-\,4¢\\\hline\end{array}$
7. $\begin{array}{r}7\\+6\\\hline\end{array}$	**8.** $\begin{array}{r}16\\-\,9\\\hline\end{array}$	**9.** $\begin{array}{r}0\\+7\\\hline\end{array}$	**10.** $\begin{array}{r}13\\-\,4\\\hline\end{array}$	**11.** $\begin{array}{r}7¢\\+9¢\\\hline\end{array}$	**12.** $\begin{array}{r}14¢\\-\,6¢\\\hline\end{array}$
13. $\begin{array}{r}15\\-\,8\\\hline\end{array}$	**14.** $\begin{array}{r}9\\+9\\\hline\end{array}$	**15.** $\begin{array}{r}11\\-\,8\\\hline\end{array}$	**16.** $\begin{array}{r}9\\+6\\\hline\end{array}$	**17.** $\begin{array}{r}18¢\\-\,9¢\\\hline\end{array}$	**18.** $\begin{array}{r}8¢\\+6¢\\\hline\end{array}$

19. 17 − 8   **20.** 6 + 6   **21.** 15 − 7   **22.** 6¢ + 7¢   **23.** 3¢ + 8¢

4

Whole Number Operations II

Related Facts

These four facts are **related facts.**
They all use the same numbers.

$6 + 5 = 11$ $11 - 5 = 6$

$5 + 6 = 11$ $11 - 6 = 5$

Study these examples.

$12 = 4 + 8$
$12 = 8 + 4$
$8 = 12 - 4$
$4 = 12 - 8$

$3 + 3 = 6$
$6 - 3 = 3$

Write the related facts for each pair.

1.

4, 6

2.

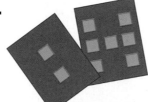

2, 7

3.

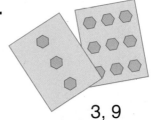

3, 9

4.

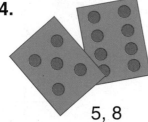

5, 8

5.

4, 5

6.

3, 7

7. 9, 5   **8.** 2, 5   **9.** 8, 8   **10.** 6, 7

Copy and complete.

11. $\underline{\ ?\ } + 7 = 13$
$7 + \underline{\ ?\ } = 13$
$13 - 7 = \underline{\ ?\ }$
$13 - \underline{\ ?\ } = 7$

12. $\underline{\ ?\ } + 9 = 17$
$9 + \underline{\ ?\ } = 17$
$17 - \underline{\ ?\ } = 9$
$17 - 9 = \underline{\ ?\ }$

13. $15 = \underline{\ ?\ } + 8$
$15 = 8 + \underline{\ ?\ }$
$8 = 15 - \underline{\ ?\ }$
$\underline{\ ?\ } = 15 - 8$

Whole Number Operations III

Adding without Regrouping

To find the total, add:
$2110 + 3022 + 1657 = \underline{\ ?\ }$

Paid Attendance	
Friday	2110
Saturday	3022
Sunday	1657

Align. Add. Start with the ones.

Add ones.	Add tens.	Add hundreds.	Add thousands.
2110	2110	2110	2110
3022	3022	3022	3022
+ 1657	+ 1657	+ 1657	+ 1657
9	89	789	6789

Study these examples.

```
  1421          251        12 + 40 + 30 + 13 = 95
    32        + 437
+  534          688
  1987
```

Find the sum.

1. 42 + 33

2. 32 + 25

3. 15 + 62

4. 53 + 42

5. 72 + 26

6. 140 + 57

7. 658 + 220

8. 128 + 820

9. 321 + 66

10. 173 + 13

11. 8317 + 1222

12. 4375 + 5014

13. 6416 + 2103

14. 1624 + 255

15. 8117 + 782

16. 12 13 + 51

17. 124 331 + 533

18. 202 260 + 132

19. 1141 4013 + 1224

20. 2145 4202 + 3031

21. 15 + 22 + 50 + 11

22. 23 + 11 + 34 + 21

23. 464 + 203 + 122

24. 300 + 240 + 159

Whole Number Operations IV

Subtracting without Regrouping

Subtract: 5867 − 4536 = __?__

Align. Subtract. Start with the ones.

Subtract ones.	Subtract tens.	Subtract hundreds.	Subtract thousands.
5867 −4536 1	5867 −4536 31	5867 −4536 331	5867 −4536 1331

Study these examples.

94 −54 40	347 −210 137	689 −632 57	2475 −321 2154

Find the difference.

1. 53
  − 21

2. 85
  − 23

3. 26
  − 12

4. 74
  − 11

5. 46
  − 25

6. 279
  − 151

7. 657
  − 242

8. 878
  − 843

9. 793
  − 243

10. 886
  − 475

11. 5986
  − 5082

12. 9929
  − 7806

13. 6495
  − 3122

14. 4819
  − 2107

15. 8576
  − 1423

16. 167
  − 35

17. 581
  − 21

18. 724
  − 12

19. 398
  − 75

20. 465
  − 23

21. 6837
  − 434

22. 7389
  − 176

23. 5677
  − 307

24. 4985
  − 562

25. 9688
  − 647

Subtract.

26. 67 − 5

27. 175 − 25

28. 438 − 16

Whole Number Operations V

Meaning of Multiplication

▶ To find how many, you can add
3 sets of 7: $7 + 7 + 7 = 21$

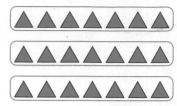

There is the *same number* in each set.
You can **multiply:**

3 sets of 7
3 sevens
3×7

number of sets	×	number in each set	=	total number

$$3 \quad \times \quad 7 \quad = \quad 21$$

or

$$\begin{array}{r} 7 \longleftarrow \textbf{factor} \\ \times\ 3 \longleftarrow \textbf{factor} \\ \hline 21 \longleftarrow \textbf{product} \end{array}$$

Remember: $3 \times 7 = 21$ is a number sentence for multiplication.

▶ Add: $2¢ + 2¢ + 2¢ + 2¢ = 8¢$

Or multiply: $4 \times 2¢ = \underline{\ ?\ }$

$$\begin{array}{r} 2¢ \\ \times 4 \\ \hline 8¢ \end{array} \qquad \text{or} \qquad 4 \times 2¢ = 8¢$$

$$\textbf{factors}\quad\textbf{product}$$

4 sets of 2¢
4 twos
$4 \times 2¢$

Write an addition sentence and
a multiplication sentence for each.

1.

2.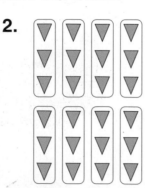

3.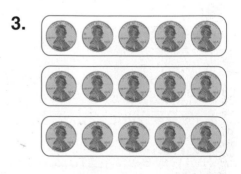

Whole Number Operations VI

Multiplying by 2, 3, 4, and 5

Add:

$$9 + 9 + 9 + 9 + 9 = 45$$

Or multiply:

$$\begin{array}{r} 9 \\ \times 5 \\ \hline 45 \end{array}$$
 or $5 \times 9 = 45$

5 sets of 9
5 nines
5×9

Find the product.

| **1.** $\begin{array}{r} 8 \\ \times 2 \end{array}$ | **2.** $\begin{array}{r} 7 \\ \times 4 \end{array}$ | **3.** $\begin{array}{r} 6 \\ \times 3 \end{array}$ | **4.** $\begin{array}{r} 5 \\ \times 5 \end{array}$ | **5.** $\begin{array}{r} 9 \\ \times 3 \end{array}$ | **6.** $\begin{array}{r} 7 \\ \times 2 \end{array}$ |

| **7.** $\begin{array}{r} 2¢ \\ \times 5 \end{array}$ | **8.** $\begin{array}{r} 8¢ \\ \times 3 \end{array}$ | **9.** $\begin{array}{r} 9¢ \\ \times 2 \end{array}$ | **10.** $\begin{array}{r} 5¢ \\ \times 4 \end{array}$ | **11.** $\begin{array}{r} 7¢ \\ \times 3 \end{array}$ | **12.** $\begin{array}{r} 8¢ \\ \times 5 \end{array}$ |

| **13.** $\begin{array}{r} 5 \\ \times 2 \end{array}$ | **14.** $\begin{array}{r} 9¢ \\ \times 4 \end{array}$ | **15.** $\begin{array}{r} 3 \\ \times 3 \end{array}$ | **16.** $\begin{array}{r} 9¢ \\ \times 5 \end{array}$ | **17.** $\begin{array}{r} 8¢ \\ \times 4 \end{array}$ | **18.** $\begin{array}{r} 6¢ \\ \times 2 \end{array}$ |

19. 4×6 **20.** 3×4 **21.** $5 \times 6¢$ **22.** $4 \times 4¢$

23. 3×1 **24.** 5×3 **25.** $2 \times 4¢$ **26.** $5 \times 4¢$

Problem Solving Write a number sentence for each.

27. One factor is 4. The product is 24. What is the other factor?

28. The product is 35. One factor is 5. What is the other factor?

29. The factors are 3 and 7. What is the product?

9

Whole Number Operations VII

Multiplying by 6, 7, 8, and 9

Multiply: $9 \times 3 =$ ___?___

$$\begin{array}{r} 3 \\ \times 9 \\ \hline 27 \end{array}$$ **or** $9 \times 3 = 27$

9 sets of 3
9 threes
9×3

Multiply.

1. $\begin{array}{r}1\\ \times 6\\ \hline\end{array}$	**2.** $\begin{array}{r}3\\ \times 8\\ \hline\end{array}$	**3.** $\begin{array}{r}5\\ \times 7\\ \hline\end{array}$	**4.** $\begin{array}{r}2\\ \times 9\\ \hline\end{array}$	**5.** $\begin{array}{r}4¢\\ \times 9\\ \hline\end{array}$	**6.** $\begin{array}{r}5¢\\ \times 6\\ \hline\end{array}$
7. $\begin{array}{r}3\\ \times 7\\ \hline\end{array}$	**8.** $\begin{array}{r}7\\ \times 9\\ \hline\end{array}$	**9.** $\begin{array}{r}6\\ \times 8\\ \hline\end{array}$	**10.** $\begin{array}{r}9\\ \times 6\\ \hline\end{array}$	**11.** $\begin{array}{r}8¢\\ \times 8\\ \hline\end{array}$	**12.** $\begin{array}{r}8¢\\ \times 6\\ \hline\end{array}$
13. $\begin{array}{r}7\\ \times 7\\ \hline\end{array}$	**14.** $\begin{array}{r}4\\ \times 6\\ \hline\end{array}$	**15.** $\begin{array}{r}5\\ \times 8\\ \hline\end{array}$	**16.** $\begin{array}{r}8\\ \times 9\\ \hline\end{array}$	**17.** $\begin{array}{r}7¢\\ \times 6\\ \hline\end{array}$	**18.** $\begin{array}{r}4¢\\ \times 8\\ \hline\end{array}$

Find the product.

19. 7×1 **20.** 9×6 **21.** $6 \times 6¢$ **22.** $9 \times 9¢$

23. 9×5 **24.** 8×9 **25.** $7 \times 4¢$ **26.** $9 \times 3¢$

27. 7×9 **28.** 6×3 **29.** $8 \times 7¢$ **30.** $7 \times 6¢$

Problem Solving Write a number sentence for each.

31. Ms. Black made 7 paper triangles for each of 8 mobiles. How many paper triangles did Ms. Black make in all?

32. There were 9 collages. On each collage students glued 7 bottle caps and 5 stickers. How many bottle caps were there? how many stickers?

Whole Number Operations VIII

Understanding Division

 Multiply when you want to find the total number.

$$3 \quad \times \quad 6 \quad = \quad 18$$

number of sets	number in each set	total number

18 in all
6 in each set
3 equal sets

▶ **Divide** when you want to find:

- the number of equal sets.

- the number in each equal set.

$$18 \quad \div \quad 6 \quad = \quad 3$$

total number	number in each set	number of sets

$$18 \quad \div \quad 3 \quad = \quad 6$$

total number	number of sets	number in each set

▶ Use multiplication facts to help you find division facts.

$4 \times 5 = 20$ $\quad$ $20 \div 5 = 4$
$5 \times 4 = 20$ $\quad$ $20 \div 4 = 5$

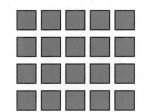

These four facts make up a **family of facts** for the numbers 4, 5, and 20.

Copy and complete each family of facts.

1. $6 \times 5 = 30$
$\underline{?} \times 6 = 30$
$30 \div 5 = \underline{?}$
$30 \div 6 = \underline{?}$

2. $9 \times 7 = 63$
$\underline{?} \times 9 = 63$
$63 \div 7 = \underline{?}$
$63 \div 9 = \underline{?}$

3. $4 \times 4 = 16$
$16 \div 4 = \underline{?}$

Write a family of facts for each set of numbers.

4. 2, 4, 8 $\qquad$ **5.** 3, 7, 21 $\qquad$ **6.** 4, 3, 12 $\qquad$ **7.** 5, 7, 35

8. 7, 6, 42 $\qquad$ **9.** 9, 1, 9 $\qquad$ **10.** 8, 3, 24 $\qquad$ **11.** 3, 2, 6

12. 8, 7, 56 $\qquad$ **13.** 9, 5, 45 $\qquad$ **14.** 5, 8, 40 $\qquad$ **15.** 6, 6, 36

Whole Number Operations IX

Dividing by 2, 3, 4, and 5

▶ Divide: $35 \div 5 =$ ___?___

Think: ___?___ $\times 5 = 35$
$7 \times 5 = 35$

So $\quad \underset{\uparrow}{35} \quad \underset{\uparrow}{\div} \quad \underset{\uparrow}{5} \quad = \quad \underset{\uparrow}{7.}$

$\quad\quad$ **dividend** $\quad$ **divisor** $\quad$ **quotient**

or

$\quad\quad\quad\quad\quad\quad\quad \overset{7}{} \leftarrow$ **quotient**
divisor $\longrightarrow 5\overline{)35} \leftarrow$ **dividend**

35 in all
5 in each set

Remember: $35 \div 5 = 7$
is a number sentence
for division.

▶ Find the quotient: $27¢ \div 3 =$ ___?___

Think: $3 \times$ ___?___ $= 27¢$
$3 \times 9¢ = 27¢$

So $27¢ \div 3 = 9¢ \quad$ or $\quad 3\overline{)27¢}^{\,9¢}$.

27¢ in all
3 equal sets

Division undoes multiplication.

Find the quotient.

1. $2\overline{)0}^{\,0}$
2. $4\overline{)24}$
3. $5\overline{)40}$
4. $3\overline{)15}$
5. $2\overline{)18¢}$
6. $5\overline{)5¢}$

7. $4\overline{)16}$
8. $3\overline{)21}$
9. $2\overline{)16}$
10. $4\overline{)36}$
11. $5\overline{)25¢}$
12. $2\overline{)12¢}$

13. $28 \div 4$
14. $45 \div 5$
15. $21 \div 3$
16. $8 \div 2$
17. $20 \div 5$

18. $3¢ \div 3$
19. $14¢ \div 2$
20. $28¢ \div 4$
21. $30¢ \div 5$

12

Whole Number Operations X

Dividing by 6, 7, 8, and 9

▶ Find the quotient: 24 ÷ 6 = __?__

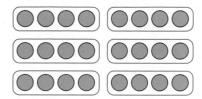

Think: $6 \times$ __?__ $= 24$
$6 \times 4 = 24$ So 24 ÷ 6 = 4.

24 in all
6 equal sets

▶ Divide: 24 ÷ 8 = __?__

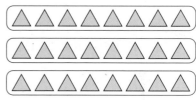

Think: __?__ $\times 8 = 24$
$3 \times 8 = 24$ So $8\overline{)24}$.

24 in all
8 in each set

Divide.

1. Dividing by 6

$6\overline{)0}$ $6\overline{)6}$ $6\overline{)12}$ $6\overline{)18}$ $6\overline{)24}$ $6\overline{)30}$ $6\overline{)36}$ $6\overline{)42}$ $6\overline{)48}$ $6\overline{)54}$

2. Dividing by 7

$7\overline{)0}$ $7\overline{)7}$ $7\overline{)14}$ $7\overline{)21}$ $7\overline{)28}$ $7\overline{)35}$ $7\overline{)42}$ $7\overline{)49}$ $7\overline{)56}$ $7\overline{)63}$

3. Dividing by 8

$8\overline{)0}$ $8\overline{)8}$ $8\overline{)16}$ $8\overline{)24}$ $8\overline{)32}$ $8\overline{)40}$ $8\overline{)48}$ $8\overline{)56}$ $8\overline{)64}$ $8\overline{)72}$

4. Dividing by 9

$9\overline{)0}$ $9\overline{)9}$ $9\overline{)18}$ $9\overline{)27}$ $9\overline{)36}$ $9\overline{)45}$ $9\overline{)54}$ $9\overline{)63}$ $9\overline{)72}$ $9\overline{)81}$

Fractions

Identifying Fractions

A fraction can name one or more *equal parts* of a whole or of a set.

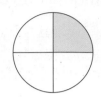

▶ $\frac{1}{4}$ of the circle is shaded.

$\frac{3}{4}$ of the circle is *not* shaded.

4 equal parts

▶ $\frac{3}{5}$ of the set of circles is shaded.

$\frac{2}{5}$ of the set of circles is *not* shaded.

5 equal parts

Write the fraction for the shaded part of each whole or set.
Then write the fraction for the part that is not shaded.

1.

2.

3.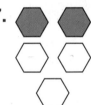

4.

5.

6.

7.

8.

Write a fraction for the red part of each set.
Then write a fraction for the yellow part.

9.

10.

11.

12.

Customary Units of Length

The **inch (in.)** is a customary unit of length.

A quarter is about 1 inch wide. You can use a quarter as a benchmark for 1 inch.

> A **benchmark** is an object of known measure that can be used to estimate the measure of other objects.

The **foot (ft)** and the **yard (yd)** are also customary units of length.

> 12 inches (in.) = 1 foot (ft)
> 3 feet (ft) = 1 yard (yd)
> 36 inches (in.) = 1 yard (yd)

1 ft

A license plate is about 1 foot long.

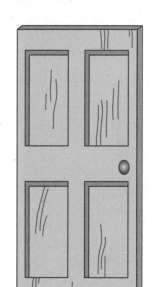

1 yd

A door is about 1 yard wide.

Write the letter of the best estimate.

1. length of a paintbrush   **a.** 9 ft   **b.** 9 yd   **c.** 9 in.

2. length of a bus   **a.** 40 in.   **b.** 40 ft   **c.** 40 yd

3. height of a wall   **a.** 3 in.   **b.** 3 yd   **c.** 3 ft

Measurement II

Cup, Pint, Quart, Gallon

The **cup (c)**, the **pint (pt)**, the **quart (qt)**, and the **gallon (gal)** are customary units of liquid capacity.

2 cups = 1 pint
2 pints = 1 quart
2 quarts = 1 half gallon
4 quarts = 1 gallon

1 cup

1 pint

1 quart

1 half gallon

1 gallon

Write *c, pt, qt,* or *gal* for the unit you would use to measure the capacity of each.

1. swimming pool

2. cereal bowl

3. can of soup

4. can of house paint

5. tanker truck

6. small container of frozen yogurt

7. large glass of juice

8. bottle of seltzer

9. family-size jar of mayonnaise

10. car tank of gasoline

Pound

The **pound (lb)** is a customary unit of weight.

A can of beans weighs about 1 pound.

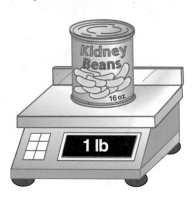

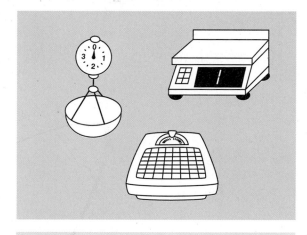

Weight is measured on a **scale.**

Does each actual object weigh more than or less than 1 pound?
Write *more than* or *less than*.

1.

2.

3.

4.

5.

6.

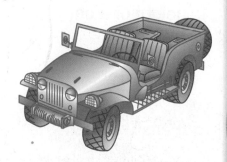

Measurement IV

Centimeter and Meter

The **centimeter (cm)** and the **meter (m)** are metric units of length.

100 centimeters (cm) = 1 meter (m)

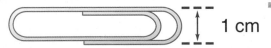
1 cm

A large paper clip is about 1 centimeter wide.

A full-size baseball bat is about 1 meter long.

Write the letter of the best estimate.

1. height of a mug   **a.** 2 cm   **b.** 9 cm   **c.** 2 m

2. width of a room   **a.** 4 m   **b.** 20 cm   **c.** 12 m

3. length of a soccer field   **a.** 10 m   **b.** 100 cm   **c.** 100 m

4. height of a cat   **a.** 99 cm   **b.** 1 m   **c.** 30 cm

5. length of a bed   **a.** 2 m   **b.** 20 cm   **c.** 20 m

Write *cm* or *m* for the unit you would use to measure each.

6. width of a dollar bill   7. height of a giraffe

Liter

The **liter (L)** is a metric unit of liquid capacity.

Seltzer and springwater are sold in bottles that hold 1 L.

Does each actual object hold more than or less than 1 liter? Write *more than* or *less than*.

1.

2.

3.

4.

5.

6.

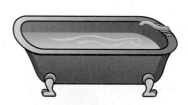

7.

8.

9.

Measurement VI

Kilogram

The **kilogram (kg)** is a metric unit of mass.

A small bag of flour has a mass
of about 1 kilogram.

Mass is measured on
a **balance**.

Does each actual object have a mass of more than or
less than 1 kilogram? Write *more than* or *less than*.

1.

2.

3.

4.

5.

6.

Perimeter

Find the perimeter of the figure.

> **Perimeter** is the distance around a figure.

To find the perimeter of a figure,
add the lengths of its sides.

$$
\begin{array}{r}
100 \text{ ft} \\
60 \text{ ft} \\
75 \text{ ft} \\
+\ 50 \text{ ft} \\
\hline
285 \text{ ft}
\end{array}
$$

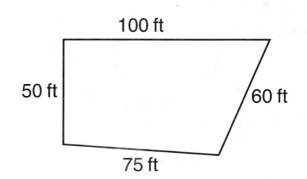

Find the perimeter of each figure.

1.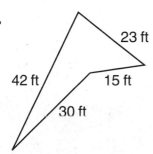

2.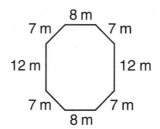

3.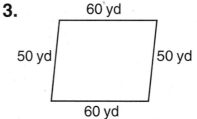

4.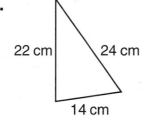

5.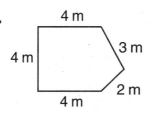

6.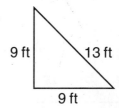

7. a polygon whose sides
measure 100 ft, 142 ft,
68 ft, and 127 ft

8. a polygon whose sides
measure 92 m, 109 m,
and 92 m

Geometry 1

Congruent Figures

Each of the patterns below was made
using congruent figures.

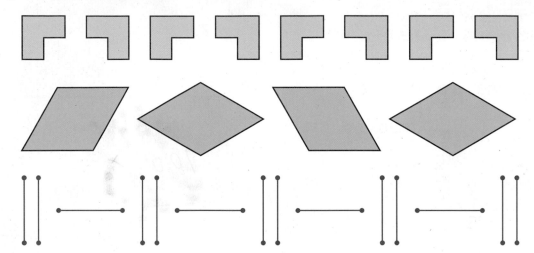

Congruent figures have exactly
the same size and the same shape.

To find whether two figures are congruent:

- Carefully trace one figure onto tracing paper.

- Lay the tracing over the other figure.

If the tracing and the figure match,
the two figures are congruent.

Are the figures congruent? Write yes or no.
You may use tracing paper.

1.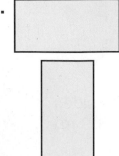

2.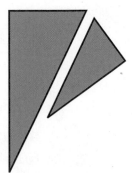

3.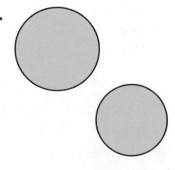

Lines of Symmetry

If you can fold a figure in half so that the two halves exactly match, the figure is **symmetrical.**

The fold line is a **line of symmetry**.

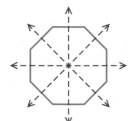 4 lines of symmetry

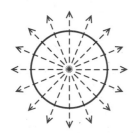 A circle has more lines of symmetry than you can count.

You can also use a **reflection** to see if the two halves exactly match.

Is each red line a line of symmetry? Write yes or no.

1.
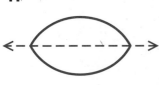

2.
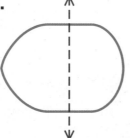

3.
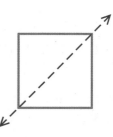

4.

5.
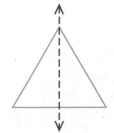

6.
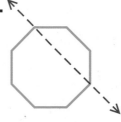

7.
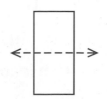

8.
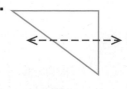

23

Statistics I

Recording and Organizing Data

▶ The **tally chart** at the right shows how many birds of different kinds came to a bird feeder one day.

Remember:
I = 1 and ЖГ = 5

Kind of Bird	Tally	Total
House Sparrow	ЖГ ЖГ ЖГ ЖГ ЖГ ЖГ II	32
House Finch	ЖГ ЖГ ЖГ ЖГ ЖГ	25
Blue Jay	ЖГ ЖГ III	13
Chickadee	ЖГ ЖГ ЖГ I	16
Nuthatch	IIII	4
Junco	ЖГ ЖГ ЖГ ЖГ III	23

Which kind of bird visited the feeder most often? least often?

▶ Organizing information in a table from least to greatest or greatest to least makes it easier to find and compare data.

House sparrows visited the feeder most often. Nuthatches visited least often.

Birds at My Feeder	Kind	Number
	House Sparrow	32
	House Finch	25
	Junco	23
	Chickadee	16
	Blue Jay	13
	Nuthatch	4

The tally chart below shows the number of farm animals Alex and Rachel saw on a trip.

Copy and complete the chart.

	Animal	Tally	Total
1.	Cows	ЖГ ЖГ ЖГ ЖГ ЖГ ЖГ ЖГ ЖГ II	?
2.	Pigs	?	11
3.	Goats	ЖГ ЖГ ЖГ III	?
4.	Horses	ЖГ ЖГ ЖГ ЖГ I	?
5.	Sheep	?	26
6.	Chickens	ЖГ ЖГ ЖГ ЖГ ЖГ ЖГ ЖГ III	?

Problem Solving Use the blue tally chart above.

7. Organize the data from least to greatest in a table.

8. What kind of animal was seen most often? least often?

Reading a Pictograph

The pictograph shows how many tulips of each color are planted in a courtyard.

Tulips	
Red	🌷 🌷 🌷 🌷 🌷 🌷 🌷
Yellow	🌷 🌷 🌷 🌷 🌷
Pink	🌷 🌷
White	🌷 🌷 🌷 🌷 🌷 🌷
Key: Each 🌷 = 10 tulips.	

Use the key to find how many of each color tulip are planted in the courtyard.

Skip count or add.

White:

🌷 🌷 🌷 🌷 🌷 🌷

10 + 10 + 10 + 10 + 10 + 10 = 60

There are 60 white tulips.

Use the pictograph to solve each problem.

1. How many red tulips are there? how many pink?

2. Of which color were there 50 tulips?

3. Of which color were there 40 more than the number of pink tulips?

4. Are there more or fewer red tulips than yellow tulips? how many more or fewer?

5. If the number of white tulips were doubled, how many more white tulips than red tulips would there be?

6. How many tulips in all are planted in the courtyard?

Statistics III

Reading a Bar Graph

The bar graph shows the lengths
of several dinosaurs.

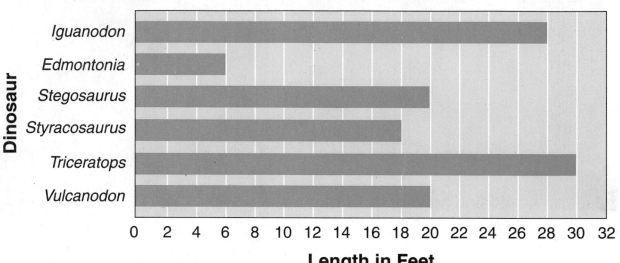

Dinosaur Lengths

How long was *Styracosaurus*?

Look at the bar labeled *Styracosaurus*.
The end of the bar is at the 18-foot mark.

So *Styracosaurus* was 18 feet long.

Use the bar graph to solve each problem.

1. Which of the dinosaurs was
   longest? shortest?

2. How much longer was
   Triceratops than *Styracosaurus*?

3. How long was *Iguanodon*?

4. Which dinosaur was 30 feet long?

5. Which two dinosaurs were the
   same length?

6. How much shorter was
   Edmontonia than *Vulcanodon*?

7. One *Apatosaurus* skeleton is 86 feet long.
   Is this more or less than triple the length
   of *Iguanodon*? by how many feet?

Key Sequences

Multiply: $4 \times \$10.95 = $ _?_

You can use a **calculator** to find the product. Use the decimal point key, ⬚ , to separate dollars and cents.

▶ To multiply $4 \times \$10.95$ on a calculator,

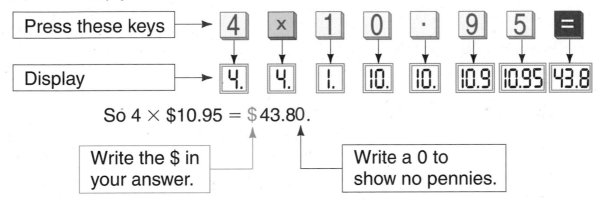

So $4 \times \$10.95 = \$43.80$.

Write the $ in your answer.

Write a 0 to show no pennies.

▶ Compute: $7)\overline{35}$

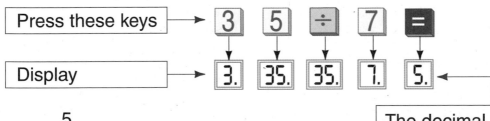

So $7)\overline{35}$.
$$\begin{array}{r} 5 \\ 7)\overline{35} \end{array}$$

The decimal point may not appear in the display.

Use a calculator to compute. Watch for +, −, ×, and ÷.

1. $5 \times \$24.50$
2. $88 \div 11$
3. $\$4.60 + \$1.05$

4. $968 - 55$
5. 41×9
6. $\$112.50 \div 15$

7. $\begin{array}{r} 415 \\ 514 \\ +145 \end{array}$
8. $\begin{array}{r} \$909.00 \\ -\ \ \ 99.99 \end{array}$
9. $9)\overline{\$164.25}$

Introduction to Problem Solving

Dear Student,

Problem solvers are super sleuths. We invite you to become a super sleuth by using these *five* steps when solving problems.

1 IMAGINE
Create a mental picture.

2 NAME
List the facts and the questions.

3 THINK
Choose and outline a plan.

4 COMPUTE
Work the plan.

5 CHECK
Test that the solution is reasonable.

Sleuths use clues to find a solution to a problem. When working together to solve a problem, you may choose to use one or more of these *strategies* as clues:

USE THESE STRATEGIES:
Extra Information
Hidden Information
Information from a Table
Working Backwards
Multi-Step Problem
Logical Reasoning
Two-Step Problem

USE THESE STRATEGIES:
Use Formulas
Make a Table/List
Interpret the Remainder
Write a Number Sentence
Make Up a Question
Logic and Analogies

USE THESE STRATEGIES:
Choose the Operation
Find a Pattern
Use a Model/Drawing
Guess and Test
More Than One Way
Use Simpler Numbers

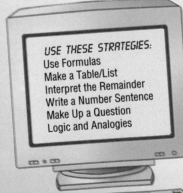

1 ▶ IMAGINE

Create a mental picture.

As you read a problem, create a picture in your mind. Make believe you are there in the problem. This will help you think about:
- what facts you will need;
- what the problem is asking;
- how you will solve the problem.

After reading the problem, it might be helpful to sketch the picture you imagined so that you can refer to it.

2 ▶ NAME

List the facts and the questions.

Name or list all the facts given in the problem. Be aware of *extra* information not needed to solve the problem. Look for *hidden* information to help solve the problem. Name the question or questions the problem asks.

3 ▶ THINK

Choose and outline a plan.

Think about how to solve the problem by:
- looking at the picture you drew;
- thinking about what you did when you solved similar problems;
- choosing a strategy or strategies for solving the problem.

4 ▶ COMPUTE

Work the plan.

Work with the listed facts and the strategy to find the solution. Sometimes a problem will require you to add, subtract, multiply, or divide. Two-step problems require more than one choice of operation or strategy. It is good to *estimate* the answer before you compute.

5 ▶ CHECK

Test that the solution is reasonable.

Ask yourself:
- "Have I answered the question?"
- "Is the answer reasonable?"

Check the answer by comparing it to your estimate. If the answer is not reasonable, check your computation. You may use a calculator.

Problem Solving

Strategy: Choose the Operation

Number Sentence	Definition
□ + □ = □	Join sets or quantities.
□ − □ = □	Separate, or take away, from a set. Compare two sets or quantities. Find part of a set. Find how many more are needed.
□ × □ = □	Join only equal sets or quantities.
□ ÷ □ = □	Separate a set into equal groups. Share a set equally.

Problem: Meg collects comic books. She puts 7 comic books into each envelope. How many envelopes does she need for 42 comic books?

 1 IMAGINE Picture yourself in the problem.

 **2 NAME**

Facts: 7 comic books in each envelope
42 comic books

Question: How many envelopes does she need?

 **3 THINK** You are separating a set into equal groups.
Divide: 42 ÷ 7 = _?_

Think: ? × 7 = 42

4 COMPUTE 42 ÷ 7 = 6
Meg needs 6 envelopes.

 **5 CHECK** Use a calculator. Multiply to check division:

The answer is reasonable.

Problem Solving

Strategy: Guess and Test

Algebra ✓

Problem: Pat's bank holds dimes and quarters. There are 4 more dimes than quarters in the bank. The value of all the coins is $2.85. How many quarters are in Pat's bank?

1 ▶ IMAGINE Create a mental picture of combinations of quarters and dimes.

2 ▶ NAME

Facts: bank hold dimes and quarters
4 more dimes than quarters
$2.85 in quarters and dimes

Question: How many quarters are in Pat's bank?

3 ▶ THINK First **guess** a number of quarters. 5 quarters

Add 4 to find the number of dimes. 9 dimes

Then **test** to find whether the value of the coins equals $2.85.

Make a table to record your guesses.

4 ▶ COMPUTE

		Quarter Value	Dime Value	Total Value	Test
Guesses	1st	5 quarters = $1.25	9 dimes = $ .90	$1.25 + $ .90 = $2.15	too low
	2nd	6 quarters = $1.50	10 dimes = $1.00	$1.50 + $1.00 = $2.50	too low
	3rd	7 quarters = $1.75	11 dimes = $1.10	$1.75 + $1.10 = $2.85	correct

5 ▶ CHECK The third guess is correct because:

- 11 dimes is 4 coins more than 7 quarters.

- 7 quarters ($1.75) and 11 dimes ($1.10) equal $2.85.

31

Problem Solving

Strategy: Information from a Table

Problem: Chris went to the beach. The air temperature was 90°F. The relative humidity was 70 percent. How hot did the air temperature feel?

Heat Index Table

Percent Relative Humidity	Temperature (°F)				
	75	80	85	**90**	95
50	75	81	88	96	107
60	76	82	90	100	114
70 ▾	77	85	93	106	124
80	78	86	97	113	136
90	79	88	102	122	
100	80	91	108		

1 IMAGINE Place yourself in the problem.

2 NAME

Facts: air temperature of 90°F
relative humidity of 70 percent

Question: How hot did the air temperature feel?

3 THINK The table shows a large amount of data, or information. Study the table carefully. Choose only the data needed to solve the problem.

Percent Relative Humidity: 70

Temperature (°F): 90

4 COMPUTE To find how hot it felt, use the Heat Index Table.

- Read *down* to 70 percent.
- Read *across* to 90°F.

The 70 percent relative humidity made the air temperature of 90°F feel as if it were 106°F.

5 CHECK Read across to 90°F and down to 70 percent relative humidity to find the same answer, 106°F.

Problem Solving

Strategy: Write a Number Sentence

Algebra ✓

Problem: A nursery donates 36 trees to a city. The city plants 4 trees in each of its parks. At most, how many parks could there be?

 **1 IMAGINE**

You are planting the trees. Draw and label a picture.

2 NAME

Facts: 36 trees donated
4 trees in each park

Question: How many parks could there be?

3 THINK

Because the 36 trees are being separated into equal sets of 4 trees each, write a number sentence for division.

$$36 \div 4 = \underline{\ ?\ }$$
$$\text{parks}$$

Think:
Number ÷ Number = Number
in all   in each   of groups
group

4 COMPUTE

Divide to find the quotient.

$$\begin{array}{r} 9 \\ 4\overline{)3\ 6} \\ -3\ 6 \\ \hline 0 \end{array}$$

Think: How many 4s are in 36? 9

There could be 9 parks.

5 CHECK

Multiply the quotient by the divisor.

$$\begin{array}{r} 9 \\ \times\ 4 \\ \hline 36 \end{array}$$

The answer checks!

33

Problem Solving

Applications

Choose a strategy from the list or use another strategy you know to solve each problem.

1. Olivia works at a zoo gift shop. She sold 6 small, 8 medium, and 4 large T-shirts. How many T-shirts did she sell?

2. Olivia sold 16 posters. Penguins were pictured on 7 of the posters. Pandas were on the rest. How many panda posters did Olivia sell?

3. Stu packed 6 ceramic animals into each small box. How many boxes does he need for 54 ceramic animals?

4. Ryan sent 22 animal buttons to three cousins. Sue received twice as many buttons as Mike and 3 more than Jill. How many buttons did each receive?

5. Lin wants to use 7 animal beads for each of 9 necklaces he is making for the zoo gift shop. How many animal beads will he need?

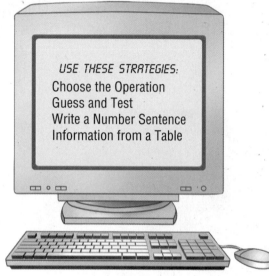

USE THESE STRATEGIES:
Choose the Operation
Guess and Test
Write a Number Sentence
Information from a Table

Use the table for problems 6–8.

6. How much would it cost to buy one of all three items on sale?

7. Which item is on sale for half price?

8. Jay spent $14 to buy 5 items on sale. Which items did Jay buy? how many of each?

Sale at the Zoo Shop		
Item	Regular Price	Sale Price
Polar Bear Key Chain	$3	$2
Toucan Shirt	$12	$10
Fish Cards	$8	$4

Place Value

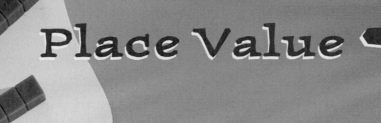

1

In this chapter you will:

Explore one million

Compare, order, and round whole
  numbers and money

Locate numbers on a number line

Make change

Use the STORE and RECALL
  calculator keys

Solve problems by making a table
  or list

Critical Thinking/Finding Together

There are 10 hundreds in 1000.
How many hundred miles are in a
twenty-six-thousand mile trip?

Willis C. Sick

There once was a young man on a ship
Who counted each pitch and each dip,
    Each roll and each yaw,
    Each sea and each saw
On a twenty-six-thousand mile trip.

John Ciardi

1-1 Thousands

A **place-value chart** makes understanding large numbers easier.

In 206,493 the value of:
  2 is 2 hundred thousands or 200,000.
  0 is 0 ten thousands or 0.
  6 is 6 thousands or 6000.
  4 is 4 hundreds or 400.
  9 is 9 tens or 90.
  3 is 3 ones or 3.

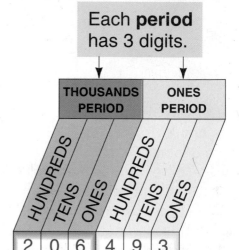

Each **period** has 3 digits.

In numbers larger than 9999, use a comma to separate the periods.

Standard Form: 206,493

Word Name: two hundred six thousand,

four hundred ninety-three

Four-digit numbers may be written with or without a comma.

**Write the place of the red digit.
Then write its value.**

1. 6,541   **2.** 7,843   **3.** 3,962   **4.** 5,034

5. 27,142   **6.** 46,359   **7.** 65,186   **8.** 92,170

9. 156,143   **10.** 983,567   **11.** 495,638   **12.** 374,826

13. 632,018   **14.** 275,941   **15.** 321,235   **16.** 176,404

17. 205,866   **18.** 652,048   **19.** 520,124   **20.** 804,397

Write the number in standard form.

21. nine hundred four

22. twelve thousand

23. six hundred thousand

24. eight thousand

25. five hundred twenty-one thousand, one hundred twelve

26. sixty-four thousand, seven hundred thirty-five

27. two hundred forty thousand, three hundred ninety-two

28. ninety thousand, four hundred eight

29. one hundred fifteen thousand, five hundred sixty

30. three hundred thousand, two

31. four hundred one thousand, eighteen

Write the word name for each number.

32. 762

33. 431

34. 605

35. 911

36. 4,918

37. 1,265

38. 7,016

39. 3,402

40. 25,461

41. 51,824

42. 90,160

43. 80,007

44. 169,818

45. 748,295

46. 300,040

47. 809,006

Critical Thinking

48. How many different four-digit numbers can you make using all the digits in each set only once?
a. 1, 2, 3, 4   **b.** 0, 1, 2, 3   **c.** 0, 0, 1, 2

What Is One Million?

The numbers from 1 to 999 are in the ones period. The numbers from 1000 to 999,999 are in the thousands period. Today you will discover the next counting number.

 Discover Together

You Will Need: calculator, paper, pencil

Use the calculator to compute exercise 1. Record each number sentence and the answer.

1. $10 \times 1 = \underline{?}$
$10 \times 10 = \underline{?}$
$10 \times 100 = \underline{?}$
$10 \times 1000 = \underline{?}$
$10 \times 10,000 = \underline{?}$
$10 \times 100,000 = \underline{?}$

2. What patterns do you notice?

The number that is $10 \times 100,000$ is *one million,* or 1,000,000. One million is the next counting number after 999,999.

3. How is 1,000,000 like 1000; 10,000; and 100,000? How is it different?

If $1,000,000 = 10$ hundred thousands, then $1,000,000 = 100$ ten thousands.

4. How many thousands is one million equal to? how many hundreds? You may use the calculator to find out.

Suppose you counted one number per second. You would take about

- <u>?</u> to count to 100.

- $16\frac{1}{2}$ minutes to count to 1000.

- 2 hours and 42 minutes to count to 10,000.

- 1 day to count to 100,000.

- $11\frac{1}{2}$ days to count to 1,000,000!

You may use the calculator to find the answers to questions 5–7.

5. If you were 100 days old, would you be older or younger than 1 year old?

6. About how many years old would you be if you were 1000 days old? 10,000 days old? (*Hint:* 1 year = 365 days)

7. About how many years old would you be if you were 100,000 days old? 1,000,000 days old?

Communicate

8. How did you discover how old you would be if you were 100 days old?

Discuss ✓

9. How did you discover how old you would be if you were 1000; 10,000; 100,000; and 1,000,000 days old?

 **Project**

Communicate ✓

10. About how much distance is equal to 1,000,000 inches? Name a place that is about 1,000,000 inches away from your school. Tell the class how you found out.

39

Millions

Recently, the population of Brazil was 158,202,019.

In the millions period of 158,202,019, the value of:
 1 is 1 hundred million, or 100,000,000.
 5 is 5 ten millions, or 50,000,000.
 8 is 8 millions, or 8,000,000.

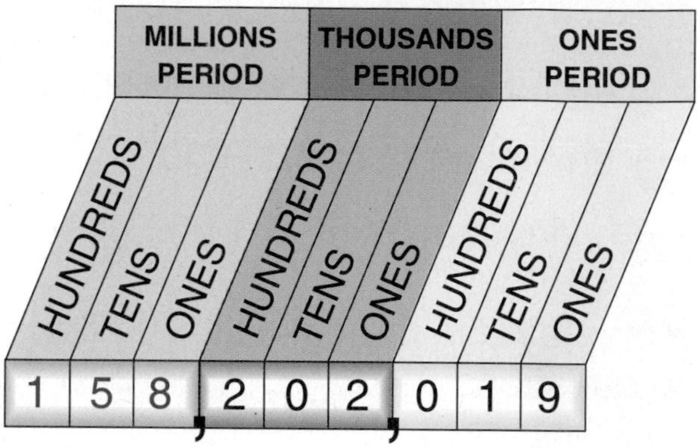

MILLIONS PERIOD			THOUSANDS PERIOD			ONES PERIOD		
HUNDREDS	TENS	ONES	HUNDREDS	TENS	ONES	HUNDREDS	TENS	ONES
1	5	8	2	0	2	0	1	9

Standard Form: 158,202,019

Word Name: one hundred fifty-eight million,

two hundred two thousand,

nineteen

Write the period of the underlined digits.

1. 45,678

2. 59,650

3. 26,545

4. 456,789

5. 567,890

6. 148,337

7. 9,456,789

8. 567,890,000

9. 617,148,337

Write in standard form.

10. thirty-one million

11. three million

12. six hundred million

13. eighty million

**Write the place of the red digit.
Then write its value.**

14. 482,165,016

15. 904,628,153

16. 617,465,089

17. 38,296,145

18. 10,692,534

19. 4,797,123

20. 412,076,531

21. 217,945,310

22. 842,005,301

23. 920,354,876

24. 105,643,129

25. 732,530,481

Write the word name for each number.

26. 5,460,000

27. 920,015,300

28. 10,300,000

29. 475,000

30. 1,006,005

31. 20,000,012

 Connections: Geography

Brazil is the largest country in South America.

32. The land area of Brazil is three million,
two hundred eighty-six thousand,
four hundred seventy square miles. How would
you write this number in standard form?

33. In Brazil there are two million,
one hundred thirty-five thousand,
six hundred thirty-seven square miles
of forest. Write this number in standard form.

34. In 1985 the Brazilian city of Rio de Janeiro had
an estimated population of 5,615,149. Write
this number in words.

Brazil

1-4 Place Value

▶ Understanding the place of each digit in a number can help you write the number in **expanded form**.

Standard Form	Expanded Form
178	$100 + 70 + 8$
25,613	$20,000 + 5,000 + 600 + 10 + 3$
4,381,256	$4,000,000 + 300,000 + 80,000 + 1000 + 200 + 50 + 6$
60,070,005	$60,000,000 + 70,000 + 5$
800,500,020	$800,000,000 + 500,000 + 20$

▶ Understanding the place of each digit in a number can help you count on and count back by 10, 100, or 1000.

Count on by 10.	Count on by 100.	Count back by 1000.
25,613	25,613	25,613
25,623	25,713	24,613
25,633	25,813	23,613
25,643	25,913	22,613

Write each number in expanded form.

1. 65

2. 38

3. 246

4. 975

5. 352

6. 810

7. 6143

8. 7924

9. 5491

10. 4035

11. 13,827

12. 62,473

13. 90,303

14. 184,001

15. 705,060

16. 350,900

17. 6,320,079

18. 19,430,600

19. 75,260,080

20. 507,104,908

21. 800,002,100

22. 300,400,050

Write each number in standard form.

23. 2000 + 400 + 90 + 6   **24.** 7000 + 100 + 80

25. 30,000 + 5000 + 800 + 20 + 9

26. 800,000 + 90,000 + 4000 + 600 + 50 + 2

27. 7,000,000 + 300,000 + 50,000 + 2000 + 90 + 4

28. 20,000,000 + 70,000 + 5000 + 8

29. 700,000,000 + 300,000 + 4000 + 5

Write the numbers that are 10 more, 100 more, and 1000 more. Then write the numbers that are 10 less, 100 less, and 1000 less.

30. 7825   **31.** 92,614   **32.** 365,829   **33.** 482,565

34. 7,342,675   **35.** 32,489,267   **36.** 107,361,072

 **Connections: Language Arts**

A **googol** is the number 1 followed by a hundred zeros. A **googolplex** is the number 1 followed by a googol of zeros.

10,000,000,000,

37. Write a googol.

38. How many commas did you need?

39. In your Math Journal, write some things that might be counted using a googol.

Math Journal

43

 **Estimation**

 Discover Together

Sometimes it is inconvenient, difficult, or even impossible to report the exact number of items in a group or set.

When you cannot report an exact number, you can use an estimate. An **estimate** is any number that tells *about* how much or *about* how many.

Here are some examples of estimates:

- So far, 1,000,000 different species of insects have been discovered.

- The worldwide population of tigers had fallen from 1,000,000 in 1990 to 7000 by 1993.

- The age of the oldest bird on record, a cockatoo, was 80 years. It was fully grown when captured in 1902 and died in the London Zoo in 1982.

Discuss these questions with your group:

1. For each example, why is each number reported as an estimate rather than an exact number?

2. How are all the estimated numbers alike? How are they different?

3. Read the examples again. Do you think estimating is the same as guessing? Why or why not?

Work with your group to estimate the number of fish in the picture below. *Do not try to count all the fish.* You may use a ruler, tracing paper, a calculator, or any other tools you think might help you.

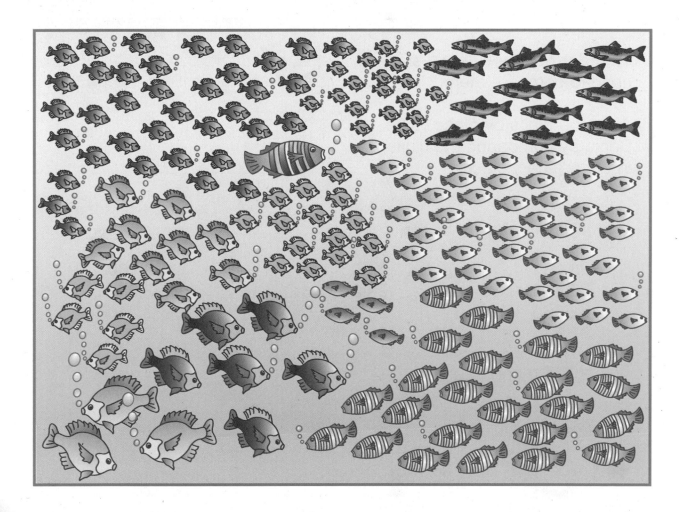

Communicate

4. What is your estimate of the number of fish in the picture?

5. How did you make your estimate?

6. Do *you* ever use estimates instead of exact numbers? When?

Update your skills. See page 2.

1-6 Comparing and Ordering Whole Numbers

Order the heights of the mountains from greatest to least.

To order numbers, you first need to compare them.

Align the digits by place value. Begin by comparing the digits in the greatest place.

Algebra ✓

Mountain	Height in Feet
Aconcagua, Argentina-Chile	23,034
Huascarán, Peru	22,198
Kangto, Tibet	23,260
Simvuo, India	22,346

Compare ten thousands.	Compare thousands. Rearrange.	Compare hundreds. Rearrange.
23,034 22,198 23,260 22,346	23,034 23,260 22,198 22,346	23,260 23,034 22,346 22,198
20,000 = 20,000	3000 > 2000	200 > 0 300 > 100

So 23,034 and 23,260 are greater than 22,198 and 22,346.

So 23,260 > 23,034 and 22,346 > 22,198.

The order from greatest to least: 23,260; 23,034; 22,346; 22,198

The order from least to greatest: 22,198; 22,346; 23,034; 23,260

Compare. Write <, =, or >.

1. 3705 ? 992   Think: No thousands.

2. 4783 ? 4378

3. 98,050 ? 98,305

4. 63,582 ? 62,975

5. 36,758 ? 36,721

Write in order from least to greatest.

6. 23; 29; 25; 21

7. 426; 505; 431; 424

8. 671; 680; 707; 679; 702

9. 843; 839; 87; 841; 836

10. 4515; 3204; 7661; 1139; 4500

11. 6714; 6783; 6756; 679; 6744

12. 24,316; 34,316; 24,416; 34,416; 24,404

13. 57,554; 58,641; 5784; 57,590; 579

Write in order from greatest to least.

14. 86; 89; 84; 82

15. 343; 349; 434; 352

16. 526; 642; 589; 538; 658

17. 295; 32; 289; 27; 281

18. 8451; 8468; 8450; 8464; 8445

19. 3605; 3679; 369; 3610; 3600

20. 46,824; 46,785; 46,804; 46,815; 46,790

21. 94,747; 9547; 95,754; 959; 94,763

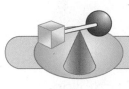

Challenge

22. List the countries in order from greatest area to least area.

23. List the countries with areas greater than four hundred thousand square miles.

Country	Area in Square Miles
France	211,208
Argentina	1,072,067
Peru	496,222
Ethiopia	472,432
Japan	143,574
India	1,229,737
Mexico	761,600

Number Sense: Using a Number Line

Halfway points can help you to find numbers
on a number line.

▶ About where on each number line is 75?

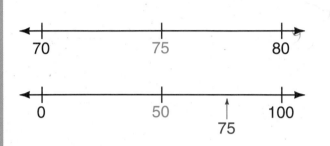

75 is exactly halfway
between 70 and 80.

50 is the halfway point.
75 is exactly halfway
between 50 and 100.

▶ About where on each number line is 142?

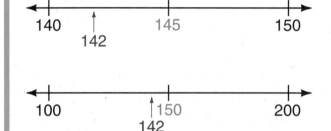

145 is the halfway point.
142 is between 140 and 145.
142 is closer to 140.

150 is the halfway point.
142 is between 100 and 150.
142 is much closer to 150.

**Write the number that is halfway between
the two numbers.**

1. 20; 30   **2.** 0; 50   **3.** 600; 700   **4.** 0; 200

5. 0; 500   **6.** 0; 80   **7.** 10; 70   **8.** 150; 200

**Draw a number line to show the halfway point
between the two numbers.**

9. 0; 10   **10.** 40; 50   **11.** 0; 60   **12.** 800; 900

13. 0; 1000   **14.** 510; 520   **15.** 1000; 2000   **16.** 0; 2000

About what number is each arrow pointing toward?

17.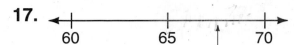

18.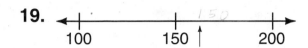

19.

20.

21.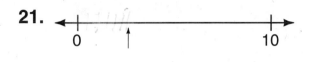

22.

23.

24.

25.

26.

Draw each number line.

27. Draw a number line from 50 to 60. Show the halfway point. Draw an arrow that points toward 53.

28. Draw a number line from 0 to 100. Show the halfway point. Draw an arrow that points toward 40.

29. Draw a number line from 0 to 500. Draw an arrow that points toward 300.

Share Your Thinking

30. Bring home the number lines you drew for exercises 27–29. Tell your family about halfway points. Then ask your family to tell which numbers the arrows are pointing toward.

1-8 | Making Change

Imagine that you are working in a music store. A customer wants to buy a CD that costs $13.88 and gives you a twenty-dollar bill. What coins and bills would you give the customer as change? What would be the value of the change?

twenty-dollar bill

$20.00

To make change:

- Count up from the cost to the amount given.

- Start with the coins that have the least value.

- Use the fewest possible coins and bills.

cost   **amount given**

$13.88 → $13.89 → $13.90 → $14.00 → $15.00 ⟶ $20.00

Arrange the money in order.
Count the change: $5.00 + $1.00 + $0.10 + $0.01 + $0.01
$5.00 → $6.00 → $6.10 → $6.11 → $6.12

You would give the customer 2 pennies, 1 dime, 1 one-dollar bill, and 1 five-dollar bill as change. The value of the change is $6.12.

Use play money. Write the fewest coins and bills you would give as change. Then write the value of the change.

1. Cost: $0.81
Amount given: $1.00

2. Cost: $2.54
Amount given: $3.00

Use play money. Write the fewest coins and bills you would receive as change. Then write the value of the change.

3.

 $4.75

Amount given: $10.00

4.

 $9.98

Amount given: $20.00

5. Cost: $3.16
Amount given: $5.00

6. Cost: $4.22
Amount given: $10.00

7. Cost: $12.99
Amount given: $15.00

8. Cost: $13.08
Amount given: $14.00

9. Cost: $13.70
Amount given: $20.00

10. Cost: $14.10
Amount given: $20.00

11. Cost: $15.46
Amount given: $20.00

12. Cost: $19.55
Amount given: $20.00

13. Cost: $10.60
Amount given: $20.00

14. Cost: $2.67
Amount given: $20.00

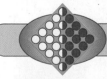 **Finding Together**

Use nickels, dimes, and quarters. List all the ways you can make each amount. You may use play money.

$ .20	
nickels	dimes
4	0
2	1
0	2

15. $0.15

16. $0.30

17. $0.25

18. $0.35

19. $0.50

20. $0.40

21. $0.60

22. $0.75

Comparing and Ordering Money

Chuck earned $25.35.
Evan earned $24.50.
Who earned more?

To find who earned more,
compare $25.35 and $24.50.

XYZ CORPORATION	0192
	Sept. 20 2002
PAY TO THE ORDER OF Evan Smith	$ 24.50
Twenty-four and 50/100	DOLLARS
	Glenn Johnson
32759 533004 8976321 883 1122856	

▶ Compare money as you compare whole numbers.
  • Line up the amounts by the pennies.
  • Compare digits. Start at the left.

Compare ten dollars.	Compare dollars.
$25.35	$25.35
$24.50	$24.50
$20.00 = $20.00	$5.00 > $4.00

So $25.35 > $24.50. Chuck earned more.

Order $7.49, $7.43, and $6.43 from least to greatest.

▶ Order money as you order whole numbers.

Compare dollars.	Compare dimes.	Compare pennies.
$7.49	$6.43	$6.43
$7.43	$7.49	$7.43
$6.43	$7.43	$7.49
$6.00 < $7.00	$0.40 = $0.40	$0.03 < $0.09

least amount

The order from least to greatest: $6.43, $7.43, $7.49

The order from greatest to least: $7.49, $7.43, $6.43

Compare. Write <, =, or >.

1. $0.07 _?_ $0.09
2. $0.76 _?_ $0.73
3. $0.52 _?_ $0.52

4. $3.49 _?_ $4.69
5. $8.03 _?_ $8.50
6. $2.81 _?_ $2.80

7. $5.38 _?_ $5.36
8. $9.75 _?_ $9.75
9. $7.63 _?_ $7.66

10. $10.30 _?_ $10.70
11. $42.25 _?_ $25.42
12. $87.95 _?_ $87.75

13. $36.99 _?_ $36.98
14. $77.07 _?_ $77.70
15. $61.18 _?_ $61.18

16. $1.95 _?_ $0.19
17. $2.67 _?_ $26.07
18. $74.50 _?_ $7.85

Write in order from least to greatest.

19. $0.76, $0.35, $0.57, $0.83

20. $0.18, $0.15, $0.19, $0.12, $0.17

21. $4.65, $4.62, $4.26, $5.24, $5.42

22. $75.39, $78.36, $7.48, $74.48, $75.93

Write in order from greatest to least.

23. $1.11, $1.10, $1.01, $1.17, $1.71

24. $24.42, $24.48, $24.24, $2.48, $2.84

25. $9.91, $9.19, $91.19, $91.91, $99.11

26. $68.50, $65.80, $68.05, $6.85, $65.08

PROBLEM SOLVING

27. Jill saved $32.40. Ed saved $34.20. Lynn saved $34.40. Who saved the most money? Who saved the least money?

28. Adam has saved $85.25. Can he buy a jacket that costs $58.82?

53

You can round numbers and money amounts to tell **about** how much or **about** how many.

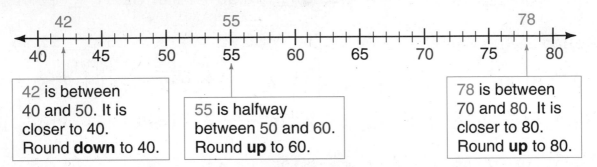

42 is between 40 and 50. It is closer to 40. Round **down** to 40.

55 is halfway between 50 and 60. Round **up** to 60.

78 is between 70 and 80. It is closer to 80. Round **up** to 80.

To round numbers:
- find the place you are rounding to.
- look at the digit to its right.

Round 65 to the nearest ten.

65
↓
70

5 = 5
Round **up** to 70.

Round $2.53 to the nearest ten cents.

$2.53
↓
$2.50

3 < 5
Round **down** to $2.50.

Round $5.86 to the nearest dollar.

$5.86
↓
$6.00

8 > 5
Round **up** to $6.00.

Round 2174 to the nearest hundred.

2174
↓
2200

7 > 5
Round **up** to 2200.

Round 8214 to the nearest thousand.

8214
↓
8000

2 < 5
Round **down** to 8000.

Round $625.95 to the nearest ten dollars.

$625.95
↓
$630.00

5 = 5
Round **up** to $630.00.

Round to the nearest ten or ten cents.

1. 16   **2.** 49   **3.** 94   **4.** 315   **5.** 871

6. $0.55   **7.** $0.83   **8.** $1.24   **9.** $8.39   **10.** $5.66

Round to the nearest hundred or dollar.

11. 285   **12.** 674   **13.** 503   **14.** 857   **15.** 449

16. 9173   **17.** 3426   **18.** 1250   **19.** 7314   **20.** 2693

21. $1.44   **22.** $6.70   **23.** $3.95   **24.** $7.56   **25.** $8.39

26. $55.20   **27.** $38.98   **28.** $27.49   **29.** $18.88   **30.** $71.53

Round to the nearest thousand or ten dollars.

31. 9437   **32.** 1878   **33.** 8564   **34.** 2946

35. 74,806   **36.** 32,521   **37.** 60,719   **38.** 45,133

39. $53.68   **40.** $15.89   **41.** $94.87   **42.** $27.95

43. $836.42   **44.** $351.25   **45.** $708.50   **46.** $484.62

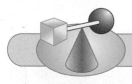

Challenge

Round to the nearest ten thousand or hundred dollars.

47. 36,455   **48.** 52,630   **49.** $654.70   **50.** $895.99

Round to the nearest hundred thousand.

51. 743,299   **52.** 250,343   **53.** 571,320

54. 462,135   **55.** 325,523   **56.** 2,704,810

1-11 Working with Money

Use the skills and strategies you have learned to solve each problem.

1. Dan buys school supplies for $8.47. He gives the cashier a twenty-dollar bill. The cashier makes change with the fewest possible coins and bills. What coins and bills does Dan receive as change?

Communicate ✓

2. A sweater is on sale for $18.89. Elena says the sweater costs about $18.00. Rita says it costs about $19.00. Who do you think is right? Why?

3. Juwon receives a total of $75.00 for his birthday. He wants to buy a pair of sneakers that costs $79.95, including tax. Will his birthday money be enough to pay for the sneakers?

4. Neither Mei nor Jaycie has pennies, but they both have $0.45. Mei has 3 coins. Jaycie has 6 coins. Which coins does each girl have?

5. The cost of Ms. Johnson's purchases at the drugstore is $7.82. She gives the clerk a ten-dollar bill and 2 pennies. Since the ten-dollar bill is more than enough to pay for Ms. Johnson's purchases, why might she give the clerk the extra 2 pennies?

6. Would you rather have 5 quarters, 15 dimes, or a one-dollar bill? Why?

7. The table at the right shows how many pennies different classes in Glenn School have collected for charity. Which class collected the fewest pennies? Which class collected the most? How much money did each class collect?

Class	Number of Pennies
4A	1430
4B	1432
4C	1342
3A	1324
3B	1483
3C	1384

Solve each problem. Then explain how you found each answer.

8. Trucker Bob's check at the diner comes to $8.55. He pays with a ten-dollar bill. The cashier has run out of quarters. How can she give him change using the fewest possible coins and bills?

9. Dominique has saved $15.00 for a birthday present for her mother. She spends $12.76 for earrings and a pin. Does she have enough money left over to buy a gift bag that costs $2.98?

10. Is $6.53 closer to $6.00 or closer to $7.00? How do you know?

11. Mr. Mackintosh hires students to pick apples in his orchard. The more apples a student picks, the more money he or she earns. Jessie earned $125.75. Zach earned $127.25. Tommy earned $125.27. Sara earned $127.17. Which student earned the most? the least? Did Jessie pick more or fewer apples than Sara?

12. Manny has 9 coins that have a value of $0.88. What coins does Manny have?

13. Alonzo buys a dog collar and a leash at a pet supply store. The dog collar and leash cost $11.56. Alonzo pays with a twenty-dollar bill. If he receives the fewest possible coins and bills as his change, what coins and bills does he receive? What is the value of his change?

14. Tom has 1 quarter, 6 dimes, 3 nickels, and 4 pennies. Dick has 2 quarters, 3 dimes, 2 nickels, and 7 pennies. Harry has 1 half dollar, 1 quarter, and 5 nickels. Whose coins have the greatest value? What is the value of these coins?

TECHNOLOGY

Store and Recall Keys

You can use a calculator to find the standard form of a number written in expanded form.

▶ Use a calculator to find the standard form of 70 + 8.

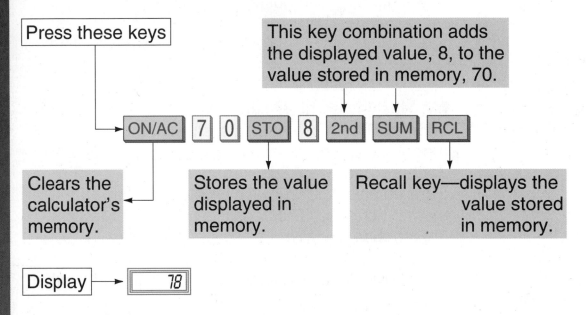

Press these keys

This key combination adds the displayed value, 8, to the value stored in memory, 70.

ON/AC · 7 · 0 · STO · 8 · 2nd · SUM · RCL

Clears the calculator's memory.

Stores the value displayed in memory.

Recall key—displays the value stored in memory.

Display → 78

The standard form of 70 + 8 is 78.

▶ Use a calculator to find the standard form of 100 + 40 + 5.

Press these keys

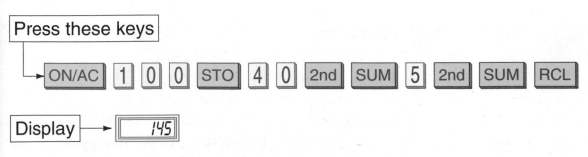

ON/AC · 1 · 0 · 0 · STO · 4 · 0 · 2nd · SUM · 5 · 2nd · SUM · RCL

Display → 145

The standard form of 100 + 40 + 5 is 145.

Enter the following key sequences into your calculator. Write each number in standard form.

1. $\boxed{4}\ \boxed{0}\ \boxed{+}\ \boxed{6}\ \boxed{=}$ 2. $\boxed{3}\ \boxed{0}\ \boxed{0}\ \boxed{+}\ \boxed{2}\ \boxed{0}\ \boxed{=}$

3. $\boxed{6}\ \boxed{0}\ \boxed{STO}\ \boxed{7}\ \boxed{2nd}\ \boxed{SUM}\ \boxed{RCL}$

4. $\boxed{8}\ \boxed{0}\ \boxed{STO}\ \boxed{1}\ \boxed{2nd}\ \boxed{SUM}\ \boxed{RCL}$

5. $\boxed{2}\ \boxed{0}\ \boxed{0}\ \boxed{STO}\ \boxed{2}\ \boxed{0}\ \boxed{2nd}\ \boxed{SUM}\ \boxed{3}\ \boxed{2nd}\ \boxed{SUM}\ \boxed{RCL}$

6. $\boxed{1}\ \boxed{0}\ \boxed{0}\ \boxed{STO}\ \boxed{9}\ \boxed{0}\ \boxed{2nd}\ \boxed{SUM}\ \boxed{9}\ \boxed{2nd}\ \boxed{SUM}\ \boxed{RCL}$

Use the $\boxed{STO}$, $\boxed{2nd}$, $\boxed{SUM}$, and $\boxed{RCL}$ keys to find each number in standard form.

7. $30 + 2$ 8. $70 + 7$ 9. $300 + 10 + 8$

10. $700 + 60 + 9$ 11. $600 + 40 + 4$ 12. $900 + 9$

13. $200 + 2$ 14. $5000 + 200 + 10 + 3$ 15. $1000 + 300 + 50 + 5$

 Finding Together

16. Work with a classmate. Choose a number from 20–30. Let this be your target number. Take turns entering a number from 1–5 into your calculator. Be sure to use the $\boxed{2nd}$ $\boxed{SUM}$ key combination. The object is to add numbers so that a sum equal to your target number will be stored in the calculator's memory. When you or your classmate think you have reached the target number, press the $\boxed{RCL}$ (recall) key. If the number displayed is the target number, you score 2 points, the activity ends, and you can play again. If the number displayed is *not* your target number, use the $\boxed{STO}$ key to store it in memory and continue to play. The activity ends when a player scores 10 points.

1-13 Problem Solving: Make a Table or List

Problem: Steve has 24 marbles. Each marble is green or red. For every green marble, Steve has 3 red marbles. How many red marbles does Steve have?

Color	Number of Marbles			
green	1	2	?	?
red	3	?	?	?
total	4	?	?	?

1 IMAGINE Make a table.

2 NAME

Facts: Steve has 24 marbles. For 1 green marble, there are 3 red marbles.

Question: How many red marbles does Steve have?

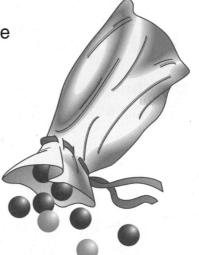

3 THINK If Steve has 1 green marble, he would have 3 red marbles. Write those numbers in the table.

If Steve has 2 green marbles, he would have 2 × 3 red marbles.

4 COMPUTE Complete the table. Multiply each number of green marbles by 3. Then add to find a column that shows 24 marbles.

¹
18 red
+   6 green
24

Color	Number of Marbles					
green	1	2	3	4	5	6
red	3	6	9	12	15	18
total	4	8	12	16	20	24

Steve has 18 red marbles.

5 CHECK Use a calculator to check your computation, or act out the problem.

Make a table or list to solve each problem.

	Blue Shirt	Gray Shirt	White Shirt
Tie Colors red brown green yellow			

1. Mr. Hoody bought 3 shirts and 4 ties. The shirts are blue, gray, and white. The ties are red, brown, green, and yellow. How many ways can he wear the shirts and ties together?

IMAGINE   Make an organized list.

NAME   *Facts:*   3 shirts: blue, gray, white
4 ties: red, brown, green, yellow

Question: How many ways can the shirts and ties be worn?

THINK   List each shirt color.
Write the ties that can be worn with each shirt.
Count the total number of combinations.

COMPUTE ⟶ **CHECK**

2. Apple juice costs 50¢. The juice machine accepts quarters, dimes, and nickels. Make a list of coin combinations that can be used to buy juice.

3. Adam and Ashlee use three 1-6 number cubes. They look for different ways to roll the sum of 12. How many ways will they find?

4. Calvin has 90 stamps. For every Mexican stamp, Calvin has 8 U.S. stamps. How many Mexican stamps does Calvin have?

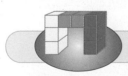

Make Up Your Own

5. Write a problem that uses a table or list. Ask a classmate to solve the problem.

1-14 Problem-Solving Applications

Solve each problem and explain the method you used.

1. The school book fair wanted to raise $1500. It raised $2500. What is the difference in the amounts?

2. Abigail bought a science fiction novel for $17.89. How much change did she receive from a twenty-dollar bill? What coins and bills could she have received as change?

3. Abigail's science fiction novel describes life one hundred thousand years from now. What will the date be one hundred thousand years from today?

4. Paperbacks sold for 50¢ each. Hardcover books sold for $1.25 each. Was it more expensive to buy 3 paperbacks or 1 hardcover book?

5. Ray sold handmade bookmarks for 75¢ each. What five coins could be used to pay for 1 bookmark?

6. One book at the sale was printed 100 years ago. In what year was that book printed?

7. The book fair sold 437 books this year. Last year it sold 327 books. In which year were more books sold? how many more?

8. Zena brought 10 dollars to the book fair. She bought 2 books about mountain climbing for $4.20 each. How much change did she get?

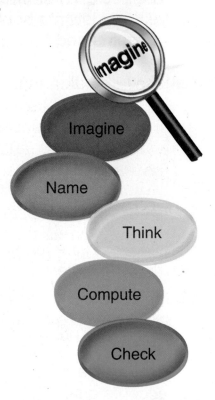

Imagine

Name

Think

Compute

Check

Choose a strategy from the list or use another strategy to solve each problem.

USE THESE STRATEGIES:
Make a Table or List
Choose the Operation
Guess and Test
Write a Number Sentence

9. There were 428 people at the book fair. Three hundred eighteen of them bought books. How many people did not buy a book?

10. Stella made a triangular book display. She put 9 books in the first row, 8 books in the second row, 7 books in the third row, and so on. How many books did Stella use in her display?

11. Hank wrote 14 poems. Julio wrote 5 more poems than Hank. How many poems did Julio write?

12. Ray's bookmarks were made of red or blue plastic with purple, white, or yellow fringe. How many different bookmarks could Ray make?

13. Sue reads adventure books. There are 11 books on her desk. She has read 7 books. How many books does Sue have left to read?

14. The book fair charged 30¢ admission. How many different ways could people give the exact amount if no pennies were allowed?

 **Share Your Thinking**

Math Journal

15. In your Math Journal, write the name of the strategy you think is the most fun to use. Explain why you think it is fun. Then write the numbers of the problems you solved by using that strategy.

Write the number in standard form.   *(See pp. 36–43.)*

1. eighty-four thousand, two hundred six   **2.** six hundred thousand, five

Write the place of the red digit. Then write its value.

3. 56,651,020   **4.** 205,640,311   **5.** 67,451

Compare. Write <, =, or >.   *(See pp. 46–47, 52–53.)*

6. $7.45 _?_ $4.75   **7.** 1450 _?_ 1450   **8.** 61,905 _?_ 61,950

Write in order from greatest to least.

9. $25.10; $52.10; $51.20   **10.** 625; 217; 451; 332

11. 3542; 3320; 4310; 5403   **12.** 46,532; 46,503; 46,330

About what number is each arrow pointing toward?   *(See pp. 48–49.)*

13.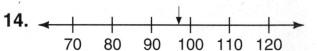
5   10   15   20   25   **14.** 70   80   90   100   110   120

Round each number.   *(See pp. 54–55.)*

To the nearest 100:   **15.** 4486   **16.** 6824

To the nearest 1000:   **17.** 76,534   **18.** 153,462

To the nearest dollar: **19.** $12.75   **20.** $57.45

Write the change you would receive.
Then write the value of the change.   *(See pp. 50–51, 56–57.)*

21. Cost: $10.72   **22.** Cost: $.93
      Amount given: $15.00   Amount given: $5.00

(See Still More Practice, p. 461.)

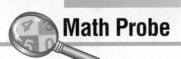

ROMAN NUMERALS

The ancient Romans used letters to write numbers.

1 = I	4 = IV	7 = VII	10 = X	40 = XL	70 = LXX
2 = II	5 = V	8 = VIII	20 = XX	50 = L	80 = LXXX
3 = III	6 = VI	9 = IX	30 = XXX	60 = LX	90 = XC
					100 = C

Use these rules to read and write Roman numerals:

- When letters that stand for smaller numerals come *after* letters that stand for larger numerals, *add*.

  III → 1 + 1 + 1 = 3
  VIII → 5 + 3 = 8
  LVIII → 50 + 8 = 58

- When a letter that stands for a smaller numeral comes *before* a letter that stands for a larger numeral, *subtract*.

  IV → 5 − 1 = 4
  IX → 10 − 1 = 9
  XL → 50 − 10 = 40

CXLVII = ?

$$\underset{\downarrow}{C} \quad \underset{\downarrow}{XL} \quad \underset{\downarrow}{VII}$$
$$100 + 40 + 7 = 147$$

Write the Roman numeral in standard form.

1. LXIV   **2.** XXXIX   **3.** LXIX   **4.** CXXVI   **5.** CCVII

Write each as a Roman numeral.

6. 17   **7.** 48   **8.** 300   **9.** 89   **10.** 56   **11.** 234

Performance Assessment

Make an organized list.

1. Robert buys a sandwich and milk. With tax the total is $2.84. He pays with a ten-dollar bill. List the greatest number of bills and coins he could receive in change. List the least number of bills and coins he could receive in change.

Write in standard form.

2. five hundred eight

3. two hundred four thousand

4. fourteen million, fifteen

5. 700 thousands + 60 tens + 8 ones

Write in expanded form.

6. 420,635,010

7. 56,431

8. 7,532,060

Write in order from least to greatest.

9. $56.20; $50.62; $52.60

10. 72,310; 72,130; 73,303

About what number is each arrow pointing toward?

11.

5 10 15 20 25

12.

70 80 90 100 110 120

Round each number.

To the nearest 10: 13. 1471

To the nearest 100: 14. 732

To the nearest dollar: 15. $24.31

16. $162.58

PROBLEM SOLVING *Use a strategy you have learned.*

17. Jared has 22 plastic animals in shapes of tigers, antelopes, and elephants. He has 2 more tigers than elephants and 3 more antelopes than tigers. How many of each does he have?

Addition and Subtraction Concepts

2

MATH MAKES ME FEEL SAFE

Math isn't just adding and subtracting.
Not for me.

Math makes me feel safe
knowing that my brother will always be
three years younger than I am,
and every day of the year will have
twenty-four hours.
That a snowflake landing on my mitten
will have exactly six points,
and that I can make new shapes
from my Tangram pieces
whenever I feel lonely.

Math isn't just adding
and subtracting,
Not for me.

Math makes me feel safe.

Betsy Franco

In this chapter you will:

Use addition properties and strategies
Learn about subtraction concepts
Estimate sums and differences
Check addition and subtraction
Add and subtract whole
  numbers and money
Learn about READ/DATA statements
Solve problems using logical reasoning

Critical Thinking/Finding Together

Suppose you are the person in the
poem. When your brother is 28 years
old, how old will you be?

2-1 Addition Properties

Algebra

The properties of addition can help you to add quickly and correctly.

- Changing the *order* of addends does not change the sum.

Think: "order."

$$5 + 6 = 11 \qquad 5 \qquad 6$$
$$6 + 5 = 11 \qquad \frac{+6}{11} \qquad \frac{+5}{11}$$

- The sum of *zero* and a number is the same as that number.

Think: "same number."

$$7 + 0 = 7 \qquad 7 \qquad 0$$
$$0 + 7 = 7 \qquad \frac{+0}{7} \qquad \frac{+7}{7}$$

- Changing the *grouping* of the addends does not change the sum.

Think: "grouping."

$$(4 + 5) + 2 = 4 + (5 + 2)$$

Always do the computation in parentheses first.

$$9 \quad + 2 = 4 + \quad 7$$
$$11 = 11$$

Use the properties to make adding a list of numbers easier.

Change the order.

Add down.   Add up.

$$
\begin{array}{ll}
4 & \\
5 & 9 \\
1 & 10 \\
+3 & 13 \\
\hline
13 &
\end{array}
\qquad
\begin{array}{ll}
4 & 13 \\
5 & 9 \\
1 & 4 \\
+3 & \\
\hline
13 &
\end{array}
$$

Change the order and the grouping.

$$
\begin{array}{l}
2 \\
3 \\
10 \begin{array}{|l} 0 \\ 7 \end{array} \\
+5 \\
\hline
17
\end{array}
$$

$$(3 + 7) + 2 + 0 + 5 = 17$$
$$10 \quad + 2 + 0 + 5 = 17$$

Add. Use the addition properties.

1.	2.	3.	4.	5.	6.	7.
3 +0	6 +3	3 +6	8 +7	7 +8	8 +0	0 +5

8.	9.	10.	11.	12.	13.	14.
5 +4	0 +6	9 +7	7 +9	4 +8	9 +0	8 +4

15.	16.	17.	18.	19.	20.	21.	22.
7 9 0 +3	2 6 1 +4	5 4 2 +5	1 2 8 +0	2 1 3 +9	1 9 7 +0	3 4 3 +6	1 2 7 +8

Add the number in the center to each number around it.

23.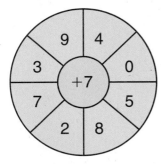

24.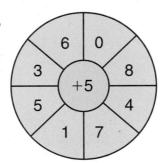

25.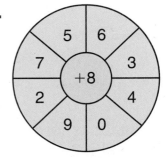

Critical Thinking

Use the scoreboard to answer the questions.

Inning	1	2	3	4	5	6	7	8	9
Bluebirds	5	1	0	0	4	0	1	3	0
Robins	0	2	1	3	0	0	3	3	4

26. Who won the game?

27. What was the final score?

28. After which inning was the score 11 to 9?

29. After which inning was there a tie score?

30. What was the score after 2 innings? 6 innings? 8 innings?

31. How many runs did the Bluebirds and Robins score in the 5th inning?

2-2 Addition Strategies

Tyrone and Maria use doubles to find 6 + 7.

doubles

Tyrone thinks: 6 + 6 = 12
6 + 7 = 13

1 more than 6 + 6

Maria thinks: 7 + 7 = 14
6 + 7 = 13

1 less than 7 + 7

James uses 10 to find 9 + 4.

James thinks: 10 + 4 = 14
So, 9 + 4 = 13.

1 more than 9 + 4

Tania looks for sums of 10 and doubles
when she adds more than two numbers.

$$\begin{array}{r} 2 \\ 5 \\ +\ 8 \\ \hline 15 \end{array}$$ ⟩10   10 + 5 = 15

$$\begin{array}{r} 6 \\ 3 \\ 6 \\ +\ 2 \\ \hline 17 \end{array}$$ ⟩12   12 + 3 + 2 = 17

Find the sum.

1. $\begin{array}{r}3\\+4\\\hline\end{array}$	**2.** $\begin{array}{r}5\\+6\\\hline\end{array}$	**3.** $\begin{array}{r}8\\+7\\\hline\end{array}$	**4.** $\begin{array}{r}6\\+7\\\hline\end{array}$	**5.** $\begin{array}{r}5\\+4\\\hline\end{array}$	**6.** $\begin{array}{r}8\\+8\\\hline\end{array}$
7. $\begin{array}{r}9\\+5\\\hline\end{array}$	**8.** $\begin{array}{r}7\\+9\\\hline\end{array}$	**9.** $\begin{array}{r}4\\+9\\\hline\end{array}$	**10.** $\begin{array}{r}9\\+9\\\hline\end{array}$	**11.** $\begin{array}{r}3\\+9\\\hline\end{array}$	**12.** $\begin{array}{r}9\\+2\\\hline\end{array}$

13. 3 + 2   **14.** 4 + 4   **15.** 8 + 9   **16.** 9 + 6

Add mentally.

17.	18.	19.	20.	21.	22.
1	3	2	4	3	5
2	3	7	5	7	6
+9	+8	+8	+4	+7	+5

23.	24.	25.	26.	27.	28.
1	4	5	3	6	1
4	2	7	3	2	8
8	3	0	3	4	1
+2	+2	+3	+3	+2	+7

29. 10 + 5   **30.** 9 + 5   **31.** 6 + 5   **32.** 6 + 7

33. 8 + 10   **34.** 8 + 9   **35.** 9 + 9   **36.** 8 + 8

37. 3 + 10   **38.** 3 + 4   **39.** 9 + 3   **40.** 4 + 9

PROBLEM SOLVING

41. Tara needs to mail 6 letters and 5 postcards. How many stamps does she need?

42. Kim has 4 Canadian stamps, 5 English stamps, and 6 French stamps in his collection. How many stamps does he have altogether?

Calculator Activity

Find the first sum. Predict the second sum.

43. 42 + 42   **44.** 16 + 16   **45.** 35 + 35   **46.** 48 + 48

    42 + 41   17 + 16   45 + 35   48 + 38

47. 50 + 50   **48.** 20 + 20   **49.** 26 + 26   **50.** 21 + 21

    50 + 65   20 + 25   27 + 27   25 + 25

2-3 Subtraction Concepts

Algebra

Subtraction has four different meanings.

▶ Take Away

Mr. Wu displayed 12 Planet Search videogames. He sold 9 of the games. How many Planet Search games does he have left?

$$12 - 9 = 3$$

He has 3 Planet Search games left.

▶ Compare

Jenny had 4 baby dolls. Inez had 8 baby dolls. How many more baby dolls did Inez have than Jenny?

$$8 - 4 = 4$$

Inez had 4 more baby dolls.

Jenny

Inez

▶ Part of a Whole Set

Lisa packed 15 cartons of model trucks. She shipped 8 of the cartons to Ohio. How many cartons were *not* shipped to Ohio?

$$15 - 8 = 7$$

Seven cartons were not shipped to Ohio.

▶ How Many More Are Needed

Manny had 6 bull's-eyes in a board game. He needed 10 bull's-eyes to win. How many more bull's-eyes did Manny need?

$$10 - 6 = 4$$

Manny needed 4 more bull's-eyes.

PROBLEM SOLVING

1. Bobby had 10 action figures. He gave 2 of them away. How many action figures does Bobby have left?

2. Mr. Wu put 5 puppets on a shelf that can hold 14 puppets. How many more puppets can fit on the shelf?

3. Cara had 12 dolls. Three of them were from Russia. How many were from other countries?

4. Mr. Wu sold 8 soft bears and 14 soft rabbits. How many more rabbits did he sell?

Rules for Subtraction

Use these rules to help you subtract quickly and correctly.

When zero is subtracted from a number, the difference is that same number.

$$\begin{array}{r} 4 \\ -0 \\ \hline 4 \end{array} \qquad 4 - 0 = 4$$

When a number is subtracted from itself, the difference is zero.

$$\begin{array}{r} 9 \\ -9 \\ \hline 0 \end{array} \qquad 9 - 9 = 0$$

Subtract.

5. $\begin{array}{r} 7 \\ -0 \\ \hline \end{array}$
6. $\begin{array}{r} 5 \\ -5 \\ \hline \end{array}$
7. $\begin{array}{r} 9 \\ -0 \\ \hline \end{array}$
8. $\begin{array}{r} 6 \\ -6 \\ \hline \end{array}$
9. $\begin{array}{r} 4 \\ -4 \\ \hline \end{array}$
10. $\begin{array}{r} 1 \\ -1 \\ \hline \end{array}$
11. $\begin{array}{r} 3 \\ -0 \\ \hline \end{array}$

12. $\begin{array}{r} 13¢ \\ -\ 6¢ \\ \hline \end{array}$
13. $\begin{array}{r} 8¢ \\ -8¢ \\ \hline \end{array}$
14. $\begin{array}{r} 3¢ \\ -3¢ \\ \hline \end{array}$
15. $\begin{array}{r} 9¢ \\ -9¢ \\ \hline \end{array}$
16. $\begin{array}{r} 10¢ \\ -\ 5¢ \\ \hline \end{array}$
17. $\begin{array}{r} 12¢ \\ -3¢ \\ \hline \end{array}$
18. $\begin{array}{r} 16¢ \\ -\ 9¢ \\ \hline \end{array}$

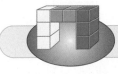

Make Up Your Own

19. Use 15 and 8 and use 13 and 5. Make up two different kinds of subtraction problems for your friends to solve.

Update your skills. See page 5.

2-4   **Addition and Subtraction Sentences** *Algebra* ✓

Nicki is making a 12-square red and blue quilt. She has cut 7 red squares. How many blue squares does she need?

To find how many blue squares, find the missing addend: $7 + \underline{?} = 12$

To find a missing number in a number sentence, think of a related fact.

Think: $12 - 7 = 5$
So $7 + 5 = 12.$

> Remember: You can write $7 + 5 = 12$ as $12 = 7 + 5.$

Nicki needs 5 blue squares.

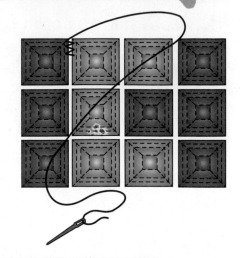

$$\begin{array}{r} \text{Minuend} \\ -\text{Subtrahend} \\ \hline \text{Difference} \end{array}$$

Study these examples.

Find the missing minuend:

$$8 = \underline{?} - 3$$
Think: $8 + 3 = 11$
So $8 = 11 - 3.$

Find the missing subtrahend:

$$15 - \underline{?} = 9$$
Think: $15 - 9 = 6$
So $15 - 6 = 9.$

Find the missing addend.

1. $5 + \underline{?} = 11$ **2.** $9 + \underline{?} = 16$ **3.** $8 + \underline{?} = 15$

4. $\underline{?} + 4 = 10$ **5.** $\underline{?} + 5 = 13$ **6.** $\underline{?} + 2 = 6$

7. $8 = 5 + \underline{?}$ **8.** $12 = 3 + \underline{?}$ **9.** $4 = 1 + \underline{?}$

10. $14 = \underline{?} + 6$ **11.** $7 = \underline{?} + 4$ **12.** $5 = \underline{?} + 0$

Find the missing minuend or subtrahend.

13. $\underline{\ ?\ } - 8 = 2$ **14.** $\underline{\ ?\ } - 6 = 6$ **15.** $\underline{\ ?\ } - 7 = 4$

16. $12 - \underline{\ ?\ } = 5$ **17.** $14 - \underline{\ ?\ } = 8$ **18.** $18 - \underline{\ ?\ } = 9$

19. $\underline{\ ?\ } - 3 = 9$ **20.** $\underline{\ ?\ } - 8 = 5$ **21.** $\underline{\ ?\ } - 9 = 4$

22. $13 - \underline{\ ?\ } = 6$ **23.** $5 - \underline{\ ?\ } = 0$ **24.** $15 - \underline{\ ?\ } = 8$

25. $2 = 9 - \underline{\ ?\ }$ **26.** $7 = \underline{\ ?\ } - 6$ **27.** $3 = \underline{\ ?\ } - 8$

28. $7 = \underline{\ ?\ } - 8$ **29.** $1 = \underline{\ ?\ } - 9$ **30.** $3 = 11 - \underline{\ ?\ }$

Find the missing number.

31.
$$\begin{array}{r} ? \\ +\,3 \\ \hline 6 \end{array}$$
32.
$$\begin{array}{r} 4 \\ +\ ? \\ \hline 12 \end{array}$$
33.
$$\begin{array}{r} ? \\ -\,6 \\ \hline 4 \end{array}$$
34.
$$\begin{array}{r} 13 \\ -\ ? \\ \hline 5 \end{array}$$
35.
$$\begin{array}{r} 8 \\ +\ ? \\ \hline 16 \end{array}$$
36.
$$\begin{array}{r} ? \\ -\,7 \\ \hline 6 \end{array}$$

37.
$$\begin{array}{r} ? \\ +\,7 \\ \hline 15 \end{array}$$
38.
$$\begin{array}{r} ? \\ -\,9 \\ \hline 5 \end{array}$$
39.
$$\begin{array}{r} 4 \\ +\ ? \\ \hline 13 \end{array}$$
40.
$$\begin{array}{r} 12 \\ -\ ? \\ \hline 3 \end{array}$$
41.
$$\begin{array}{r} ? \\ +\,5 \\ \hline 12 \end{array}$$
42.
$$\begin{array}{r} ? \\ -\,8 \\ \hline 9 \end{array}$$

 Connections: Art

Communicate

The art of quilting has been practiced for centuries all over the world. In the United States we are most familiar with patchwork quilts. The tops of these quilts are sewn together from smaller pieces of fabric.

43. Use grid paper and crayons or colored pencils to design a quilt. Make up number sentences like the ones in this lesson about your quilt. Share your design and number sentences with a classmate.

Mental Math

Algebra ✓

Here are some methods to help you add and subtract mentally.

Think of tens or hundreds.

Add: 120 + 30 = __?__

Think: 120 = 12 tens

$$\begin{array}{r} 12 \text{ tens} \\ + \ \ 3 \text{ tens} \\ \hline 15 \text{ tens} = 150 \end{array}$$

Subtract: 6500 − 400 = __?__

Think: 6500 = 65 hundreds

$$\begin{array}{r} 65 \text{ hundreds} \\ - \ \ 4 \text{ hundreds} \\ \hline 61 \text{ hundreds} = 6100 \end{array}$$

Look for patterns.

$$\begin{array}{r} 375 \\ - \ 10 \\ \hline 365 \end{array} \quad \begin{array}{r} 375 \\ - \ 20 \\ \hline 355 \end{array} \quad \begin{array}{r} 375 \\ - \ 30 \\ \hline 345 \end{array} \qquad \begin{array}{r} 500 \\ + \ 26 \\ \hline 526 \end{array} \quad \begin{array}{r} 500 \\ + \ 36 \\ \hline 536 \end{array} \quad \begin{array}{r} 500 \\ + \ 46 \\ \hline 546 \end{array}$$

Look for pairs of numbers that add to 10 or 100.

Add: 5 + 7 + 5 = __?__

$$\begin{array}{r} 5 \\ 7 \\ + \ 5 \\ \hline 17 \end{array}$$

10 + 7 = 17

Add: 26 + 70 + 30 = __?__

$$\begin{array}{r} 26 \\ 70 \\ + \ 30 \\ \hline 126 \end{array}$$

100 + 26 = 126

Add or subtract mentally.

1. 40 + 50   **2.** 60 − 10   **3.** 80 − 80   **4.** 50 + 70

5. 900 − 700   **6.** 300 + 800   **7.** 400 + 700   **8.** 690 − 80

9. 250 + 20   **10.** 160 + 30   **11.** 5700 − 200   **12.** 7400 + 500

Use patterns to add or subtract mentally.

13.	267	267	267	267	267	267
	− 10	− 20	− 30	− 40	− 50	− 60

14.	915	915	915	915	915	915
	+ 20	+ 30	+ 40	+ 50	+ 60	+ 70

15.	300	300	300	400	400	400
	+ 32	+ 42	+ 52	+ 62	+ 72	+ 82

Add mentally.

16.	4	17.	3	18.	5	19.	5	20.	9	21.	6
	6		4		2		5		6		6
	+ 8		+ 7		+ 8		+ 3		+ 1		+ 4

22.	50	23.	76	24.	20	25.	70	26.	20	27.	10
	87		40		53		30		80		97
	+ 50		+ 60		+ 80		+ 62		+ 28		+ 90

 **Skills to Remember**

Round to the nearest ten or ten cents.

28. 35 **29.** 452 **30.** 928 **31.** $.83 **32.** $3.95

Round to the nearest hundred or dollar.

33. 734 **34.** 3946 **35.** $4.75 **36.** $7.28 **37.** $54.62

Round to the nearest thousand or ten dollars.

38. 6789 **39.** 5412 **40.** $62.43 **41.** $17.29 **42.** $25.20

2-6 Estimating Sums and Differences

Rounding is one way to estimate sums and differences.

- Round each number to the greatest place of the smaller number.

4360 → 4400

- Add or subtract the rounded numbers.

Estimate: 4360 + 254 + 1207

Round to hundreds.

$$
\begin{array}{rcl}
4360 & \longrightarrow & 4400 \\
254 & \longrightarrow & 300 \\
+\ 1207 & \longrightarrow & +\ 1200 \\
\hline
 & \text{about} & 5900
\end{array}
$$

Estimate: 6924 − 3123

Round to thousands.

$$
\begin{array}{rcl}
6924 & \longrightarrow & 7000 \\
-\ 3123 & \longrightarrow & -\ 3000 \\
\hline
 & \text{about} & 4000
\end{array}
$$

Study these examples.

Round to dollars.

$$
\begin{array}{rcl}
\$56.39 & \longrightarrow & \$56.00 \\
-\ 4.25 & \longrightarrow & -\ 4.00 \\
\hline
 & \text{about} & \$52.00
\end{array}
$$

Round to ten cents.

$$
\begin{array}{rcl}
\$.27 & \longrightarrow & \$.30 \\
.12 & \longrightarrow & .10 \\
+\ .39 & \longrightarrow & +\ .40 \\
\hline
 & \text{about} & \$.80
\end{array}
$$

Estimate the sum.

1. 53
    + 76

2. $.25
    + .14

3. 632
    + 149

4. $5.25
    + 2.30

5. $37.47
    + 42.58

6. 1432
    4290
    + 134

7. 625
    38
    + 707

8. $1.52
    .18
    + .13

9. $21.07
    14.95
    + 42.78

10. $61.35
    2.75
    + 14.38

11. 42 + 25 + 22

12. 243 + 627 + 139

13. 2163 + 155 + 547

78

Estimate the difference.

14. 54
  − 23

15. $.38
  − .16

16. 932
  − 629

17. $8.57
  − 5.08

18. $42.34
  − 15.75

19. 6152
  − 2830

20. 4819
  − 592

21. $7.29
  − .11

22. $84.88
  − 16.27

23. $29.13
  − 6.58

Using Estimation to Check

Use estimation to check addition or subtraction
to see if your answer is reasonable.

Estimated Sum		Estimated Difference	
3 5 6 $\longrightarrow$	3 6 0	6 7 8 $\longrightarrow$	7 0 0
+ 4 3 $\longrightarrow$	+ 4 0	− 4 5 1 $\longrightarrow$	− 5 0 0
3 9 9	4 0 0	2 2 7	2 0 0

399 is close to 400.
The answer is reasonable.

227 is close to 200.
The answer is reasonable.

**Is the answer reasonable? Estimate to check.
Then write *yes* or *no*.**

24. $34 + 15 = 49$

25. $61 + 30 = 201$

26. $56 - 22 = 34$

27. $43 - 21 = 22$

28. $121 + 405 = 426$

29. $2.61 + $3.28 = 5.89

30. $302 + 517 + 160 = 979$

31. $49.95 - $36.20 = 13.75

32. $9428 - 207 = 9221$

33. $.17 + $.20 + $3.51 = 7.92

34. $859 - 33 = 526$

35. $2462 + 1301 + 5234 = 8997$

36. $88.89 - $6.27 = 82.62

37. $21.46 + $3.98 + $32.54 = 87.98

2-7 Adding and Subtracting Money

Suppose you bought a racquet and a pair of tennis shoes. How much money would you spend in all? How much more would you pay for the racquet than the shoes?

Estimate the sum and the difference:

$50.00 + $40.00 = $90.00
$50.00 − $40.00 = $10.00

To find how much in all,
add: $54.59 + $42.40 = ?

To find how much more,
subtract: $54.59 − $42.40 = ?

$$\begin{array}{r} \$54.59 \\ +\ \ 42.40 \\ \hline \$96.99 \end{array}$$

Adding and subtracting money is like adding and subtracting whole numbers. Just write the $ and . in the answer.

$$\begin{array}{r} \$54.59 \\ -\ \ 42.40 \\ \hline \$12.19 \end{array}$$

You would spend $96.99 in all.

You would pay $12.19 more for the racquet.

$96.99 is close to $90.00.
$12.19 is close to $10.00.
The answers are reasonable.

Study these examples.

$$\begin{array}{r} \$32.50 \\ +\ \ 6.27 \\ \hline \$38.77 \end{array} \qquad \begin{array}{r} \$9.98 \\ -\ \ .41 \\ \hline \$9.57 \end{array} \qquad \begin{array}{r} \$7.24 \\ +\ \ .05 \\ \hline \$7.29 \end{array} \qquad \begin{array}{r} \$.65 \\ -\ \ .62 \\ \hline \$.03 \end{array}$$

This 0 must be written.

Estimate. Then add.

1. $.18
   + .20

2. $.24
   + .34

3. $.50
   + .38

4. $.51
   + .25

5. $7.23
   + 2.55

6. $4.21
   + 1.75

7. $2.22
   + 6.37

8. $17.26
   + 12.73

9. $50.62
   + 24.15

10. $71.40
    + 26.48

11. $9.13
    + .82

12. $8.52
    + .35

13. $43.77
    + 6.02

14. $50.33
    + 8.24

15. $67.91
    + .08

16. $32.13 + $4.75

17. $5.23 + $.06

18. $24.08 + $1.91

Estimate. Then subtract.

19. $.84
    − .62

20. $.66
    − .44

21. $.39
    − .19

22. $8.95
    − 4.51

23. $7.55
    − 2.10

24. $4.67
    − 2.64

25. $3.95
    − 1.85

26. $78.89
    − 74.13

27. $56.39
    − 15.25

28. $99.98
    − 67.50

29. $6.26
    − .24

30. $9.58
    − .46

31. $29.99
    − 8.75

32. $75.83
    − 4.40

33. $86.37
    − .05

34. $49.86 − $7.21

35. $2.86 − $.05

36. $57.92 − $.32

PROBLEM SOLVING

37. Lauren had $15.95. She bought a pedometer for $4.75. How much money did she have left?

38. Ana bought a bike helmet for $32.25 and elbow pads for $15.60. How much did she spend in all?

81

2-8 Checking Addition and Subtraction

Suni did the addition problems at the right. How can she check her answers?

▶ To check addition with more than two addends, add up.

```
  → 1899

    1427 ↑
     352 |
   + 120 |
  → 1899        The answer checks.
```

1427
352
$+ 120$
1899

$\$2.36$
$+ 3.02$
$\$5.38$

▶ To check addition with two addends, subtract one addend from the sum. The answer is the other addend.

```
  $2.36 ←       $5.38
 + 3.02       − 3.02
  $5.38   → $2.36        The answer checks.
```

Joaquim subtracted $2.12 from $8.37. How can he check his answer?

$\$8.37$
$- 2.12$
$\$6.25$

▶ To check subtraction, add the difference and the subtrahend. The answer is the minuend.

```
  $8.37 ←       $6.25
 − 2.12       + 2.12
  $6.25   → $8.37        The answer checks.
```

Add or subtract. Then check the answer.

1. 153 + 516
2. $4.95 − $1.74
3. 762 − 250

Check each answer. If an answer is incorrect, write the correct answer.

4.	5.	6.	7.	8.
2410	$6.22	8050	204	$12.01
132	1.34	27	361	.84
+1225	+ 1.02	+ 810	+ 413	+ 3.12
3757	$8.58	8887	978	$42.97

9.	10.	11.	12.	13.
$5.06	2413	$8.17	$6.22	8251
+ 4.91	+ 5062	+ .62	+ 3.56	+ 543
$9.97	7485	$8.77	$9.78	8894

14.	15.	16.	17.	18.
$.99	$7.95	9388	6975	$38.46
− .36	− 2.62	− 8072	− 733	− 16.25
$.63	$4.33	1116	6248	$22.21

Add or subtract. Then check the answer.

19.	20.	21.	22.	23.
$11.46	252	$42.01	813	6040
3.21	314	10.25	42	122
+ 24.30	+ 321	+ 4.52	+ 33	+ 36

24.	25.	26.	27.	28.
7411	6359	$75.59	9300	$8.88
+ 1505	− 144	− 13.25	+ 458	− 7.37

29. 2301 + 5090   **30.** 7799 − 626   **31.** $24.41 + $11.44

 **Mental Math**   Algebra

Add or subtract mentally. Look for patterns.

32.				
8912	8912	8912	8912	8912
− 701	− 601	− 501	− 401	− 301

33.				
1042	1042	1042	1042	1042
+ 111	+ 212	+ 313	+ 414	+ 515

2-9   Addition and Subtraction Applications

Use the skills and strategies you have learned to solve each problem.

1. All fourth graders in Glenecho must complete a project about either space exploration, dinosaurs, or oceanography. Out of 879 fourth graders, 452 students chose space exploration or oceanography. How many students chose dinosaurs?

2. From its fossil footprints, scientists estimate that the dinosaur *Breviparopus* was as much as 157 feet long. How much longer was *Breviparopus* than the 25-foot-long *Corythosaurus*?

3. There were 7897 fossils in the geology room of a museum. Of these fossils, 735 were chosen for a special exhibit in the museum lobby. How many fossils were left in the geology room?

4. Prehistoric elephants and mammoths had two tusks. The heaviest single fossil tusk ever found weighed 330 pounds. If the other tusk weighed the same, what would have been their combined weight?

5. Marshall had $87.65 in his checking account. If he writes a check for $24.45 to buy a dinosaur model, how much money will be left in his account?

6. The dinosaur *Iguanodon* grew to be 28 feet long. *Diplodocus* was 61 feet longer. How long was *Diplodocus*?

Read each problem carefully before you solve it.

7. The first discovery of dinosaur bones was made in Windsor, Connecticut, in 1818. A complete skeleton of the dinosaur *Herrerasaurus* was found in Argentina in 1989. How many years passed between the two discoveries?

8. The record number of points scored in a dinosaur video game at an arcade is 7695. So far, Wanda has earned 6252 points. How many more points does she need to earn to tie the record?

9. The longest prehistoric snake measured about 444 inches. The modern indigo snake of the United States is about 102 inches long. Is the indigo snake longer or shorter than the longest prehistoric snake? by how many inches?

10. A scientist discovered a cluster of 147 fossilized dinosaur eggs in May. From June to December he found 542 more dinosaur eggs. Did he discover at least 600 dinosaur eggs?

11. At a fossil and mineral show, Trevor bought a fossil shell for $2.30, a fossil field guide for $4.15, and a package of polished stones for $1.42. He said that he spent $7.75. Was Trevor correct? How do you know?

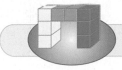 **Make Up Your Own**

12. Use the numbers 3576 and 3321 to make up an addition or subtraction problem. Give it to a classmate to solve.

TECHNOLOGY

READ/DATA Statements

A **computer program** is a set of statements written in a computer language that tells the computer what to do. Line numbers are used in programs to tell the computer the order in which to process the statements.

▶ The program below shows different ways the PRINT statement can be used.

PROGRAM	OUTPUT
10 PRINT "10 + 10 = "	10 + 10 =
20 PRINT 10 + 10	20
30 PRINT "10 + 10 = " 10 + 10	10 + 10 = 20
40 END	

Remember: Use a PRINT statement *with* quotation marks to display exact information.
Use a PRINT statement *without* quotation marks to display an answer.

▶ The program below introduces the REM, READ, and DATA statements.

A remark statement. It explains what the program does. This statement is not processed.

10 REM This program finds the sum of three numbers.

20 READ A, B, C ◀──────── Reads information stored in a DATA statement and assigns it to a letter.

30 PRINT A+B+C

40 GOTO 20

50 DATA 123, 45, 210, 90, 108, 310 ◀── Stores information used by the program.

60 END

Output: 378
     508
    Out of DATA in 20

The program will run until all the information in the DATA statement is used.

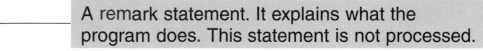

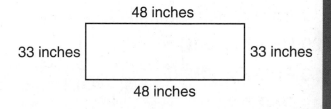

Write the output for each program.

1. 10 READ A,B
20 PRINT "A+B= " A+B
30 GOTO 10
40 DATA 453, 187
50 END

2. 10 READ A,B
20 PRINT "A−B= " A−B
30 GOTO 10
40 DATA 605, 232
50 END

3. 10 READ A,B,C
20 PRINT "A+B+C= " A+B+C
30 GOTO 10
40 DATA 48, 25, 100, 32, 51, 49
50 END

4. 10 READ A,B
20 PRINT "A−B= " A−B
30 GOTO 10
40 DATA 99, 20, 185, 49
50 END

PROBLEM SOLVING

5. Complete the program below so that it
will find the sum of 327 and 194 and of 809 and 36.
What will be the output?

50 DATA _?_ , _?_ , _?_ , _?_
30 PRINT _?_ + _?_
10 _?_ This program finds the sum of two numbers.
60 END
20 _?_ A,B
40 GOTO 20

6. Write a program to find the
distance around the rectangle
at the right. Include REM,
READ, and DATA statements.
What is the distance?

48 inches

33 inches ▢ 33 inches

48 inches

7. Write a program that will
find the sum of all the
even numbers from 0 to 10.
Include REM, READ, and DATA
statements. What is the sum?

8. Write a program that will find
the sum, difference, product,
and quotient of 256 and 8.
Include REM, READ, and DATA
statements. What is the output?

2-11 Problem Solving: Logical Reasoning

Problem: Lee, Hoshi, and Yori have their hats and scarves mixed up. Each boy puts on another boy's cap and a different boy's scarf. Hoshi wears Yori's cap. Whose cap and scarf does each boy wear?

1 IMAGINE Create a mental picture.

2 NAME *Facts:* Hoshi wears Yori's cap. Each wears another boy's cap and a different boy's scarf.

Question: Whose cap and scarf is each boy wearing?

3 THINK Draw and label a table.
Fill in the facts you know.
Consider the possible answers.

	Lee	**Hoshi**	**Yori**
cap	Hoshi's	Yori's	Lee's
scarf	Yori's	Lee's	Hoshi's

4 COMPUTE Hoshi wears Yori's cap, so he must wear Lee's scarf.

Lee didn't wear his own cap, so he must wear Hoshi's cap and Yori's scarf.

That means that Yori wears Lee's cap and Hoshi's scarf.

5 CHECK Are the answers reasonable?
Is each boy wearing another boy's cap and a different boy's scarf? Yes.

Use logical reasoning to solve each problem.

1. Mimi, Pedro, and Martin live in three houses in a row on Mountain Lane. Mimi does not live next to Pedro. Pedro lives on a corner. Who lives in the middle house?

Mountain Lane

IMAGINE   Create a mental picture.

NAME   *Facts:*   Mimi, Pedro, and Martin live on Mountain Lane. Mimi does not live next to Pedro. Pedro lives on a corner.

Question: Who lives in the middle house?

THINK   Pedro cannot live in the middle house.

COMPUTE ──→ **CHECK**

2. What number would you move from one box to another to make the sums in each box equal?

1	2	3
4	5	6
7	8	9

3. Van has six coins that are worth 57¢ in all. Only one coin is a quarter. What are the other coins?

4. Rudy was born in the month whose name has the most letters. The date is an even 2-digit number. The sum of the digits is 5. What is Rudy's birthday?

5. Mary, Anne, and Rose spent $43.51, $47.46, and $50.44. Rose spent the least and did not buy a blazer. Anne's skirt did not cost the most. How much money did each girl spend? Who bought a sweater?

Make Up Your Own

6. Write a problem modeled on problem 3 above. Have a classmate solve it.

2-12 Problem-Solving Applications

Solve each problem and explain the method you used.

1. Deirdre needs 14 yards of white fabric to make costumes for a play. She has 3 yards. How many yards of fabric does she need to buy?

2. Glenn brought home 50 tickets to sell for the school play. He sold 20 tickets. How many does he have left to sell?

3. The theater has 100 seats on the first level and 55 seats in the balcony. How many seats does the theater have?

4. The first act of the play is 69 minutes long. The second act is 54 minutes long. How much longer is the first act?

5. Gini plays the ice queen. She buys a plastic crown for $5.78 and a jar of silver glitter for $1.20. How much does she spend?

6. The director bought 12 boxes of plastic snowflakes and has 7 boxes left. How many boxes has he already used?

7. Bill paints the ice castle door, which is 70 inches tall. The top of the castle is 80 inches higher. How tall is the ice castle?

8. There are 58 penguin puppets in the last scene of the play. Ida has finished making 42 of them. How many does she still have to make?

Choose a strategy from the list or use another strategy you know to solve each problem.

9. The two-act play is 84 pages long. The first act is 43 pages long. How long is the second act?

10. The play was performed on Thursday, Friday, and Saturday. Ben, Sue, and Dana went on different nights. Sue went after Dana. Ben missed the first night, so he went the next night. When did Sue and Dana see the play?

11. The cast received 3 curtain calls on Thursday and double that on Friday. On Saturday there were 2 more than on Thursday. What was the total number of curtain calls?

12. There were 142 people in the audience on Thursday night. Forty of them were adults. How many were children?

13. There are 3 bears and 2 penguins in the animal dance line. In how many different ways can the animals be arranged?

14. Jake, Kyle, and Lou play the jester, the king, and the leopard. No one plays a part that begins with the same letter as his name. Kyle decided not to play the jester. Who plays the king?

USE THESE STRATEGIES:
Logical Reasoning
Choose the Operation
Make a Table or List
Guess and Test
Write a Number Sentence

Use the sign for problems 15 and 16.

15. Mr. Mendez bought tickets for 2 adults and 2 children. How much more than ten dollars did he spend?

16. Ms. Shapiro spent $14.40 on tickets. What tickets did she buy?

Tickets	
Adults	**Children**
$3.20	$2.00

Find the sum. *(See pp. 68–71.)*

1. $8 + 0$ **2.** $9 + 6$ **3.** $7 + 8$ **4.** $6 + 5$

5. $6 + 7 + 4$ **6.** $8 + 0 + 3 + 8$ **7.** $3 + 7 + 8$

Add or subtract. Then check the answer. *(See pp. 80–83.)*

8.	**9.**	**10.**	**11.**	**12.**
153	205	3051	$23.74	$42.04
412	381	$+ 1738$	1.12	3.41
$+ 323$	$+ 413$		$+ \quad .13$	$+ \; 10.22$

13.	**14.**	**15.**	**16.**	**17.**
56	549	798	$94.36	6759
$- 16$	$- 427$	$- 55$	$- 40.13$	$- 542$

Estimate. *(See pp. 78–79.)*

18.	**19.**	**20.**	**21.**	**22.**
42	$5.49	85	568	$8.39
$+ 38$	$+ \quad .23$	$- 23$	$- 399$	$- \quad .21$

Find the missing number. *(See pp. 74–75.)*

23. $3 + \underline{\ ?\ } = 11$ **24.** $\underline{\ ?\ } + 7 = 15$ **25.** $7 = \underline{\ ?\ } + 0$

26. $12 - \underline{\ ?\ } = 7$ **27.** $\underline{\ ?\ } - 9 = 8$ **28.** $9 - \underline{\ ?\ } = 0$

PROBLEM SOLVING *(See pp. 84–85, 90–91.)*

29. The Madison Arts and Crafts Fair had 33 art exhibits and 49 craft exhibits. About how many exhibits were at the Fair?

30. Does $5 + (2 + 3) + 8$ equal $(8 + 3) + (2 + 5)$? Explain.

(See Still More Practice, p. 462.)

THE ABACUS

The ancient Greeks and Romans used an **abacus** to make computations. The abacus is still used today in Asian cultures.

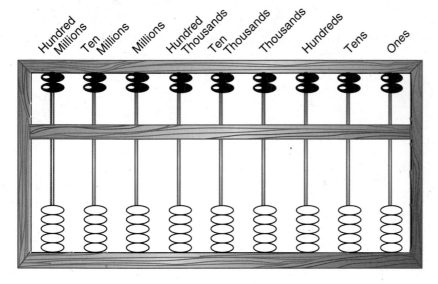

Each black bead stands for 5 units.

Each white bead stands for 1 unit.

A number is shown by moving the appropriate beads to the crossbar.

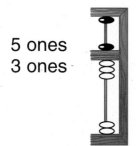

5 ones
3 ones

5 tens
3 tens
4 ones

2 hundreds
5 tens
1 ten
5 ones
2 ones

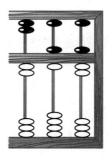

$5 + 3 = 8$ $50 + 30 + 4 = 84$ $200 + 50 + 10 + 5 + 2 = 267$

Make your own abacus. Use buttons, beads, or counters.

Show each number on your abacus.

1. 39   **2.** 326   **3.** 26   **4.** 681   **5.** 78   **6.** 589

Performance Assessment

Write the missing output. Then write a rule for each table.

1.
Input	25	40	80
Output	47	62	?

2.
Input	$1.25	$2.10	$3.30
Output	$1.79	$2.64	?

Add.

3. $5 + 9$

4. $0 + 7$

5. $9 + 8$

6. $7 + 1 + 7 + 3$

Find the sum or difference. Then check.

7.
```
  172
  205
+  22
```

8.
```
 1583
  112
+ 204
```

9.
```
 $42.63
   5.12
+   .14
```

10.
```
 $32.71
+ 46.28
```

11.
```
  846
- 230
```

12.
```
  497
-  43
```

13.
```
 $24.98
-  3.05
```

14.
```
 $58.49
- 22.41
```

Find the missing number.

15. $5 + \underline{\ ?\ } = 13$

16. $\underline{\ ?\ } - 4 = 6$

17. $8 = \underline{\ ?\ } + 8$

PROBLEM SOLVING *Use a strategy you have learned.*

18. Maria went to the grocery store. She bought items that cost $3.09, $1.39, $.20, and $4.15. Did Maria pay more than $10 for the items? Estimate to find the answer.

19. At the City Zoo there are more zebras than lions and more monkeys than zebras. Are there more monkeys or lions?

Addition and Subtraction

3

A LOT OF KIDS

There are a lot of kids
Living in my apartment building
And a lot of apartment buildings on my street
And a lot of streets in this city
And cities in this country
And a lot of countries in the world.
So I wonder if somewhere there's a kid I've never met
Living in some building on some street
In some city and country I'll never know—
And I wonder if that kid and I might be best friends
If we ever met.

Jeff Moss

In this chapter you will:

Learn about front-end estimation
Add and subtract larger numbers
with regrouping
Add three or more addends
Solve problems that have
extra information

Critical Thinking/Finding Together
Mary visited a friend. She drove 126
miles from New York to New Jersey
and 140 miles from New Jersey to
Pennsylvania. If she traveled a total
distance of 425 miles, how far is it
from Pennsylvania to New York?

95

3-1 Front-End Estimation

Students in the Hilldale elementary schools held a Read-a-Thon in October. About how many books did they read altogether?

School	Books Read
Central	2534
North	2496
South	3875

To find about how many, estimate:
2534 + 2496 + 3875

▶ **Add the front digits.**   **Write 0s for the other digits.**

```
   2534          2534
   2496          2496
 + 3875        + 3875
 ───────       ───────
    7          about 7000
```

Rough estimate: 7000

▶ To get a closer estimate, make groups of about 1000 from the other digits.

2534 + 2496 + 3875

about 1000   about 1000

Think: 7000 + 1000 + 1000 = 9000

Adjusted estimate: 9000

Altogether, the students read about 9000 books.

Study these examples.

```
  $5.26
   1.52  } about $1
   3.78  } about $1
 + 2.45
 ─────────
 about $11.00
```

$11 + $1 + $1 = $13
Rough estimate: $11
Adjusted estimate: $13

```
   738
   223  } about 100
 + 569
 ──────
 about 1400
```

1400 + 100 = 1500
Rough estimate: 1400
Adjusted estimate: 1500

Make a rough estimate. Then adjust.

1. 212 672 +827	**2.** 358 143 +796	**3.** 588 419 +622	**4.** $3.47 1.30 + 9.65	**5.** $6.98 4.25 + 6.10
6. 3235 5871 +1886	**7.** 9139 2584 +4475	**8.** 5405 1679 +2961	**9.** $67.99 73.46 + 36.49	**10.** $78.65 18.98 + 21.49

11. 635 + 198 + 474 + 360

12. $5.32 + $7.12 + $3.69 + $1.95

13. 283 + 722 + 542 + 156

14. $6.58 + $1.40 + $2.56 + $4.61

PROBLEM SOLVING Use the table on page 96.

15. Hilldale Middle School students read 4073 books.
About how many books did all the students read?

Estimating Differences

To estimate differences using front-end estimation:

- Subtract the front digits.
- Write 0s for the other digits.

$73.45 − 26.50 about $50.00	5736 −1775 about 4000	$8.21 − 7.35 about $1.00	963 −315 about 600

Estimate the difference. Use front-end estimation.

16. 646 −519	**17.** 441 −193	**18.** 938 −256	**19.** $8.98 − 3.50	**20.** $2.56 − 1.48
21. 7149 −3861	**22.** 5460 −1509	**23.** 8432 −5954	**24.** $49.90 − 24.95	**25.** $37.21 − 18.88

3-2 Adding with Regrouping

What was the total membership of the
U.S. House of Representatives in 1987?

To find the total,
add: 258 + 177 = __?__

| 1987 U.S. House of Representatives ||
Democrats	Republicans
258	177

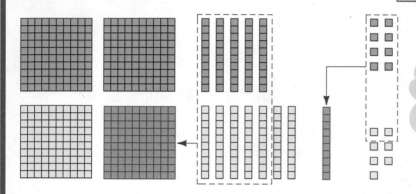

Remember:
10 ones = 1 ten
10 tens = 1 hundred

Add the ones. Regroup.

```
  h t o
      1
  2 5 8
+ 1 7 7
      5
```

15 ones =
1 ten 5 ones

Add the tens. Regroup.

```
  h t o
  1 1
  2 5 8
+ 1 7 7
    3 5
```

13 tens =
1 hundred 3 tens

Add the hundreds.

```
  h t o
  1 1
  2 5 8
+ 1 7 7
  4 3 5
```

The total membership was 435 representatives.

Study these examples.

```
    1
  $.4 7
+  .2 9
  $.7 6
```

```
    1
  5 2 6
+ 2 3 7
  7 6 3
```

```
    1
  $1.7 3
+ 6.5 4
  $8.2 7
```

```
    1
  6 5
+    7
  7 2
```

Estimate mentally. Then add.

1. 48
    + 46

2. 37
    + 16

3. 58
    + 22

4. 85
    + 8

5. 73
    + 9

6. 329
    + 543

7. 480
    + 253

8. 675
    + 162

9. 781
    + 47

10. 909
    + 64

11. 168
    + 743

12. 643
    + 259

13. 345
    + 469

14. 877
    + 95

15. 768
    + 99

16. $.78
    + .06

17. $.46
    + .28

18. $1.75
    + 3.61

19. $5.28
    + 2.49

20. $4.65
    + 4.99

21. 75 + 18

22. 389 + 276

23. 581 + 229

24. $.19 + $.66

25. $6.19 + $2.32

26. $3.97 + $4.33

PROBLEM SOLVING

27. There were 156 Democrats and 137 Republicans in the U.S. House of Representatives in 1878. How many members of the House were there?

28. In 1925 the U.S. Congress was made up of 435 Representatives and 96 Senators. How many members of Congress were there in 1925?

 **Challenge**

29. Find two 3-digit addends with the same digits in each number whose sum is 404.

30. What are the largest and the smallest possible addends of two 3-digit numbers whose sum is 555? 999?

99

3-3 Four-Digit Addition

How many votes were cast in the School Board election?

To find how many votes, add: 1279 + 2355 = ?

School Board Election Results	
Green	Myers
1279	2355

Add the ones. Regroup.

th	h	t	o
		1	
1	2	7	9
+ 2	3	5	5
			4

14 ones = 1 ten 4 ones

Add the tens. Regroup.

th	h	t	o
	1	1	
1	2	7	9
+ 2	3	5	5
		3	4

13 tens = 1 hundred 3 tens

Add the hundreds.

th	h	t	o
	1	1	
1	2	7	9
+ 2	3	5	5
	6	3	4

Add the thousands.

th	h	t	o
	1	1	
1	2	7	9
+ 2	3	5	5
3	6	3	4

In the School Board election, 3634 votes were cast.

Study these examples.

```
  1 1
  4 7 8 0
+ 2 9 5 6
  7 7 3 6
```

```
    1     1
$6 4.3 8
+  1 7.2 5
$8 1.6 3
```

```
      1 1
$5 0.6 7
+    5.4 5
$5 6.1 2
```

Estimate. Then add.

1. 3165
    + 2917

2. 4227
    + 1905

3. 2774
    + 6407

4. 5538
    + 614

5. 4168
    + 3454

6. 6075
    + 2845

7. 8264
    + 1349

8. 9438
    + 395

Find the sum.

9. 3670 + 3458	**10.** 5891 + 2768	**11.** 6655 + 1563	**12.** 8492 + 945
13. 5329 + 1398	**14.** 4921 + 3486	**15.** 6482 + 1843	**16.** 7560 + 488
17. $34.27 + 46.17	**18.** $65.05 + 13.98	**19.** $87.98 + 10.75	**20.** $51.75 + 9.15

Align and add.

21. 6414 + 979

22. 495 + 1272

23. 8067 + 86

24. $28.95 + $56.60

25. $69.75 + $8.94

26. $4.35 + $24.89

27. $6.08 + $44.56

PROBLEM SOLVING

28. Ms. Davis and Mr. Brown ran for mayor of Newton. Ms. Davis received 2365 votes and Mr. Brown received 4915 votes. How many people voted in the election?

29. A campaign worker spent $23.96 on phone calls and $57.32 for posters. How much did she spend?

30. Three people ran for town manager. Mr. Miller received 4286 votes. Mr. Rush received 3907 votes. Ms. Adams received 7454 votes. Did Mr. Miller and Mr. Rush together receive more or fewer votes than Ms. Adams?

31. Mr. Jones received 2487 votes for sheriff. Mr. Long received double that number. How many votes did Mr. Long receive?

3-4 Three- and Four-Digit Addition

The Botanical Gardens held a two-day open house. What was the total attendance at the open house?

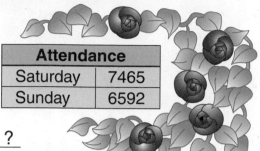

Attendance	
Saturday	7465
Sunday	6592

To find the total, add: 7465 + 6592 = __?__

Add the ones.	Add the tens. Regroup.	Add the hundreds. Regroup.
| ```
  7 4 6 5
+ 6 5 9 2
────────
        7
``` | ```
    1
  7 4 6 5
+ 6 5 9 2
────────
      5 7
``` | ```
  1 1
  7 4 6 5
+ 6 5 9 2
────────
    0 5 7
``` |
| | 15 tens = 1 hundred 5 tens | 10 hundreds = 1 thousand 0 hundreds |

Add the thousands.

```
    1 1
    7 4 6 5
+     6 5 9 2
──────────
  1 4,0 5 7
```

The total attendance was 14,057.

Add.

1.
```
    1 1
  $  8.79
+     4.46
────────
  $13.25
```

2.
```
   $6.39
+  6.21
──────
```

3.
```
   $41.75
+  54.50
───────
```

4.
```
   $65.49
+  82.90
───────
```
```

Estimate mentally. Then find the sum.

5.	951 + 735	6.	694 + 508	7.	873 + 456	8.	298 + 769
9.	3742 + 8616	10.	5390 + 6475	11.	2615 + 9218	12.	3099 + 1908
13.	$6.75 + 4.37	14.	$5.92 + 7.35	15.	$2.36 + 8.84	16.	$9.06 + 4.85
17.	$47.29 + 72.36	18.	$20.82 + 91.33	19.	$38.91 + 64.07	20.	$57.35 + 52.72

Align and add.

21. 742 + 381 **22.** 397 + 721 **23.** 876 + 950

24. 6344 + 5812 **25.** 1423 + 9182 **26.** 7227 + 5090

27. $2.89 + $7.56 **28.** $8.25 + $6.96 **29.** $98.99 + $9.62

PROBLEM SOLVING

30. Visitors to the Botanical Gardens bought 8429 flowering plants and 4872 vegetable plants. How many plants did they buy?

 **Skills to Remember**

Add.

31.	32.	33.	34.	35.	36.
4 6 7 + 2	3 9 3 + 4	8 2 4 + 4	9 0 1 + 6	5 4 3 + 4	9 7 4 + 8

103

Algebra

Ms. Pei drove from Chicago to Kansas City. Then she drove to Indianapolis and Pittsburgh before returning to Chicago. How many miles did she travel?

First estimate:

499 + 485 + 353 + 452

400 + 400 + 300 + 400 + 100 + 100 + 100 = 1800

To find how many miles, add: 499 + 485 + 353 + 452 = __?__

Add the ones. Regroup.	Add the tens. Regroup.	Add the hundreds.

    1
  4 9 9
  4 8 5
  3 5 3
+ 4 5 2
     9
	2 1
	  4 9 9
	  4 8 5
	  3 5 3
	+ 4 5 2
	   8 9
		2 1
		  4 9 9
		  4 8 5
		  3 5 3
		+ 4 5 2
		1 7 8 9

19 ones = 1 ten 9 ones	28 tens = 2 hundreds 8 tens	17 hundreds = 1 thousand 7 hundreds

Ms. Pei traveled 1789 miles.

Think: 1789 is close to 1800. The answer is reasonable.

Study these examples.

```
  1 2
$1.2 7
  .3 2
  .0 5
+  .9 7
$2.6 1
```

```
 1 1 1
 4 1 6 1
 3 7 2 8
+  7 5 6 2
1 5,4 5 1
```

```
  1 2
6 5 7
    8
  3 4
+ 1 2 4
8 2 3
```

Estimate mentally. Then find the sum.

1.	27 34 61 + 58	**2.**	15 70 46 + 22	**3.**	93 9 56 + 82	**4.**	$.33 .12 .68 + .71	**5.**	$.84 .07 .55 + .06
6.	247 191 322 + 423	**7.**	539 293 612 + 109	**8.**	316 875 26 + 6	**9.**	$7.32 5.17 1.97 + 3.28	**10.**	$5.51 .99 6.37 + .03

11.	3219 8604 + 6154	**12.**	9002 2756 + 4321	**13.**	5806 275 + 1888	**14.**	$41.55 97.60 + 3.28

Align and add.

15. 63 + 147 + 735 + 8

16. 2905 + 324 + 55

17. $.51 + $2.76 + $4.29 + $.77

18. $4.26 + $22.79 + $56.07

PROBLEM SOLVING

19. One week, Mr. Mills made business trips of 16 miles, 29 miles, 9 miles, and 42 miles. How many miles did he travel?

20. Ms. Sims spent $13.48, $19.76, and $9.88 on gasoline last month. How much money did she spend on gasoline?

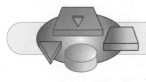

Critical Thinking

Three of the four addends have a sum of 1000.
Write the addend that does *not* belong.

21. 421, 391, 198, 381

22. 510, 237, 253, 233

23. 173, 125, 225, 602

24. 345, 352, 303, 355

3-6 Subtracting with Regrouping

How much taller is the
First Interstate Tower than
the Gas Company Tower?

First estimate:
900 − 700 = 200

To find how much taller,
subtract: 858 − 749 = ?

Height of Tall Buildings in Los Angeles, California	
The Liberty Tower	1018 feet
First Interstate Tower	858 feet
Gas Company Tower	749 feet
Arco Tower	743 feet
Wells Fargo Tower	740 feet
Sanwa Bank Plaza	717 feet

See if there are enough ones to subtract.

```
 h t o
 8 5 8
-7 4 9
```

More ones are needed. Regroup the tens to get more ones.

58 = 5 tens 8 ones
= 4 tens 18 ones

Subtract the ones.

```
 h  t  o
    4 18
 8  5  8
-7  4  9
       9
```

Subtract the tens.

```
 h  t  o
    4 18
 8  5  8
-7  4  9
    0  9
```

Subtract the hundreds.

```
 h  t  o
    4 18
 8  5  8
-7  4  9
 1  0  9
```

Check by adding.

```
   1
  1 0 9
 +7 4 9
  8 5 8
```

The First Interstate Tower is
109 feet taller.

Think: 109 is close to 200.
The answer is reasonable.

Study these examples.

```
 6 14
 7 4
-4 6
 2 8
```

```
  7 13
$8.3 8
-7.9 5
$ .4 3
```

```
 5 11
 6 1 2
-  9 1
 5 2 1
```

106

Estimate mentally. Then find the difference.

1.	2.	3.	4.	5.
82 − 17	60 − 34	72 − 25	$.94 − .58	$.43 − .29

6.	7.	8.	9.	10.
572 − 143	720 − 418	886 − 249	$3.61 − 2.25	$6.84 − 4.19

11.	12.	13.	14.	15.
927 − 692	435 − 172	228 − 147	$5.43 − 2.83	$9.69 − 5.90

16.	17.	18.	19.	20.
23 − 9	132 − 28	429 − 75	$.52 − .06	$2.75 − .08

Align and subtract.

21. 75 − 9   **22.** 32 − 8   **23.** 480 − 36

24. $6.21 − $.16   **25.** $8.19 − $.54   **26.** $5.33 − $.07

PROBLEM SOLVING
Use the table on page 106.

27. How much shorter is Sanwa Bank Plaza than the Arco Tower?

28. How much taller is Wells Fargo Tower than the Sanwa Bank Plaza?

29. The First Interstate Tower has 62 stories. 777 Tower has 55 stories. How many more stories does the First Interstate Tower have?

30. The 777 Tower in Los Angeles is 725 feet tall. Is it taller or shorter than the Wells Fargo Tower? by how much?

Subtraction: Regrouping Twice

How many more home runs did Babe Ruth hit than Frank Robinson?

To find how many more, subtract: 714 − 586 = __?__

All-Time Home Run Leaders			
H. Aaron	755	H. Killebrew	573
B. Ruth	714	R. Jackson	563
W. Mays	660	M. Schmidt	548
F. Robinson	586	M. Mantle	536

More ones are needed. Regroup. Subtract ones.	More tens are needed. Regroup. Subtract tens.	Subtract hundreds.

```
    0  14              10                    10
  7 7̸ 4̸            6  0̸  14           6  0̸  14
                    7̸ 7̸ 4̸             7̸ 7̸ 4̸
- 5 8 6            - 5 8 6             - 5 8 6
      8                2 8             1 2 8
```

1 ten 4 ones = 0 tens 14 ones

7 hundreds 0 tens = 6 hundreds 10 tens

Babe Ruth hit 128 more home runs than Frank Robinson.

Check.

```
   1 1
   1 2 8
 + 5 8 6
   7 1 4
```

Study these examples.

```
      12                  11
   7  2̸  11           3  1̸  16
  $8̸.3̸ 1̸            4̸ 2̸ 6̸
 - 7.8 4                8 0
  $ .4 7              3 3 7
```

Estimate. Then subtract.

1.	624 − 137	**2.**	930 − 452	**3.**	846 − 669	**4.**	561 − 265	**5.**	734 − 587

Find the difference.

6.	7.	8.	9.	10.
452 − 378	360 − 185	922 − 734	712 − 499	653 − 578

11.	12.	13.	14.	15.
835 − 79	561 − 94	454 − 65	946 − 58	137 − 48

16.	17.	18.	19.
$3.25 − 1.58	$5.37 − 2.49	$8.64 − 4.87	$9.52 − .99

Align and subtract.

20. 456 − 179   **21.** 837 − 488   **22.** 671 − 95

23. $9.36 − $7.59   **24.** $2.91 − $1.97   **25.** $5.42 − $.67

PROBLEM SOLVING Use the table on page 108.

26. How many more home runs did Henry Aaron hit than Frank Robinson?

27. How many fewer home runs did Harmon Killebrew hit than Willie Mays?

28. Did Babe Ruth and Reggie Jackson combined hit more or fewer home runs than Henry Aaron and Mickey Mantle combined? How many more or fewer?

29. Which is the greater difference: between the number of home runs hit by Willie Mays and Mike Schmidt or between the number of home runs hit by Henry Aaron and Willie Mays?

30. Babe Ruth, Mickey Mantle, and Reggie Jackson were home run leaders. What was their combined home run total?

Three- and Four-Digit Subtraction

The Mississippi is the longest river in the United States. How much longer than the Colorado River is it?

U.S. Rivers Length in Miles			
Colorado	862	Mississippi	2348
Porcupine	569	Missouri	2315
Rio Grande	1760	Tennessee	652

To find how much longer, subtract: 2348 − 862

Subtract the ones.	More tens are needed. Regroup. Subtract.	More hundreds are needed. Regroup. Subtract.

```
                      2 14                    12
                                            1 ⁄2 14
    2 3 4 8          2 ⁄3 ⁄4 8              ⁄2 ⁄3 ⁄4 8
  −   8 6 2        −     8 6 2            −     8 6 2
          6                8 6                  4 8 6
```

3 hundreds 4 tens = 2 hundreds 14 tens	2 thousands 2 hundreds = 1 thousand 12 hundreds

Subtract the thousands.	Check.

```
        12
      1 ⁄2 14
    ⁄2 ⁄3 ⁄4 8          1 1
  −     8 6 2         1 4 8 6
    1 4 8 6        +     8 6 2
                     2 3 4 8
```

The Mississippi is 1486 miles longer than the Colorado River.

Study these examples.

```
      13              14              12
    5 ⁄3 10         5 ⁄4 11         0 ⁄2 15         5 12 8 13
  7 ⁄6 4 ⁄0         ⁄6 ⁄5 ⁄1       $⁄1 .⁄3 ⁄5      $⁄6 ⁄2.⁄9 ⁄3
− 4 1 9 5         − 3 7 5         −   .8 9        − 3 9.0 5
  3 4 4 5           2 7 6         $  .4 6         $2 3.8 8
```

Estimate mentally. Then find the difference.

1. 521
   − 347

2. 825
   − 169

3. 612
   − 248

4. $2.79
   − 1.89

5. $9.91
   − 1.97

6. 6218
   − 5354

7. 9743
   − 4467

8. $74.36
   − 46.72

9. $84.23
   − 23.47

10. 634
    − 58

11. 1274
    − 990

12. $4.22
    − 3.48

13. $63.35
    − 8.16

Align and subtract.

14. 142 − 69

15. 360 − 74

16. $4.21 − $1.38

17. 7218 − 533

18. 6182 − 2804

19. $55.47 − $8.95

PROBLEM SOLVING Use the table on page 110.

20. Is the difference in length of the Missouri and Rio Grande rivers greater or less than the length of the Porcupine River?

21. How much shorter is the Tennessee River than the Missouri River?

 **Choose a Computation Method**

Compute. Use mental math or paper and pencil.
In your Math Journal, write when you would use each method.

Math Journal

22. 650
    200
    + 15

23. 8264
    − 736

24. 467
    + 728

25. 939
    − 400

26. 53
    + 42

27. 15 + 22 + 50 + 11

28. 2007 + 1031 + 6749

3-9 Zeros in Subtraction

When there are zeros in the minuend, you may need to regroup *more than once* before you subtract.

Subtract: 300 − 158 = ?

You can use base ten blocks to help you subtract.

Remember:
300 = 3 hundreds =
2 hundreds 10 tens

2 hundreds 10 tens =
2 hundreds 9 tens 10 ones

Hands-On Understanding

You Will Need: base ten blocks, paper, pencil

Step 1 Model 300.

$$\begin{array}{r} 3\,0\,0 \\ -\ 1\,5\,8 \\ \hline \end{array}$$

Can you take away 1 hundred 5 tens 8 ones?
Why or why not?

Step 2 Regroup 1 hundred to get more tens.

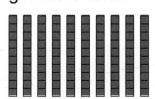

$$\begin{array}{r} {\scriptstyle 2\ 10} \\ \cancel{3}\,\cancel{0}\,0 \\ -\ 1\,5\,8 \\ \hline \end{array}$$

Step 3 Regroup 1 ten to get more ones.

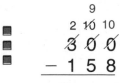

$$\begin{array}{r} {\scriptstyle 9} \\ {\scriptstyle 2\ 10\ 10} \\ \cancel{3}\,\cancel{0}\,\cancel{0} \\ -\ 1\,5\,8 \\ \hline \end{array}$$

Now can you take away 1 hundred 5 tens 8 ones?
Explain.

Step 4   First, take away 8 ones. Next, take away 5 tens.
Third, take away 1 hundred.

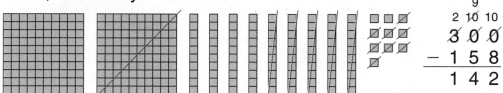

$$\begin{array}{r} 9 \\ 2\ \cancel{10}\ 10 \\ \cancel{3}\ \cancel{0}\ \cancel{0} \\ -\ 1\ 5\ 8 \\ \hline 1\ 4\ 2 \end{array}$$

What is the difference when you subtract
158 from 300?

You can also use paper and pencil to subtract.

Find the difference: $8.00 − $2.46 = __?__

Regroup to get more dimes.	Regroup to get more pennies.	Subtract.	Check.
7 10 $\cancel{\$8}.\cancel{0}\ 0$ − 2.4 6	9 7 1̶0̶ 10 $\cancel{\$8}.\cancel{0}\ \cancel{0}$ − 2.4 6	9 7 1̶0̶ 10 $\cancel{\$8}.\cancel{0}\ \cancel{0}$ − 2.4 6 $5.5 4	1 1 $5.5 4 + 2.4 6 $8.0 0

8 dollars 0 dimes = 7 dollars 10 dimes	10 dimes 0 pennies = 9 dimes 10 pennies

Subtract. You may use models.

1. 500 − 329   **2.** 806 − 447   **3.** $9.00 − $5.41   **4.** $1.05 − $.88

5. 9002 − 7865   **6.** 5000 − 718   **7.** $40.00 − $16.95

Communicate   Discuss ✓

8.   What do you think is hardest about subtracting
from a number that has zeros in the minuend?
How can you make this subtraction easier?

Larger Sums and Differences

To add or subtract larger numbers:

- Start by adding or subtracting at the right.
- Regroup as necessary.

Add: $567.86 + $341.95 = ?   Subtract: 87,731 − 65,954 = ?

```
        1    1 1
     $5 6 7.8 6
   +  3 4 1.9 5
     $9 0 9.8 1
```

```
              16 12
          6  6  2  11
       8  7, 7  3  1
    −  6  5, 9  5  4
       2  1, 7  7  7
```

Add: 36,428 + 83,985 + 759 = ?   Subtract: $490.00 − $478.81 = ?

```
      1 2 1 2
     3 6,4 2 8
     8 3,9 8 5
   +       7 5 9
   1 2 1,1 7 2
```

```
                9  9
            8  10 10 10
      $4  9  0. 0  0
    −    4  7  8. 8  1
      $     1 1. 1 9
```

Add or subtract. Watch for + or − .

1. 42,937 + 11,426	**2.** 32,864 + 94,828	**3.** 85,963 + 28,279	**4.** $562.43 + 680.79
5. 94,361 − 22,087	**6.** 75,937 − 12,649	**7.** 82,616 − 51,499	**8.** $262.71 − 140.99
9. 13,584 41,592 + 26,437	**10.** 64,205 39,811 + 52,406	**11.** 82,099 4,157 + 79,862	**12.** $902.67 51.81 + 235.27

114

Find the sum or the difference.

| 13. | 53,007
− 21,979 | 14. | 70,064
− 19,155 | 15. | 80,102
− 9,516 | 16. | $600.08
−   59.99 |

| 17. | 98,694
287
+ 5,148 | 18. | 675
44,526
+   67 | 19. | 75,628
8,073
+   48 | 20. | $ 4.97
826.13
+   65.39 |

| 21. | 81,000
− 19,625 | 22. | 94,000
− 67,887 | 23. | 70,000
− 36,678 | 24. | $600.00
−   47.89 |

Align and add or subtract.

25. 21,863 + 2,684 + 1,326

26. 82,010 + 395 + 13,692

27. 65,600 − 1,592

28. $200.00 − $126.74

29. 90,506 − 3,729

30. $645.16 + $8.88 + $.56

PROBLEM SOLVING

31. Lambeau Field in Green Bay, WI, could seat 56,189 people in 1990. Seating for 4601 people was added later. How many people can Lambeau Field seat today?

32. Rich Stadium in Buffalo, NY, can seat 80,024 people. The Superdome in New Orleans, LA, can seat 64,992. How many more people can be seated in Rich Stadium?

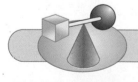 **Challenge** Algebra

Find the missing digits.

| 33. | 923
− 14▢
776 | 34. | 629
− ▢8▢
441 | 35. | 231
− ▢▢▢
85 | 36. | 856
− 4▢8
▢6▢ |

115

3-11 Problem Solving: Extra Information

Problem: The Keep Fit Shop ordered 487 pairs of high-tops. The factory has 1000 pairs in stock. The prices range from $30 to $85. How many pairs of high-tops will the factory have after they fill the order?

1 IMAGINE Create a mental picture.

2 NAME *Facts:* 487 pairs ordered
1000 pairs in stock at the factory
Pairs cost $30 to $85.

Question: How many pairs will be left after the order is filled?

3 THINK What information do you need?
- the number of pairs ordered
- the number of pairs in stock

What information is unnecessary?
- the price range

4 COMPUTE Estimate the difference.
$1000 - 500 = 500$
Then subtract.

```
  0 9 9 10
  1̸ 0̸ 0̸ 0̸
 -  4 8 7
 ─────────
    5 1 3
```

The factory will have 513 pairs of high-tops left.

5 CHECK The answer is close to the estimate. It is reasonable.

Add to check subtraction. $513 + 487 = 1000$

Look for extra information before you solve each problem.

1. Rock climbing shoes cost $82.55.
   Running shoes are on sale for $62.79.
   The regular price is $8.55 more.
   What is the regular price for the running shoes?

IMAGINE Create a mental picture.

NAME *Facts:* $82.55 rock climbing shoes
   $62.79 running shoes on sale
   Regular price is $8.55 more.

   Question: What is the regular price for
   running shoes?

THINK What information do you need?
   • sale price—$62.79
   • regular price—$8.55 more

   What information is unnecessary?
   • the price of rock climbing shoes

COMPUTE ⟶ **CHECK**

2. The Keep Fit Catalog sells 376 clothing items, 29 books, and
   107 trail maps. There are 6 order clerks and 2 managers.
   How many different items does the catalog sell?

3. Shipping costs $3 for orders under $10 and $5.50 for orders over
   $10. Delivery takes 6 to 9 days. What is the total cost of a
   $14.98 order?

4. The company sent 4850 catalogs in April and 5782 catalogs
   in May. They received 853 orders in April and 118 more than
   that in May. How many orders did they receive in May?

5. The Keep Fit Shop has sponsored a charity bike race for 15 years.
   It is 35 miles long. There are rest stops every 5 miles, including at
   the finish line. How many rest stops are there?

Problem-Solving Applications

Solve each problem and explain the method you used.

1. Jan and Kelly built a giant chain of 1378 plastic dominoes and 2267 wood dominoes. How many dominoes did they use?

2. The chain was 300 feet long. The first 127 feet were plastic dominoes. How many feet of chain were wood dominoes?

3. They set up the chain on the gym floor, which is 10,000 square feet in area. The chain took up 6341 square feet. How much of the gym floor was not covered?

4. Jan and Kelly spent 192 minutes on Friday setting up the dominoes. They worked for 218 minutes on Saturday. How long did it take them to set up the chain?

5. Their project raised $1070. They paid $318 for the dominoes. They gave the rest to charity. How much money did Jan and Kelly donate?

6. Use the graph at the right. How many people were in the audience that saw the domino chain?

7. The plastic dominoes fell in 109 seconds. Then the wood dominoes fell in 189 seconds. How long did it take the entire chain to fall down?

8. Jan and Kelly are planning next year's chain. They will use 2567 plastic dominoes and 3271 wood dominoes. How many dominoes will they use?

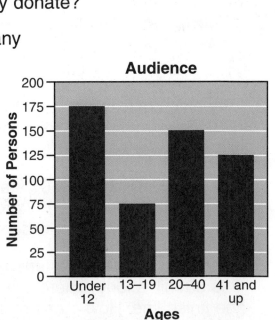

Choose a strategy from the list or use another strategy to solve each problem.

9. A class held a jump rope contest for charity. The winner jumped 9278 times without missing. The second prize went to someone who jumped 8765 times. How many more times did the winner jump?

10. There were 108 people in the contest. They each paid $2 to enter. The winner won $25. Only 27 jumpers made it to the second round. How many jumpers were eliminated after round one?

11. Paula hopped on her right foot 876 times and then on her left foot 954 times. Then she switched back to her right foot and hopped 212 times before tripping. How many times did she hop in all?

12. Paul, Maria, Gail, and Leroy play Double Dutch. Two people hold ropes and 2 jump. How many different ways could the friends play?

13. Marcia jumped for 47 minutes. How many minutes less than an hour did Marcia jump?

14. Asa, Max, and Jemma came in first, second, and third in the jump rope contest. Max did not win, but he jumped more times than Asa. Who came in first, second, and third?

15. Of the 108 contestants, the number of girls was double the number of boys. How many girls were there? how many boys?

 **Make Up Your Own**

16. Write a problem modeled on problem 12. Have a classmate solve it.

Chapter Review and Practice

Estimate. Use front-end estimation. *(See pp. 96–97.)*

1.	382 +216	**2.**	648 +175	**3.**	7060 −2955	**4.**	$63.49 − 19.79

Add. *(See pp. 98–105, 114–115.)*

5. 392 + 26   **6.** 276 + 477   **7.** 4234 + 477

8.	258 314 +126	**9.**	527 198 + 56	**10.**	$32.38 4.43 + 20.37	**11.**	12,476 9,830 +31,579

Subtract. *(See pp. 106–115.)*

12. 982 − 54   **13.** 2816 − 129   **14.** 17,150 − 3594

15.	6000 −1406	**16.**	2603 − 186	**17.**	$54.93 − 16.17	**18.**	15,168 − 7,619

PROBLEM SOLVING

(See pp. 118–119.)

19. Memorial School has 630 students. If 437 students are girls, how many are boys?

20. The new stadium has 60,000 seats. The old stadium had 45,500 seats. How many more seats does the new stadium have?

21. There are 127 roses, 416 daisies, and 216 lilies in the flower shop. How many flowers are in the shop?

(See Still More Practice, p. 463.)

Algebra

VARIABLES

▶ You can let symbols stand for numbers. Then you can decide whether a number sentence is true or false.

Let ■ = 25.

true or false? 5 + ■ = 30

> Think: 5 + 25 = 30?
> **Yes**.

So 5 + ■ = 30 is true
when ■ = 25.

Let ● = 16.

true or false? 20 − ● = 10

> Think: 20 − 16 = 10?
> **No**. 20 − 16 = 4

So 20 − ● = 10 is false
when ● = 16.

▶ You can let letters stand for numbers.

Decide whether each number sentence is true or false.

Let n = 21.

true or false? n − 8 = 15

> Think: 21 − 8 = 15?
> **No**. 21 − 8 = 13

So n − 8 = 15 is false
when n = 21.

Let a = 10.

true or false? a + 85 = 95

> Think: 10 + 85 = 95?
> **Yes**.

So a + 85 = 95 is true
when a = 10.

Write *true* or *false* for each number sentence.

1. Let ■ = 9.
   a. 6 + ■ = 18
   b. 15 − ■ = 6
   c. 93 − ■ = 85
   d. ■ + 63 = 72

2. Let ● = 36.
   a. ● − 29 = 7
   b. 74 − ● = 34
   c. 59 + ● = 85
   d. ● + 99 = 135

3. Let x = 78.
   a. 328 − x = 250
   b. x + 99 = 177
   c. x − 19 = 59
   d. 145 + x = 225

4. Let n = 208.
   a. 395 + n = 603
   b. n − 179 = 29
   c. 812 − n = 404
   d. n + n = 808

Check Your Mastery

Performance Assessment

1. Estimate and then list which three money amounts in the boxes have a sum of about $73.

$63.50	$1.35	$3.75	$8.25	$58.50

Find the sum.

2. 509 + 45

3. 283 + 179

4. 8059 + 397

5.
```
  176
  205
+ 387
```

6.
```
  374
  162
+  51
```

7.
```
$78.50
   .99
+ 5.38
```

8.
```
23,154
    96
+ 4,129
```

Find the difference.

9. 750 − 29

10. 5123 − 99

11. 56,150 − 3777

12.
```
 5430
- 298
```

13.
```
 3000
-2951
```

14.
```
$29.39
- 18.42
```

15.
```
29,126
- 8,437
```

PROBLEM SOLVING   *Use a strategy you have learned.*

16. Cindy's car cost $15,935. She paid $8599. How much more does she owe?

17. Joe is 8 years old. He has $67.19 In his bank account. Lori is 11 years old and has $96.95 in her bank account. Together how much do they have in the bank?

18. Mrs. Grant bought gifts for her 4 children. The gifts cost $29.99, $17.59, $35.79, and $49.99. How much did Mrs. Grant pay for the gifts? If she gave the cashier $135.00, how much change did she receive?

Cumulative Review II

Choose the best answer.

1. Choose the standard form of the number.

   ninety-one thousand, four hundred sixty

   a. 9146
   b. 91,046
   c. 91,460
   d. 91,460,000

2. Choose the standard form of the number.

   30,000,000 + 80,000 + 5000 + 6

   a. 30,085,006
   b. 30,085,060
   c. 30,805,006
   d. 30,850,006

3. Which is greatest to least?

   a. 5718; 57,180; 56,032; 57,099
   b. 57,099; 57,180; 56,032; 5718
   c. 57,180; 57,099; 56,032; 5718
   d. 57,180; 57,099; 5718; 56,032

4. Round 5638 to the nearest thousand.

   a. 5000
   b. 5600
   c. 5700
   d. 6000

5. About what number is the arrow pointing toward?

   250 ↑ 260

   a. 225
   b. 250
   c. 255
   d. 270

6. What is the value of the change?

   Cost: $14.52
   Amount given: $20.00

   a. $5.48
   b. $6.48
   c. $6.52
   d. $6.58

7. Find the missing addend.

   $12 = \underline{?} + 4$

   a. 5
   b. 8
   c. 12
   d. 16

8. Find the missing subtrahend.

   $17 - \underline{?} = 9$

   a. 6
   b. 7
   c. 8
   d. 9

9. $8502 - 647$

   a. 7865
   b. 7855
   c. 9149
   d. not given

10. $\begin{array}{r} \$89.60 \\ .88 \\ + \quad 6.49 \\ \hline \end{array}$

    a. $95.97
    b. $96.87
    c. $150.48
    d. not given

11. Which is more than 20,000 but less than 28,000?

    a. 15,987 + 13,162
    b. 29,000 − 900
    c. 30,255 − 11,065
    d. 23,154 + 96 + 4129

12. Which statement is true?

    a. $6 + (3 + 4) = (6 + 3) + 1$
    b. $(6 + 3) + 4 = 6 + (3 + 4)$
    c. $(4 + 3) + 1 = (6 + 3) + 1$
    d. none of these

Ongoing Assessment I

For Your Portfolio

Solve each problem. Explain the steps and the strategy or strategies you used for each. Then choose one from problems 1–4 for your Portfolio.

1. Lynette caught 224 crabs. She sold 195 crabs to a fish store and kept the rest. How many crabs did Lynette keep?

2. Dan's Deli sold 134 tuna sandwiches, 246 turkey sandwiches, and 371 ham sandwiches. How many sandwiches is that altogether?

3. Milk is sold at $.99 per quart and yogurt at $3.89 per quart. Adela bought a quart of yogurt. How much change did she receive from $10.00?

4. Ty has 45 posters. Some of the posters are of cars and the others are of trains. For each train poster, he has 8 car posters. How many train posters does Ty have?

Tell about it.

5. Identify the extra information in problem 3. How can you change problem 3 so that the extra information is needed to answer the question?

Communicate

6. Did you make a table to help you solve problem 4? What would your answer be if the problem asked for the number of car posters?

For Rubric Scoring

Listen for information on how your work will be scored.

7. Which statement below is *not* true? Explain why.
   - The sum of two odd numbers is always even.
   - The difference between an even number and an odd number is always even.

8. Write two other true statements about even or odd numbers. You may use +, −, ×, or ÷.

Multiplying By One and Two Digits

4

Is Six Times One a Lot of Fun?

Is six times one a lot of fun?
Or eight times two?
Perhaps for you.
But five times three
Unhinges me,
While six and seven and eight times eight
Put me in an awful state
And four and six and nine times nine
Make me want to cry and whine
So when I get to twelve times ten
I begin to wonder when
I can take a vacation from multiplication
And go out
And start playing again.

Karla Kuskin

8 x 2
12 x 10
3 x 5
5 x 3
12 x 10
3 x 5

In this chapter you will:

Use multiplication properties
Learn about missing factors,
 special factors, and patterns
Explore multiplication models
Estimate and multiply whole
 numbers and money
Study computer spreadsheets
Solve problems by working
 backwards

Critical Thinking/Finding Together

Use base ten blocks to model
and find the product for each
multiplication the girl is thinking of.

125

4-1 | **Multiplication Properties**

Algebra ✓

The properties of multiplication can help you
to multiply quickly and correctly.

- Changing the *order* of the factors
  does not change the product.

  Think: "order"

$4 \times 5 = 20$ 5   4
$5 \times 4 = 20$ $\times 4$ $\times 5$
           20   20

- Changing the *grouping* of the factors
  does not change the product.

  Think: "grouping"

$(2 \times 4) \times 1 = 2 \times (4 \times 1)$

    8    $\times 1 = 2 \times$ 4
        $8 = 8$

- The product of *one* and a number
  is the same as that number.

  Think: "same number"

$1 \times 6 = 6$ 6   1
$6 \times 1 = 6$ $\times 1$ $\times 6$
           6   6

- The product of *zero* and a number
  is 0.

  Think: "zero"

$0 \times 3 = 0$ 3   0
$3 \times 0 = 0$ $\times 0$ $\times 3$
           0   0

Find the product.

1.   5   2   **2.**   6   3   **3.**   7   9   **4.**   4   0
    $\times 2$ $\times 5$ $\times 3$ $\times 6$ $\times 9$ $\times 7$ $\times 0$ $\times 4$

5.   1   8   **6.**   9   4   **7.**   2   6   **8.**   7   5
    $\times 8$ $\times 1$ $\times 4$ $\times 9$ $\times 6$ $\times 2$ $\times 5$ $\times 7$

9. 4×8 **10.** 7×3 **11.** 5×9 **12.** 1×0

13. 6×7 **14.** 8×5 **15.** 6×0 **16.** 9×1

Copy and complete.

17. $2 \times (3 \times 1) = (2 \times 3) \times 1$
$2 \times \underline{\ ?\ } = \underline{\ ?\ } \times 1$
$\underline{\ ?\ } = \underline{\ ?\ }$

18. $(3 \times 2) \times 2 = 3 \times (2 \times 2)$
$\underline{\ ?\ } \times \underline{\ ?\ } = \underline{\ ?\ } \times \underline{\ ?\ }$
$\underline{\ ?\ } = \underline{\ ?\ }$

19. $2 \times (5 \times 0) = (2 \times \underline{\ ?\ }) \times \underline{\ ?\ }$
$\underline{\ ?\ } \times \underline{\ ?\ } = \underline{\ ?\ } \times \underline{\ ?\ }$
$\underline{\ ?\ } = \underline{\ ?\ }$

20. $(1 \times 6) \times 2 = \underline{\ ?\ } \times (\underline{\ ?\ } \times \underline{\ ?\ })$
$\underline{\ ?\ } \times \underline{\ ?\ } = \underline{\ ?\ } \times \underline{\ ?\ }$
$\underline{\ ?\ } = \underline{\ ?\ }$

PROBLEM SOLVING

21. The product is 8. One factor is 8. What is the other factor?

22. The product is 9. One factor is 1. What is the other factor?

23. If $8 \times 12 = 96$, what is the product of 12×8?

24. Write two multiplication facts using 5 and 0.

Multiplying Sums

The product of a number and the sum of two addends is the same as the sum of the two products.

Think: "sum"

$5 \times (2 + 1) = (5 \times 2) + (5 \times 1)$
$5 \times \quad 3 \quad = \quad 10 \quad + \quad 5$
$\qquad 15 \quad = \quad 15$

Copy and complete.

25. $2 \times (3 + 2) = (2 \times 3) + (2 \times 2)$
$2 \times \underline{\ ?\ } = \underline{\ ?\ } + \underline{\ ?\ }$
$\underline{\ ?\ } = \underline{\ ?\ }$

26. $3 \times (5 + 4) = (3 \times \underline{\ ?\ }) + (3 \times \underline{\ ?\ })$
$3 \times \underline{\ ?\ } = \underline{\ ?\ } + \underline{\ ?\ }$
$\underline{\ ?\ } = \underline{\ ?\ }$

27. $5 \times (6 + 3) = (\underline{\ ?\ } \times \underline{\ ?\ }) + (\underline{\ ?\ } \times \underline{\ ?\ })$
$\underline{\ ?\ } \times \underline{\ ?\ } = \underline{\ ?\ } + \underline{\ ?\ }$
$\underline{\ ?\ } = \underline{\ ?\ }$

4-2 **Missing Factors**

Jill designed 24 greeting cards.
She wants to box them in sets
of 6 cards each. How many sets
of cards can she make?

To find how many sets, think:
  $\underline{\ ?\ }$ sets of 6 cards = 24 cards
      $\underline{\ ?\ }$ sixes = 24
        $\underline{\ ?\ } \times 6 = 24$

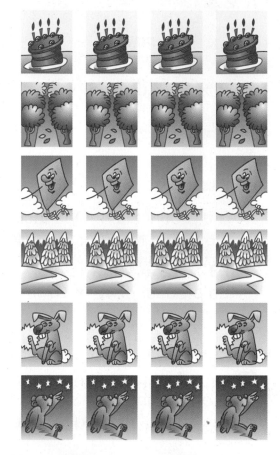

To find the missing factor, think:
What number times 6 is 24?

$2 \times 6 = 12$ too small
$3 \times 6 = 18$ too small
$4 \times 6 = 24$ just right!

She can make 4 sets of cards.

Study these examples.

$3 \times \underline{\ ?\ } = 21$ $\underline{\ ?\ } \times 7 = 42$
$3 \times \ 5\ = 15$ too small $8 \times 7 = 56$ too large
$3 \times \ 6\ = 18$ too small $7 \times 7 = 49$ too large
$3 \times \ 7\ = 21$ just right! $6 \times 7 = 42$ just right!

Find the missing factor.

1. $\underline{\ ?\ } \times 3 = 6$ **2.** $\underline{\ ?\ } \times 5 = 15$ **3.** $\underline{\ ?\ } \times 4 = 20$

4. $\underline{\ ?\ } \times 6 = 36$ **5.** $\underline{\ ?\ } \times 7 = 56$ **6.** $\underline{\ ?\ } \times 8 = 72$

7. $\underline{\ ?\ } \times 2 = 2$ **8.** $\underline{\ ?\ } \times 9 = 36$ **9.** $\underline{\ ?\ } \times 3 = 9$

10. $\underline{\ ?\ } \times 1 = 8$ **11.** $\underline{\ ?\ } \times 4 = 28$ **12.** $\underline{\ ?\ } \times 6 = 42$

Copy and complete.

13. $9 \times \underline{?} = 18$

14. $4 \times \underline{?} = 16$

15. $7 \times \underline{?} = 21$

16. $2 \times \underline{?} = 16$

17. $8 \times \underline{?} = 40$

18. $5 \times \underline{?} = 25$

19. $54 = 9 \times \underline{?}$

20. $24 = 3 \times \underline{?}$

21. $49 = 7 \times \underline{?}$

22. $63 = \underline{?} \times 9$

23. $64 = \underline{?} \times 8$

24. $48 = \underline{?} \times 6$

25.
$$\begin{array}{r} ? \\ \times\ 5 \\ \hline 30 \end{array}$$

26.
$$\begin{array}{r} ? \\ \times\ 2 \\ \hline 12 \end{array}$$

27.
$$\begin{array}{r} ? \\ \times\ 6 \\ \hline 0 \end{array}$$

28.
$$\begin{array}{r} ? \\ \times\ 3 \\ \hline 27 \end{array}$$

29.
$$\begin{array}{r} ? \\ \times\ 7 \\ \hline 35 \end{array}$$

30.
$$\begin{array}{r} ? \\ \times\ 4 \\ \hline 32 \end{array}$$

31.
$$\begin{array}{r} 1 \\ \times\ ? \\ \hline 9 \end{array}$$

32.
$$\begin{array}{r} 5 \\ \times\ ? \\ \hline 10 \end{array}$$

33.
$$\begin{array}{r} 4 \\ \times\ ? \\ \hline 12 \end{array}$$

34.
$$\begin{array}{r} 9 \\ \times\ ? \\ \hline 45 \end{array}$$

35.
$$\begin{array}{r} 3 \\ \times\ ? \\ \hline 18 \end{array}$$

36.
$$\begin{array}{r} 6 \\ \times\ ? \\ \hline 24 \end{array}$$

PROBLEM SOLVING

37. The product of 8 and another factor is 72. What is the other factor?

38. How many threes are equal to 18? How many nines are equal to 18?

39. What number times 7 is 35?

40. The product is 81. One factor is 9. What is the other factor?

Critical Thinking

Communicate

Complete the pattern. Describe the pattern you see.

41.
$$\begin{array}{r} 11 \\ \times\ 2 \\ \hline 22 \end{array} \quad \begin{array}{r} 11 \\ \times\ 3 \\ \hline 33 \end{array} \quad \begin{array}{r} 11 \\ \times\ 4 \\ \hline 44 \end{array} \quad \begin{array}{r} 11 \\ \times\ 5 \\ \hline \end{array} \quad \begin{array}{r} 11 \\ \times\ 6 \\ \hline \end{array} \quad \begin{array}{r} 11 \\ \times\ 7 \\ \hline \end{array} \quad \begin{array}{r} 11 \\ \times\ 8 \\ \hline \end{array} \quad \begin{array}{r} 11 \\ \times\ 9 \\ \hline \end{array}$$

42.
$$\begin{array}{r} 11 \\ \times\ 10 \\ \hline 110 \end{array} \quad \begin{array}{r} 11 \\ \times\ 11 \\ \hline 121 \end{array} \quad \begin{array}{r} 11 \\ \times\ 12 \\ \hline 132 \end{array} \quad \begin{array}{r} 11 \\ \times\ 13 \\ \hline \end{array} \quad \begin{array}{r} 11 \\ \times\ 14 \\ \hline \end{array} \quad \begin{array}{r} 11 \\ \times\ 15 \\ \hline \end{array} \quad \begin{array}{r} 11 \\ \times\ 16 \\ \hline \end{array} \quad \begin{array}{r} 11 \\ \times\ 17 \\ \hline \end{array}$$

Multiplication Models

▶ Sharon uses 16 paper clips to make a necklace. How many paper clips will she need to make 2 necklaces?

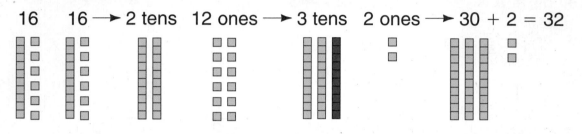

16   16 ⟶ 2 tens   12 ones ⟶ 3 tens   2 ones ⟶ 30 + 2 = 32

$2 \times 16 = 32$ Sharon needs 32 paper clips.

▶ Holly makes a bracelet with 3 rows of beads. Each row has 15 beads. How many beads are in the bracelet?

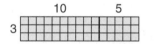

$15 + 15 + 15 = 45$

$3 \times 15 = 45$ There are 45 beads in the bracelet.

▶ Ito wants to make 4 headbands. Each headband uses 34 rubber bands. How many rubber bands does he need?

4×3 tens $= 12$ tens $= 120$ 4×4 ones $= 16$ ones $= 16$

$120 + 16 = 136$

$4 \times 34 = 136$ Ito needs 136 rubber bands.

Write a multiplication sentence for each model.

1.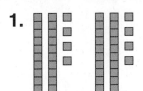

2.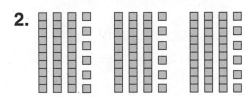

3.

4.

5.

6.

PROBLEM SOLVING You may use models.

7. Raul has 3 paper-clip chains. Each one is 55 paper clips long. He connects them. How long is the new chain?

8. Monica paints T-shirts. She paints 17 dots on each T-shirt. How many dots does she paint on 9 T-shirts?

9. Marva makes stained-glass designs. One design has 5 rows of squares. There are 10 squares in each row. How many squares does Marva use?

10. Paul builds model boats. He uses 25 craft sticks to build a rowboat. How many craft sticks does he need to build 4 rowboats?

11. Peter uses 52 toothpicks to build a model house. How many toothpicks does he need for 6 model houses?

4-4 Special Factors

Look for a pattern when you multiply tens.

4 × 1 ten = 4 tens
4 × 10 = 40

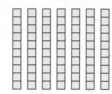
7 × 1 ten = 7 tens
7 × 10 = 70

2 × 3 tens = 6 tens
2 × 30 = 60

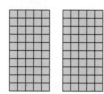
2 × 5 tens = 10 tens
2 × 50 = 100

To multiply tens, hundreds, or thousands:

- Multiply the nonzero digits.
- Write 1, 2, or 3 zeros in the product.

$$\begin{array}{r} 50 \\ \times\ \ 9 \\ \hline 450 \end{array} \quad \text{1 zero}$$

$$\begin{array}{r} 700 \\ \times\ \ \ \ 4 \\ \hline 2800 \end{array} \quad \text{2 zeros}$$

$$\begin{array}{r} 5000 \\ \times\ \ \ \ \ 6 \\ \hline 30{,}000 \end{array} \quad \text{3 zeros}$$

Algebra ✓

Write a number sentence for each.

1.

2.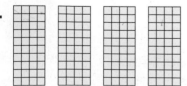

3.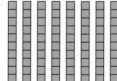

4.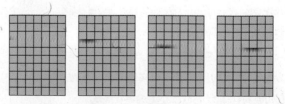

5.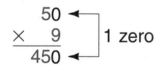

6.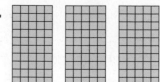

132

Copy and complete.

7. 6×3 tens $= 6 \times 30 = 180$ **8.** 5×4 tens $= 5 \times 40 = \underline{?}$

9. 9×1 hundred $= 9 \times 100 = \underline{?}$ **10.** 2×4 hundreds $= 2 \times \underline{?} = \underline{?}$

11. 4×1 thousand $= 4 \times 1000 = \underline{?}$

12. 7×3 thousands $= \underline{?} \times \underline{?} = \underline{?}$

Find the product.

13. 4×1 ten

14. 7×3 tens

15. 8×1 hundred

16. 9×6 tens

17. 2×5 hundreds

18. 4×7 hundreds

19. 7×1 thousand

20. 6×3 thousands

21. 5×8 thousands

Multiply. Use mental math or paper and pencil.

22. $\begin{array}{r} 90 \\ \times\ 3 \\ \hline \end{array}$
23. $\begin{array}{r} 70 \\ \times\ 2 \\ \hline \end{array}$
24. $\begin{array}{r} 80 \\ \times\ 4 \\ \hline \end{array}$
25. $\begin{array}{r} 50 \\ \times\ 5 \\ \hline \end{array}$

26. $\begin{array}{r} 500 \\ \times\ 3 \\ \hline \end{array}$
27. $\begin{array}{r} 900 \\ \times\ 9 \\ \hline \end{array}$
28. $\begin{array}{r} 400 \\ \times\ 5 \\ \hline \end{array}$
29. $\begin{array}{r} 300 \\ \times\ 8 \\ \hline \end{array}$

30. $\begin{array}{r} 1000 \\ \times\ 6 \\ \hline \end{array}$
31. $\begin{array}{r} 6000 \\ \times\ 4 \\ \hline \end{array}$
32. $\begin{array}{r} 5000 \\ \times\ 7 \\ \hline \end{array}$
33. $\begin{array}{r} 9000 \\ \times\ 6 \\ \hline \end{array}$

PROBLEM SOLVING

34. There are 5000 seats at Carver Stadium. Baseball games are played there 4 nights a week. How many tickets can the stadium sell each week?

35. Glen runs the 50-yard dash 8 times. How many yards does he run in all?

36. Ms. Spero swims 8 laps every day. How many laps does she swim in September?

4-5 Multiplying by One-Digit Numbers

There are 24 tropical fish in each of 2 fish tanks at Fish World. How many tropical fish are there at Fish World?

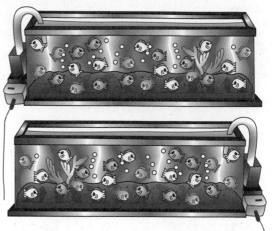

▶ To find how many, join 2 sets of 24.

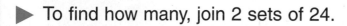

Think: 24 = 2 tens 4 ones

```
   2 tens 4 ones
+  2 tens 4 ones
   4 tens 8 ones = 48
```

▶ You can join 2 sets of 24 by multiplying and adding.

- Join 2 sets of 4 ones. $2 \times 4 = 8$
- Join 2 sets of 2 tens. $2 \times 20 = 40$
- Add. $8 + 40 = 48$

▶ You can multiply: $2 \times 24 = \underline{\ ?\ }$

Multiply the ones.	Multiply the tens.
$\begin{array}{r} 24 \\ \times\ 2 \\ \hline 8 \end{array}$	$\begin{array}{r} 24 \\ \times\ 2 \\ \hline 48 \end{array}$
2×4 ones $= 8$ ones	2×2 tens $= 4$ tens

There are 48 tropical fish at Fish World.

Multiply.

1.	**2.**	**3.**	**4.**	**5.**	**6.**
$\begin{array}{r} 12 \\ \times\ 2 \end{array}$	$\begin{array}{r} 22 \\ \times\ 3 \end{array}$	$\begin{array}{r} 13 \\ \times\ 3 \end{array}$	$\begin{array}{r} 11 \\ \times\ 5 \end{array}$	$\begin{array}{r} 14 \\ \times\ 2 \end{array}$	$\begin{array}{r} 12 \\ \times\ 4 \end{array}$

Find the product. Use mental math or paper and pencil.

7. $\begin{array}{r} 11 \\ \times\ 9 \\ \hline \end{array}$
8. $\begin{array}{r} 34 \\ \times\ 2 \\ \hline \end{array}$
9. $\begin{array}{r} 22 \\ \times\ 4 \\ \hline \end{array}$
10. $\begin{array}{r} 44 \\ \times\ 2 \\ \hline \end{array}$
11. $\begin{array}{r} 12 \\ \times\ 3 \\ \hline \end{array}$
12. $\begin{array}{r} 33 \\ \times\ 3 \\ \hline \end{array}$

13. $\begin{array}{r} 32 \\ \times\ 3 \\ \hline \end{array}$
14. $\begin{array}{r} 13 \\ \times\ 2 \\ \hline \end{array}$
15. $\begin{array}{r} 43 \\ \times\ 2 \\ \hline \end{array}$
16. $\begin{array}{r} 31 \\ \times\ 3 \\ \hline \end{array}$
17. $\begin{array}{r} 26 \\ \times\ 1 \\ \hline \end{array}$
18. $\begin{array}{r} 41 \\ \times\ 2 \\ \hline \end{array}$

19. 2×23
20. 4×22
21. 2×33
22. 3×31

23. 2×42
24. 3×21
25. 4×21
26. 8×11

27. 2×14
28. 2×31
29. 7×11
30. 2×32

PROBLEM SOLVING

31. Fish World received 3 cartons of fish food. There were 12 boxes of food in each carton. How many boxes of fish food did Fish World receive?

32. Niqui displayed 22 fish care booklets on each of 4 shelves. How many fish care booklets did Niqui display on the shelves?

33. Greg filled each of 2 fish tanks with 14 gallons of water. How much water did Greg use to fill the tanks?

34. There were 2 shipments of 42 goldfish each to Fish World. How many goldfish were there in both shipments?

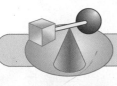

Challenge

Algebra

Find the next three numbers.

35. 10, 15, 25, 30, 40, _?_, _?_, _?_
36. 1, 3, 2, 4, 3, 5, _?_, _?_, _?_

37. 24, 30, 28, 34, 32, _?_, _?_, _?_
38. 1, 1, 3, 3, 5, 5, 7, _?_, _?_, _?_

39. 1, 2, 3, 6, 7, 14, _?_, _?_, _?_
40. 2, 4, 6, 12, 14, _?_, _?_, _?_

4-6 Products: Front-End Estimation

Will 5 games cost more or less than $100?

To find if the games will cost more or less than $100, estimate.

| Multiply the front digit of the greater factor. | | Write 0s for the other digits. |

$$\begin{array}{r} \$25.95 \\ \times \quad 5 \\ \hline 10 \end{array}$$

Write $ and . in the product.

$$\begin{array}{r} \$25.95 \\ \times \quad 5 \\ \hline \text{about } \$100.00 \end{array}$$

Think:
$$\begin{array}{r} \$20.00 \\ \times \quad 5 \\ \hline \$100.00 \end{array}$$

Since $25.95 is greater than $20, the actual cost is close to but greater than $100.

The 5 games will cost more than $100.

Study these examples.

$$\begin{array}{r} 62 \\ \times \quad 6 \\ \hline \text{about } 360 \end{array}$$

$$\begin{array}{r} \$5.28 \\ \times \quad 7 \\ \hline \text{about } \$35.00 \end{array}$$

$$\begin{array}{r} 8406 \\ \times \quad 8 \\ \hline \text{about } 64{,}000 \end{array}$$

Estimate the product.

1. $\begin{array}{r} 82 \\ \times \ 6 \\ \hline \end{array}$
2. $\begin{array}{r} 98 \\ \times \ 1 \\ \hline \end{array}$
3. $\begin{array}{r} 46 \\ \times \ 5 \\ \hline \end{array}$
4. $\begin{array}{r} \$.73 \\ \times \ 3 \\ \hline \end{array}$
5. $\begin{array}{r} \$.57 \\ \times \ 2 \\ \hline \end{array}$

6. $\begin{array}{r} 473 \\ \times \ 8 \\ \hline \end{array}$
7. $\begin{array}{r} \$9.01 \\ \times \ 4 \\ \hline \end{array}$
8. $\begin{array}{r} 5125 \\ \times \ 9 \\ \hline \end{array}$
9. $\begin{array}{r} 1070 \\ \times \ 6 \\ \hline \end{array}$
10. $\begin{array}{r} \$32.95 \\ \times \ 7 \\ \hline \end{array}$

11. $\begin{array}{r} 849 \\ \times \ 4 \\ \hline \end{array}$
12. $\begin{array}{r} \$6.53 \\ \times \ 3 \\ \hline \end{array}$
13. $\begin{array}{r} \$46.73 \\ \times \ 8 \\ \hline \end{array}$
14. $\begin{array}{r} 7211 \\ \times \ 5 \\ \hline \end{array}$
15. $\begin{array}{r} \$32.24 \\ \times \ 9 \\ \hline \end{array}$

Estimate.

16.	55 $\times$ 2	17.	49 $\times$ 9	18.	31 $\times$ 7	19.	64 $\times$ 6	20.	78 $\times$ 3

21.	437 $\times$ 9	22.	622 $\times$ 5	23.	145 $\times$ 4	24.	744 $\times$ 7	25.	609 $\times$ 8

26.	7832 $\times$ 6	27.	8209 $\times$ 5	28.	9848 $\times$ 4	29.	4633 $\times$ 2

30.	$.65 $\times$ 9	31.	$8.33 $\times$ 7	32.	$34.72 $\times$ 5	33.	$21.24 $\times$ 6

34. 4 $\times$ $7.10   **35.** 2 $\times$ $9.67   **36.** 9 $\times$ $37.55

PROBLEM SOLVING Use estimation.

37. Will 3 controls cost more or less than $60?

38. About how much would a set of 2 speakers cost?

39. Will 7 controls cost more than 2 game systems?

40. Jenique wants to buy 1 game system, 2 speakers, and 3 controls. About how much will she spend?

Calculator Activity

Estimate. Then use a calculator to multiply.

41.	788 $\times$ 8	42.	9237 $\times$ 4	43.	$6.97 $\times$ 7	44.	$89.09 $\times$ 9

Multiplying with Regrouping

Cody needs 102 pushpins for a social studies project. There are 35 pushpins in each packet. Will Cody have enough pushpins if he buys 3 packets?

To find whether Cody will have enough pushpins, first find the product: $3 \times 35 = \underline{\ ?\ }$

Amarillo
Dallas
Austin
Houston
San Antonio

Hands-On Understanding

You Will Need: base ten blocks, paper, pencils

Record as you model.

| **Step 1** | Model 3 sets of 35. |

$$\begin{array}{r} 35 \\ \times\ 3 \\ \hline \end{array}$$

How many ones are there? how many tens?

| **Step 2** | Regroup as many ones as you can. |

$$\begin{array}{r} {\scriptstyle 1}\ \ \\ 35 \\ \times\ 3 \\ \hline 5 \end{array}$$

How many ones are there now? how many tens?
What is the product when you multiply 35 by 3?

Will Cody have enough pushpins for his project? Explain your answer.

$$\begin{array}{r} {\scriptstyle 1}\ \ \\ 35 \\ \times\ 3 \\ \hline 105 \end{array}$$

Here is a way to find the product of 3 × 35 without using base ten blocks.

Multiply the ones. Regroup.	Multiply the tens. Add the regrouped tens.

$$\begin{array}{r} \overset{1}{3}\ 5 \\ \times\quad 3 \\ \hline 5 \end{array} \qquad\qquad \begin{array}{r} \overset{1}{3}\ 5 \\ \times\quad\ 3 \\ \hline 1\ 0\ 5 \end{array}$$

3 × 5 ones = 15 ones 15 ones = 1 ten 5 ones	3 × 3 tens = 9 tens 9 tens + 1 ten = 10 tens

Multiply. You may use base ten blocks.

1. $\begin{array}{r}18\\ \times\ 3\\ \hline\end{array}$	**2.** $\begin{array}{r}16\\ \times\ 5\\ \hline\end{array}$	**3.** $\begin{array}{r}38\\ \times\ 2\\ \hline\end{array}$	**4.** $\begin{array}{r}24\\ \times\ 3\\ \hline\end{array}$	**5.** $\begin{array}{r}16\\ \times\ 4\\ \hline\end{array}$	**6.** $\begin{array}{r}25\\ \times\ 2\\ \hline\end{array}$
7. $\begin{array}{r}28\\ \times\ 2\\ \hline\end{array}$	**8.** $\begin{array}{r}17\\ \times\ 3\\ \hline\end{array}$	**9.** $\begin{array}{r}24\\ \times\ 4\\ \hline\end{array}$	**10.** $\begin{array}{r}19\\ \times\ 3\\ \hline\end{array}$	**11.** $\begin{array}{r}17\\ \times\ 5\\ \hline\end{array}$	**12.** $\begin{array}{r}19\\ \times\ 4\\ \hline\end{array}$
13. $\begin{array}{r}24\\ \times\ 6\\ \hline\end{array}$	**14.** $\begin{array}{r}46\\ \times\ 4\\ \hline\end{array}$	**15.** $\begin{array}{r}68\\ \times\ 5\\ \hline\end{array}$	**16.** $\begin{array}{r}78\\ \times\ 2\\ \hline\end{array}$	**17.** $\begin{array}{r}36\\ \times\ 3\\ \hline\end{array}$	**18.** $\begin{array}{r}86\\ \times\ 9\\ \hline\end{array}$

19. 3 × 27   **20.** 4 × 63   **21.** 5 × 84   **22.** 6 × 77

23. 9 × 58   **24.** 7 × 45   **25.** 8 × 67   **26.** 9 × 99

Communicate

Discuss ✓

27. How is multiplication like addition? How is it different?

28. Which of the exercises above were easier to do without using base ten blocks? Why?

Multiplying Three-Digit Numbers

Each of the 8 families on Pine Road receives a newspaper delivery each day of the year. How many newspapers are delivered on Pine Road each year?

To find how many, multiply: 8 × 365 = _?_

Multiply the ones. Regroup.

$$\begin{array}{r} \overset{4}{} \\ 3\ 6\ 5 \\ \times\ \ \ \ 8 \\ \hline 0 \end{array}$$

8 × 5 ones = 40 ones
40 ones = 4 tens 0 ones

Multiply the tens. Add the regrouped tens. Regroup.

$$\begin{array}{r} \overset{5\ \ 4}{} \\ 3\ 6\ 5 \\ \times\ \ \ \ 8 \\ \hline 2\ 0 \end{array}$$

8 × 6 tens = 48 tens
48 tens + 4 tens = 52 tens
52 tens = 5 hundreds 2 tens

Multiply the hundreds. Add the regrouped hundreds.

$$\begin{array}{r} \overset{5\ \ 4}{} \\ 3\ 6\ 5 \\ \times\ \ \ \ \ 8 \\ \hline 2\ 9\ 2\ 0 \end{array}$$

8 × 3 hundreds = 24 hundreds
24 hundreds + 5 hundreds = 29 hundreds
29 hundreds = 2 thousands 9 hundreds

Each year, 2920 newspapers are delivered on Pine Road.

Study these examples.

$$\begin{array}{r} \overset{1}{} \\ 504 \\ \times\ \ \ 4 \\ \hline 2016 \end{array} \qquad \begin{array}{r} 321 \\ \times\ \ \ 2 \\ \hline 642 \end{array} \qquad \begin{array}{r} \overset{1}{} \\ 621 \\ \times\ \ \ 6 \\ \hline 3726 \end{array}$$

Estimate. Then multiply.

1. 214		**2.** 101		**3.** 210		**4.** 323		**5.** 223	
× 2		× 3		× 4		× 2		× 3	

6. 308		**7.** 410		**8.** 271		**9.** 505		**10.** 192	
× 5		× 8		× 7		× 9		× 4	

11. 634		**12.** 279		**13.** 844		**14.** 575		**15.** 397	
× 6		× 9		× 7		× 8		× 5	

Find the product. Use mental math or paper and pencil.

16. 2 × 304   **17.** 3 × 131   **18.** 2 × 642

19. 5 × 160   **20.** 6 × 702   **21.** 4 × 261

22. 8 × 625   **23.** 7 × 444   **24.** 9 × 368

PROBLEM SOLVING

25. The Ecology Club brought 6 bundles of junk mail to the recycling center. Each bundle weighed 275 pounds. How many pounds of junk mail did the Ecology Club recycle?

26. Troop 42 collected 8 bins of newspaper for recycling. Four of the bins held 325 pounds of newspaper each. The other 4 bins held 450 pounds of newspaper each. How many pounds of newspaper did Troop 42 collect?

27. Six of the families on Pine Road each recycled at least 8 aluminum cans each week last year. There are 52 weeks in a year. Altogether, did these families recycle more or less than 2000 aluminum cans last year?

Multiplying Money

Cesar buys 8 notebooks for the Detective Club. Each notebook costs $3.39. What is the total cost of the notebooks?

Walkie-Talkies	$8.59 each
Periscopes	$4.58 each
Magnifying Glasses	$7.50 each
Decoder Rings	$ .99 each
Invisible Ink Markers	$1.56 each
Notebooks	$3.39 each

First estimate:
```
      $3.39
    ×     8
about $24.00
```

To find the total cost, multiply: 8 × $3.39 = _?_

To multiply money:

- Multiply the same way you multiply whole numbers.

- Write a decimal point in the product two places from the right.

- Write the dollar sign.

```
   3 7
 $3.3 9
×      8
$2 7.1 2
```

The total cost is $27.12.

$27.12 is close to $24.00. The answer is reasonable.

Study this example.

```
     1
  $.6 4
×     3
$1.9 2
```

Estimate. Then multiply.

1.
```
 $.46
×   6
```

2.
```
 $.38
×   8
```

3.
```
 $.52
×   7
```

4.
```
 $.74
×   9
```

5.
```
 $.25
×   3
```

6.
```
$1.05
×   2
```

7.
```
$5.73
×   5
```

8.
```
$6.26
×   4
```

9.
```
$8.30
×   7
```

10.
```
$4.52
×   9
```

Estimate. Then find the product.

11.	$.42 × 8	12.	$.95 × 2	13.	$.79 × 3	14.	$.12 × 5

15.	$8.31 × 4	16.	$7.95 × 9	17.	$4.36 × 2	18.	$8.95 × 6

19.	$7.50 × 8	20.	$4.31 × 7	21.	$6.08 × 5	22.	$9.49 × 3

23. 4 × $.53   **24.** 6 × $.87   **25.** 8 × $.19

26. 7 × $4.03   **27.** 9 × $1.71   **28.** 3 × $7.47

29. 2 × $9.76   **30.** 5 × $5.98   **31.** 4 × $6.61

PROBLEM SOLVING Use the sign on page 142.

32. How much would 5 periscopes cost? 2 walkie-talkies?

33. What is the cost of 4 walkie-talkies and 6 invisible ink markers?

34. How much would 2 periscopes and 3 pairs of walkie-talkies cost?

Calculator Activity

35. Find a 1-digit factor and a 2-digit factor whose product is 396.

36. Find a 1-digit factor and a 2-digit factor whose product is 220.

37. Find the missing factors in this number sentence:
_?_ × _?_ × 61 = 488

38. Find the missing factors in this number sentence:
_?_ × _?_ × 99 = 891

4-10 Multiplying Four-Digit Numbers

Mr. Carter built houses on 6 neighboring plots of land. Each plot is 6875 square feet. On how many square feet of land did he build the houses?

First estimate:
```
      6875
   ×     6
about 36,000
```

6875 > 6000 So the answer is close to, but greater than, 36,000.

To find how many square feet, multiply: 6 × 6875 = ?

```
  5 4 3
  6 8 7 5
×       6
4 1,2 5 0
```

41,250 is close to and greater than 36,000. The answer is reasonable.

He built the houses on 41,250 square feet of land.

Study these examples.

```
  1   2
  1 4 0 6
×       4
  5 6 2 4
```

```
    3 1
  $2 5.2 0
×        7
$1 7 6.4 0
```

Estimate. Then multiply. Use mental math when you can.

1.
```
  2221
×    3
```

2.
```
  1022
×    4
```

3.
```
  2432
×    2
```

4.
```
  3123
×    3
```

5.
```
  1035
×    7
```

6.
```
  2164
×    4
```

7.
```
  1146
×    6
```

8.
```
  3257
×    3
```

Estimate. Then find the product.

9. 1415 × 9	10. 6423 × 7	11. 7536 × 5	12. 3341 × 8

13. 4372 × 6	14. 5279 × 4	15. 2523 × 9	16. 8119 × 9

17. $34.68 × 7	18. $94.12 × 5	19. $21.77 × 6	20. $74.41 × 3

21. 2 × 9455   22. 5 × 3408   23. 4 × 6472

24. 6 × $36.75   25. 8 × $42.56   26. 7 × $22.95

PROBLEM SOLVING

27. Each ranch house in Shady Acres has 1256 square feet of floor space. How many square feet of flooring were used for the 8 ranch houses in Shady Acres?

28. There are 4 miles of roads through Shady Acres. One mile is equal to 5280 feet. How many feet long are all the roads through Shady Acres?

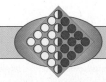 **Finding Together**

Math Journal

Predict which product is greater. Multiply to check.

29. 7 × 6321   or   6 × 7321   30. 5 × 3451   or   3 × 5451

31. 8 × 9310   or   9 × 8310   32. 4 × 9999   or   9 × 4999

33. In your Math Journal, write how you made your predictions.

Patterns in Multiplication

▶ Look for patterns to help you multiply by 10.

$1 \times 35 = 35$	$1 \times 50 = 50$	$1 \times 457 = 457$
$10 \times 35 = 350$	$10 \times 50 = 500$	$10 \times 457 = 4570$
$10 \times 350 = 3500$	$10 \times 500 = 5000$	

To multiply a number by 10:

• Write the number in the product.
• Place a 0 to its right.

$$\begin{array}{r} 35 \\ \times\ 10 \\ \hline 350 \end{array} \qquad \begin{array}{r} 500 \\ \times\ \ 10 \\ \hline 5000 \end{array} \qquad \begin{array}{r} 457 \\ \times\ \ 10 \\ \hline 4570 \end{array}$$

▶ Look for patterns to help you multiply by tens.

$9 \times 31 = 279$	$8 \times 40 = 320$
$90 \times 31 = 2790$	$80 \times 40 = 3200$
$90 \times 310 = 27{,}900$	$80 \times 400 = 32{,}000$

To multiply a number by tens:

• Multiply the number by the tens digit.
• Write 0 in the ones place in the product.

$$\begin{array}{r} 31 \\ \times\ 90 \\ \hline 2790 \end{array} \qquad \begin{array}{r} 310 \\ \times\ \ 90 \\ \hline 27{,}900 \end{array} \qquad \begin{array}{r} 40 \\ \times\ 80 \\ \hline 3200 \end{array} \qquad \begin{array}{r} 400 \\ \times\ \ 80 \\ \hline 32{,}000 \end{array}$$

Multiply mentally.

1. $\begin{array}{r}18\\\times 10\end{array}$	**2.** $\begin{array}{r}24\\\times 10\end{array}$	**3.** $\begin{array}{r}57\\\times 10\end{array}$	**4.** $\begin{array}{r}61\\\times 10\end{array}$	**5.** $\begin{array}{r}50\\\times 10\end{array}$
6. $\begin{array}{r}345\\\times\ 10\end{array}$	**7.** $\begin{array}{r}638\\\times\ 10\end{array}$	**8.** $\begin{array}{r}999\\\times\ 10\end{array}$	**9.** $\begin{array}{r}450\\\times\ 10\end{array}$	**10.** $\begin{array}{r}690\\\times\ 10\end{array}$

Find the product.

11. 23 × 20	**12.** 42 × 60	**13.** 61 × 30	**14.** 70 × 40	**15.** 60 × 50
16. 230 × 20	**17.** 420 × 60	**18.** 610 × 30	**19.** 700 × 40	**20.** 600 × 50
21. 52 × 80	**22.** 25 × 90	**23.** 19 × 70	**24.** 80 × 80	**25.** 40 × 90
26. 520 × 80	**27.** 250 × 90	**28.** 190 × 70	**29.** 800 × 80	**30.** 400 × 90

Copy and complete each pattern.

31.
$1 \times 78 = \underline{?}$
$10 \times 78 = \underline{?}$
$10 \times 780 = \underline{?}$

32.
$9 \times 60 = \underline{?}$
$90 \times 60 = \underline{?}$
$90 \times 600 = \underline{?}$

33.
$2 \times 78 = \underline{?}$
$20 \times 78 = \underline{?}$
$20 \times 780 = \underline{?}$

34.
$7 \times 60 = \underline{?}$
$70 \times 60 = \underline{?}$
$70 \times 600 = \underline{?}$

35.
$8 \times 50 = \underline{?}$
$80 \times 50 = \underline{?}$
$80 \times 500 = \underline{?}$

36.
$6 \times 35 = \underline{?}$
$60 \times 35 = \underline{?}$
$60 \times 350 = \underline{?}$

Compute mentally.

37. How many zeros are in the product when you multiply 10×670?

38. How many zeros are in the product when you multiply 40×500?

Mental Math

Complete each pattern mentally.

39.
$1 \times 56 = \underline{?}$
$10 \times 56 = \underline{?}$
$100 \times 56 = \underline{?}$
$100 \times 560 = \underline{?}$
$100 \times 5600 = \underline{?}$

40.
$7 \times 41 = 287$
$70 \times 41 = \underline{?}$
$700 \times 41 = \underline{?}$
$700 \times 410 = \underline{?}$
$700 \times 4100 = \underline{?}$

Estimating Products

A school bought 28 cans of paint for a special school project. The school bought the paint at a discounted price of $5.25 per can. About how much money did the school spend on paint?

To find about how much the school spent, estimate: 28 × $5.25

Rounding is one way to estimate products:

- Round each factor to its greatest place.
- Multiply.

$$
\begin{array}{r}
\$5.25 \longrightarrow \$5.00 \\
\times\ \ \ \ 28 \longrightarrow \times\ \ \ \ 30 \\
\hline
\text{about } \$150.00
\end{array}
$$

You can write $150.00 as $150.

The school spent about $150 on 28 cans of paint.

Study these examples.

$$
\begin{array}{r}
43 \longrightarrow 40 \\
\times 62 \longrightarrow \times 60 \\
\hline
\text{about } 2400
\end{array}
\qquad
\begin{array}{r}
586 \longrightarrow 600 \\
\times\ \ 55 \longrightarrow \times\ \ 60 \\
\hline
\text{about } 36{,}000
\end{array}
\qquad
\begin{array}{r}
\$.48 \longrightarrow \$.50 \\
\times\ \ \ 32 \longrightarrow \times\ \ \ 30 \\
\hline
\text{about } \$15.00
\end{array}
$$

Estimate the product.

1. 52 ×75	**2.** 68 ×41	**3.** 91 ×22	**4.** 86 ×57	**5.** 47 ×33
6. 19 ×62	**7.** 53 ×78	**8.** 29 ×58	**9.** 34 ×92	**10.** 85 ×38
11. $.17 × 27	**12.** $.36 × 81	**13.** $.42 × 74	**14.** $.66 × 65	**15.** $.26 × 57

Estimate.

16.	348	17.	551	18.	619	19.	809	20.	748
	× 23		× 66		× 72		× 94		× 88

21.	315	22.	754	23.	449	24.	938	25.	656
	× 38		× 24		× 57		× 46		× 53

26.	$4.59	27.	$6.53	28.	$7.24	29.	$5.39	30.	$8.57
	× 34		× 76		× 83		× 24		× 79

31. 27 × 426   **32.** 14 × 643   **33.** 36 × 338

34. 27 × $2.04   **35.** 54 × $7.15   **36.** 68 × $7.46

PROBLEM SOLVING

37. There were 24 gallons of white paint in each of 17 cartons in the storeroom. About how many gallons of white paint were in the storeroom?

38. Each sheet of maple wall paneling covers 48 square feet. Mr. Troc sold 22 sheets of the paneling. About how many square feet of paneling did he sell?

39. Each sheet of maple paneling sells for $152. Were the total sales of the 22 sheets of paneling between $2000 and $3000, between $3000 and $4000, or between $4000 and $5000?

 **Skills to Remember**

Align and add.

40. 94 + 360   **41.** 78 + 645   **42.** 65 + 940   **43.** 26 + 392

Multiplying by Two-Digit Numbers

James baked 24 dozen crescent rolls. There are 12 rolls to a dozen. How many crescent rolls did James bake?

First estimate: 24×12

$20 \times 10 = 200$

To find how many rolls, multiply: $24 \times 12 = \underline{?}$

▶ Here is one way to multiply 24×12.

12	Think:	12	12	12
×24	24 = 20 + 4	× 4	×20	×24
		48 +	240 =	288

▶ Here is another way to multiply 24×12.

Multiply by the ones.	Multiply by the tens.	Add the partial products.

12	12	12
×24	×24	×24
48	48	48 ◄ partial
	240	+ 240 ◄ products
		288

4×12 20×12

James baked 288 crescent rolls.

Estimate. Then multiply.

1.	2.	3.	4.	5.
33	23	42	24	32
×22	×11	×12	×21	×13

Find the product.

6.	7.	8.	9.	10.
12 × 44	42 × 24	32 × 32	14 × 12	22 × 21

Multiplying Money

To multiply money by a 2-digit number:

- Multiply the same way you multiply whole numbers.
- Write a decimal point in the product two places from the right.
- Write the dollar sign in the product.

```
  $.2 3          $8.00
  × 1 3          ×   40
    6 9         $320.00
+ 2 3 0
$ 2.9 9
```

You can use a calculator to multiply money.
Multiply: 10 × $.64 = _?_

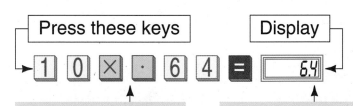

Press these keys	Display

There is no $ key on the calculator.

The display does not show final zeros to the right of the decimal point.

Write 6.4 as $6.40.

Multiply. You may use a calculator.

11.	12.	13.	14.
$.52 × 10	$6.00 × 50	$.23 × 23	$.41 × 21

15.	16.	17.	18.
$.43 × 20	$4.00 × 30	$.11 × 85	$.12 × 14

19. 40 × $.21   **20.** 12 × $.43   **21.** 32 × $3.00

More Multiplying by Two-Digit Numbers

Kara packed 24 pieces of fruit into each of 58 fruit baskets. How many pieces of fruit did Kara pack into the baskets?

First estimate:

$$\begin{array}{r} 24 \longrightarrow 20 \\ \times 58 \longrightarrow \times 60 \\ \hline \text{about } 1200 \end{array}$$

To find how many, multiply: $58 \times 24 = \underline{\ ?\ }$

Think:

$$\begin{array}{r} 2\ 4 \\ \times 5\ 8 \end{array}$$

58 = 50 + 8

$$\begin{array}{r} 3 \\ 2\ 4 \\ \times\ \ \ 8 \\ \hline 1\ 9\ 2 \end{array} + \begin{array}{r} 2 \\ 2\ 4 \\ \times 5\ 0 \\ \hline 1\ 2\ 0\ 0 \end{array} = \begin{array}{r} 2\ 4 \\ \times 5\ 8 \\ \hline 1\ 3\ 9\ 2 \end{array}$$

Multiply by the ones.	Multiply by the tens.	Add the partial products.

$$\begin{array}{r} 3 \\ 2\ 4 \\ \times 5\ 8 \\ \hline 1\ 9\ 2 \end{array}$$

$\boxed{8 \times 24}$

$$\begin{array}{r} 2 \\ \not{8} \\ 2\ 4 \\ \times 5\ 8 \\ \hline 1\ 9\ 2 \\ 1\ 2\ 0\ 0 \end{array}$$

$\boxed{50 \times 24}$

$$\begin{array}{r} 2 \\ \not{8} \\ 2\ 4 \\ \times 5\ 8 \\ \hline 1\ 9\ 2 \\ +\ 1\ 2\ 0\ 0 \\ \hline 1\ 3\ 9\ 2 \end{array}$$

1392 is close to 1200. The answer is reasonable.

Kara packed 1392 pieces of fruit.

Study these examples.

$\boxed{6 \times 22} \longrightarrow$
$\boxed{40 \times 22} \longrightarrow$

$$\begin{array}{r} \not{1} \\ 2\ 2 \\ \times 4\ 6 \\ \hline 1\ 3\ 2 \\ +\ 8\ 8\ 0 \\ \hline 1\ 0\ 1\ 2 \end{array}$$

This zero does not have to be written.

$$\begin{array}{r} 3 \\ \not{3} \\ \$.3\ 5 \\ \times\ \ 6\ 7 \\ \hline 2\ 4\ 5 \\ +\ 2\ 1\ 0\ 0 \\ \hline \$2\ 3.4\ 5 \end{array}$$

$\longleftarrow \boxed{7 \times 35}$
$\longleftarrow \boxed{60 \times 35}$

Estimate. Then multiply.

1. 21
    × 46

2. 36
    × 18

3. 42
    × 62

4. 57
    × 19

5. 73
    × 31

6. 64
    × 39

7. 83
    × 44

8. 56
    × 92

9. 29
    × 75

10. 48
    × 99

11. $.49
    ×   32

12. $.67
    ×   58

13. $.99
    ×   64

14. $.53
    ×   28

15. $.35
    ×   76

16. 95 × 76

17. 39 × 55

18. 47 × 63

19. 25 × 92

20. 16 × 52

21. 28 × 82

22. 34 × 93

23. 71 × 37

24. 15 × $.94

25. 34 × $.92

26. 85 × $.55

27. 26 × $.78

Find each product. Describe any pattern you see.

28. 12   12   12   12 •••   12   12   12
    × 2   × 3   × 4   × 5 ••• × 10   × 11   × 12
                                 •••

PROBLEM SOLVING

29. Tyrone put together 62 boxes of canned food. There were 45 cans in each box. How many cans of food were there?

30. Mill Farms donated 85 turkeys to soup kitchens. Each turkey weighed 25 pounds. How many pounds of turkey were donated?

Share Your Thinking

Communicate ✓

31. How does knowing how to multiply by tens help you multiply a 2-digit number by another 2-digit number? Discuss this with a classmate. Then tell your teacher what you have decided.

4-15 Multiplying with Three-Digit Numbers

Letisha, Marc, Robin, and Tim all made beaded wall hangings. How many beads did Robin use?

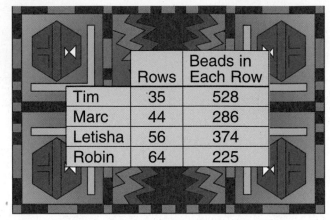

	Rows	Beads in Each Row
Tim	35	528
Marc	44	286
Letisha	56	374
Robin	64	225

First estimate: 64 × 225

$$
\begin{array}{r}
225 \longrightarrow 200 \\
\times\ 64 \longrightarrow \times\ 60 \\
\hline
\text{about } 12{,}000
\end{array}
$$

To find how many, multiply: 64 × 225 = __?__

Multiply by the ones.

```
  1 2
  2 2 5
×   6 4
-------
  9 0 0
```
→ 4 × 225

Multiply by the tens.

```
  1 3
  X 2
  2 2 5
×   6 4
-------
  9 0 0
1 3 5 0 0
```
→ 60 × 225

Add the partial products.

```
   1 3
   X 2
   2 2 5
×    6 4
--------
   9 0 0
+1 3 5 0 0
--------
 1 4,4 0 0
```

14,400 is close to 12,000. The answer is reasonable.

Robin used 14,400 beads.

Study these examples.

```
    6
    A
  3 0 9
×   7 5
-------
  1 5 4 5
+2 1 6 3 0
---------
 2 3,1 7 5
```

```
  1 3 2
×   3 1
-------
  1 3 2
+3 9 6 0
-------
  4 0 9 2
```

```
    4
    2
  $6.5 1
×    8 4
-------
  2 6 0 4
+5 2 0 8 0
---------
 $5 4 6.8 4
```

154

Estimate mentally. Then find the product.

1. 201 $\times$ 44	**2.** 132 $\times$ 23	**3.** 312 $\times$ 11	**4.** 402 $\times$ 31	**5.** 611 $\times$ 43
6. 242 $\times$ 33	**7.** 404 $\times$ 32	**8.** 723 $\times$ 24	**9.** 312 $\times$ 42	**10.** 841 $\times$ 56
11. 492 $\times$ 67	**12.** 387 $\times$ 75	**13.** 525 $\times$ 98	**14.** 906 $\times$ 86	**15.** 759 $\times$ 52
16. \$2.37 $\times$ 45	**17.** \$4.99 $\times$ 68	**18.** \$8.17 $\times$ 39	**19.** \$6.30 $\times$ 53	**20.** \$7.88 $\times$ 47

Multiply.

21. 84 $\times$ 634

22. 52 $\times$ 928

23. 79 $\times$ 837

24. 24 $\times$ \$5.09

25. 59 $\times$ \$3.25

26. 46 $\times$ \$9.72

 **Choose a Computation Method**  Communicate

When you do not need an exact answer, you may
be able to estimate to solve a problem.

**Use the table on page 154 to solve each problem.
Estimate or find an exact answer. Then explain
how you solved each.**

27. How many beads did Tim use?

28. Did Letisha use more or fewer beads than Tim?

29. Suppose Marc and Robin had the same number of rows. Who would use more beads?

30. Who used the most beads? How many beads did that person use?

TECHNOLOGY

Computer Spreadsheets

A **spreadsheet** organizes information into *columns* and *rows*. The columns of a spreadsheet are labeled with letters. The rows are labeled with numbers. Each section of a spreadsheet is called a *cell*.

The spreadsheet below shows how information can be organized.

column B

	A	B	C	D
1	Item	Price	Number Sold	Total Price
2	.table	$69.95	1	$ 69.95
3	chair	$29.95	4	$119.80
4	lamp	$45.99	2	$ 91.98
5		cell	Total Spent	$281.73

row 3 → 3

A cell is identified by its column and row.

In which cell is the price for 1 chair?

▶ To find the cell, first locate the column labeled *Price*. The label *Price* is in column B.

Then locate the row labeled *chair*. The label *chair* is in row 3.

So the price for one chair is in cell B3.

What information is in cell C4?

▶ To find the information, locate the cell C4. Cell C4 is in column C, row 4. The number of lamps sold is in cell C4.

156

PROBLEM SOLVING

1. Freddy's car needs new tires, a muffler, and brakes. He called 4 garages to get prices for the work. Complete the spreadsheet below by using the list Freddy made.

	A	B	C	D	E
1	Garage	Tires	Muffler	Brakes	Total Cost
2		$208.00			
3	Louie's				
4				$157.59	
5			$28.00		

2. Which columns contain money amounts?

3. Which row contains labels?

4. In which cell is the cost for a muffler at Louie's?

5. What information is in cell A4?

6. Which garage has the best buy on tires?

7. How would you find the amount in cell E2?

8. Find the total cost for the work at each garage. Enter the amounts in the spreadsheet.

9. Which garage has the best buy for all the work Freddy needs done?

10. Create a spreadsheet using the information below.

	Week 1	Week 2	Week 3	Total Miles
Marty	15 mi	9 mi	8 mi	?
Sarah	10 mi	12 mi	14 mi	?
Josh	20 mi	18 mi	11 mi	?

4-17 | Problem Solving: Working Backwards

Problem: Karl bought some guppies in March. He had four times as many guppies by the end of May. He had 46 fish by the end of June, which was 10 more than at the end of May. How many guppies did he buy in March?

1 IMAGINE Create a mental picture.

2 NAME

Facts: Karl bought fish in March.
4 times as many in May
10 more than that in June
46 fish in June

Question: How many fish did Karl buy in March?

3 THINK Work backwards. Use the opposite operation.

- First find the number of guppies at the end of May: Subtract 10 from the number of guppies he had in June.

  $46 - 10 = \underline{\ ?\ }$ number in May

- Then to find the number of guppies he had in March: Divide the number of guppies in May by 4.

  number in May $\div$ 4 = number in March

4 COMPUTE $46 - 10 = 36$ number in May
$36 \div 4 = 9$ number in March
Karl bought 9 guppies in March.

5 CHECK Start with 9. Use the opposite operation.
9 guppies in March
$9 \times 4 = 36$ in May
$36 + 10 = 46$ in June The answer checks.

Work backwards to solve each problem.

1. The Torres family came home from
   the matinee at 5:00 P.M. The trip
   to and from the cinema was 15 minutes
   each way. They spent 1 hour and 45 minutes
   at the cinema. What time did they leave home?

IMAGINE	Draw a clock.

NAME	*Facts:*	5:00 P.M. arrived home 15 minutes travel time each way 1 hour 45 minutes at the cinema
	Question:	What time did they leave home?

THINK	Count back each time that was added. 5:00 − 15 minutes − 15 minutes − 1 hour 45 minutes = __?__ *time to* *time from* *at the cinema*

→ **COMPUTE** → **CHECK**

2. Kari had $4.25 left after shopping. She spent
   $11.80 for party favors and $22.55 for a giant
   party pizza. How much money did Kari have
   when she began shopping?

3. Bev, Ruth, and Lisa are sisters. Bev is 8 years
   older than Ruth. Ruth is 5 years older than Lisa,
   who is 16 years old. How old is Bev?

4. Don bought a vase for $36 and a lamp for $78. He
   received $6 change. How much money did he give the cashier?

5. After lunch there were 2 pizzas left over. Grades 1, 2, and 3
   finished 6 pizzas. Grades 4 and 5 each finished 7 pizzas.
   If the teachers finished 2 pizzas, how many pizzas had
   been ordered?

Problem-Solving Applications

Solve each problem and explain the method you used.

1. Oscar's Orchard has 28 Macintosh apple trees. A tree produces about 115 pounds of fruit each year. About how many pounds of apples do the trees produce each year?

2. The orchard has 17 rows of peach trees. There are 16 trees in each row. Does the orchard have more than 300 peach trees?

3. Sonal works for 5 hours every day during harvest. How many hours does she work in thirty days?

4. A fence around the orchard is 894 feet long. Every foot of fencing has three posts. How many posts are in the fence?

5. Customers can pick raspberries for $1.75 per quart. How much would 12 quarts of berries cost?

6. The pick-your-own price at Oscar's Orchard is $3.25 per bushel of apples. Mr. Ennis picked 8 bushels. How much did he spend?

7. Mr. Ennis uses 3 pounds of apples to make 1 pint of apple butter. How many pounds of apples does he need to make 14 pints of apple butter?

8. Each pot of strawberry plants produces about 8 dozen berries. There are 58 pots of plants. About how many strawberries do 58 pots of plants produce?

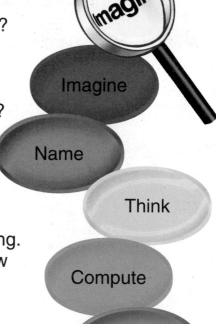

Imagine

Name

Think

Compute

Check

Choose a strategy from the list or use another strategy you know to solve each problem.

9. Emily picked 34 apples. Half of the apples were golden delicious. How many were not golden delicious?

10. Tia gave 5 apples to Ms. Lu and half of what she had left to her grandmother. She used the remaining 6 apples in a pie. How many apples had she brought home?

11. Mia, Nate, and Rob each picked either apples, pears, or grapes. Mia did not pick pears, and Rob did not pick grapes. Nate shared his apples. Which fruit did Rob pick?

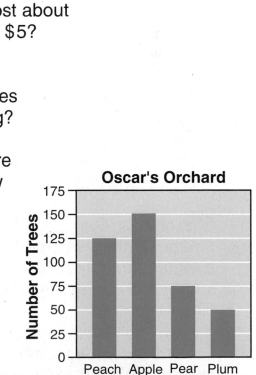

USE THESE STRATEGIES:
Working Backwards
Choose the Operation
Extra Information
Logical Reasoning
Guess and Test

12. Liam picked 124 apples and Cleo picked 152. The pick-your-own apples cost about 4 cents each. Did Cleo spend more than $5?

13. Chad stopped picking fruit at 2:30 P.M. He had picked pears for 1 hour and apples for 45 minutes. When did he start picking?

14. One apple has about 25 seeds. There are about 160 apples in a bushel. About how many seeds are in a bushel of apples?

Use the graph for problems 15 and 16.

15. How many more peach than plum trees were planted in Oscar's Orchard?

16. What kind of trees are double the number of pear trees?

Oscar's Orchard

161

Find the product. *(See pp. 130–135, 138–147, 150–155.)*

1. 2 × 34   **2.** 6 × 30   **3.** 5 × 68   **4.** 4 × 77

5. 3 × 323   **6.** 3 × 450   **7.** 9 × $5.37   **8.** 8 × 6124

9. 10 × 43   **10.** 50 × 30   **11.** 60 × 94   **12.** 10 × 364

13.   53   **14.**   74   **15.**   84   **16.**   30
     × 55   × 38   × 46   × 27

17.   524   **18.**   608   **19.**   735   **20.**   450
     ×   5   × 54   × 46   × 76

21. $6.30   **22.** $4.50   **23.** $9.40   **24.** $5.09
     ×   26   ×   34   ×   17   ×   46

Estimate the product. *(See pp. 136–137, 148–149.)*

25. 8 × 35   **26.** 6 × 736   **27.** 5 × 612   **28.** 9 × $27.50

29. 61 × 54   **30.** 86 × 91   **31.** 32 × $.17   **32.** 16 × 307

PROBLEM SOLVING *(See pp. 126–129, 160–161.)*

33. The product is zero. One factor is 8.
 What is the other factor?

34. Sharon has tiles that are 1 inch
 square. If she uses them to make
 a rectangle that is 14 inches
 long and 5 inches wide, how many
 tiles will she use?

35. Jamal bicycles 18 kilometers each day.
 How far does he bicycle in 12 days?

          (See *Still More Practice*, p. 464.)

CLUSTERING

Tommy kept a record of his family's daily mileage on a car trip to Mexico. About how many miles long was the trip?

On each of the 4 days of the trip, Tommy's family traveled about 400 miles. So you can use 400 to estimate by **clustering**.

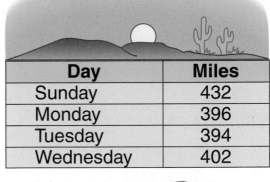

Day	Miles
Sunday	432
Monday	396
Tuesday	394
Wednesday	402

```
    400
  ×   4
  1600
```

The trip was about 1600 miles long.

Estimate the total by clustering.

1. 37 + 41 + 43 + 35

2. 85 + 98 + 87 + 88

3. 105 + 98 + 96

4. 510 + 483 + 503

5. 326 + 289 + 301 + 313

6. 740 + 675 + 690 + 727

7. 2943 + 3201 + 3065

8. 5624 + 4875 + 5133

PROBLEM SOLVING

9. In Elmsford's schools, East has 489 students, Central has 535 students, and West has 492 students. About how many students are in Elmsford?

10. VideoLand rented out 199 movies on Friday, 248 movies on Saturday, and 218 movies on Sunday. About how many movies was this?

Check Your Mastery

Performance Assessment

Use mental math to find each product. Then draw base ten blocks to check each answer.

1. $8 \times 10 = \underline{\ ?\ }$ **2.** $3 \times 50 = \underline{\ ?\ }$ **3.** $2 \times 20 = \underline{\ ?\ }$

Find the product.

4. 3×21 **5.** 7×20 **6.** 4×59 **7.** 8×47

8. 6×101 **9.** 5×360 **10.** $3 \times \$2.29$ **11.** 9×5473

12. 10×77 **13.** 90×80 **14.** 50×26 **15.** 10×133

16. $\begin{array}{r} 16 \\ \times\ 39 \\ \hline \end{array}$ **17.** $\begin{array}{r} 54 \\ \times\ 97 \\ \hline \end{array}$ **18.** $\begin{array}{r} 43 \\ \times\ 21 \\ \hline \end{array}$ **19.** $\begin{array}{r} 60 \\ \times\ 39 \\ \hline \end{array}$

20. $\begin{array}{r} 858 \\ \times\ \ \ 4 \\ \hline \end{array}$ **21.** $\begin{array}{r} 307 \\ \times\ \ 85 \\ \hline \end{array}$ **22.** $\begin{array}{r} 442 \\ \times\ \ 36 \\ \hline \end{array}$ **23.** $\begin{array}{r} 590 \\ \times\ \ 73 \\ \hline \end{array}$

24. $\begin{array}{r} \$5.50 \\ \times\ \ \ \ 18 \\ \hline \end{array}$ **25.** $\begin{array}{r} \$3.25 \\ \times\ \ \ \ 37 \\ \hline \end{array}$ **26.** $\begin{array}{r} \$1.52 \\ \times\ \ \ \ 20 \\ \hline \end{array}$ **27.** $\begin{array}{r} \$3.07 \\ \times\ \ \ \ 45 \\ \hline \end{array}$

PROBLEM SOLVING *Use a strategy you have learned.*

28. The product is 6. One factor is 6. What is the other factor?

29. Cassettes are $9.95 each. How much will Jean pay for 10 cassettes?

30. What is the product of 0 and any number?

Dividing by One Digit

5

6 ÷ 2

A Remainder of One

The story of Joe might just well explain
what happens to numbers when they must remain
after division, and they're left behind
as lonesome remainders. It seems so unkind!

*From A Remainder of One
by Elinor J. Pinczes.*

12 ÷ 9

In this chapter you will:

Study the meanings and rules
  of division
Investigate patterns, missing
  numbers, and divisibility
Estimate and divide whole
  numbers and money
Explore zeros in division
Learn about the order of
  operations and averages
Solve problems by interpreting
  the remainder

**Critical Thinking/
Finding Together**
Use counters to find the quotient
and the remainder, the number
left over, for each division on
the page.

5-1 Division Concepts

▶ You **divide** when you want to:

- **separate** a set into equal parts.

- **share** a set equally.

Cal has 12 pears. He puts 4 pears into each bag. How many bags does he use?

Jo, Meg, and Cara share 12 pears equally. How many pears does each girl get?

$$12 \div 4 = 3$$

$$12 \div 3 = 4$$

He uses 3 bags.

Each girl gets 4 pears.

▶ Here are some rules that can help you to divide quickly and correctly.

- When you divide a number by one, the quotient is the same as the dividend.

$$1\overline{)8}\ ^{8}$$

$$8 \div 1 = 8$$

- When you divide a number other than zero by itself, the quotient is 1.

$$5\overline{)5}\ ^{1}$$

$$5 \div 5 = 1$$

- When you divide zero by any other number, the quotient is 0.

$$6\overline{)0}\ ^{0}$$

$$0 \div 6 = 0$$

- It is *impossible* to divide a number by 0.

Divide.

1. $6\overline{)6}$ 2. $5\overline{)0}$ 3. $1\overline{)7}$ 4. $3\overline{)3}$ 5. $2\overline{)0}$ 6. $9\overline{)9}$

7. $4\overline{)0}$ 8. $1\overline{)5}$ 9. $1\overline{)0}$ 10. $4\overline{)4}$ 11. $1\overline{)2}$ 12. $1\overline{)6}$

Find the quotient.

13. $2 \div 2$ **14.** $9 \div 1$ **15.** $0 \div 7$ **16.** $8 \div 8$ **17.** $3 \div 1$

18. $0 \div 8$ **19.** $7 \div 7$ **20.** $4 \div 1$ **21.** $0 \div 9$ **22.** $1 \div 1$

23. $5 \div 5$ **24.** $8 \div 1$ **25.** $0 \div 3$ **26.** $0 \div 6$ **27.** $9 \div 9$

PROBLEM SOLVING

28. The dividend is 7.
The quotient is 1.
What is the divisor?

29. The divisor is 4.
The quotient is 1.
What is the dividend?

30. The divisor is 5.
The quotient is 5.
What is the dividend?

31. The quotient is 2.
The dividend is 2.
What is the divisor?

32. The dividend is 1.
The quotient is 1.
What is the divisor?

33. The quotient is 0.
What is the dividend?

34. How should 4 friends share 24 apples equally?

35. Dal and 4 friends share 15 oranges. What is a fair share for each child?

36. Sara has 64 plums. She packs them 8 to a basket. How many baskets does she use?

37. Ty packs 8 baskets with 5 peaches to a basket. How many peaches are there?

 **Mental Math**

Use the rules to divide mentally.

38. $0 \div 15$ **39.** $26 \div 26$ **40.** $49 \div 1$ **41.** $0 \div 99$ **42.** $75 \div 1$

43. $429 \div 429$ **44.** $867 \div 1$ **45.** $0 \div 539$ **46.** $938 \div 938$

5-2 Missing Numbers in Division

 Algebra

You can use multiplication facts
to find missing numbers in division.

▶ Find the missing divisor:
63 ÷ _?_ = 9

Think: 9 × _?_ = 63
          9 × 7 = 63

So   63 ÷ 7 = 9.

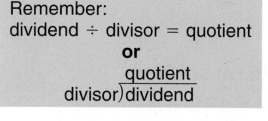

Remember:
dividend ÷ divisor = quotient
or
$$\text{divisor)}\overline{\text{dividend}}^{\text{quotient}}$$

▶ Find the missing dividend: 7)?̄ ⁶

Think: 6 × 7 = _?_
          6 × 7 = 42

So 7)42̄ ⁶.

Study these examples.

9
?)36̄

Think: 9 × _?_ = 36
          9 × 4 = 36

9
So 4)36̄.

_?_ ÷ 6 = 1

Think: 1 × _?_ = 6
          1 × 6 = 6
So   6 ÷ 6 = 1.

Find the missing divisor.

1.   6
   ?)12̄

2.   5
   ?)30̄

3.   8
   ?)32̄

4.   6
   ?)54̄

5.   7
   ?)49̄

6.   4
   ?)36̄

7.   7
   ?)56̄

8.   2
   ?)14̄

9.   3
   ?)15̄

10.   6
    ?)48̄

Find the missing dividend.

11. $9\overline{)?}$ 5

12. $4\overline{)?}$ 6

13. $5\overline{)?}$ 8

14. $9\overline{)?}$ 3

15. $8\overline{)?}$ 4

16. $6\overline{)?}$ 9

17. $2\overline{)?}$ 0

18. $8\overline{)?}$ 2

19. $5\overline{)?}$ 4

20. $3\overline{)?}$ 1

21. $7\overline{)?}$ 6

22. $5\overline{)?}$ 9

23. $6\overline{)?}$ 5

24. $9\overline{)?}$ 2

25. $4\overline{)?}$ 4

Find the missing number.

26. $35 \div \underline{\ ?\ } = 5$

27. $42 \div \underline{\ ?\ } = 6$

28. $72 \div \underline{\ ?\ } = 9$

29. $\underline{\ ?\ } \div 3 = 4$

30. $\underline{\ ?\ } \div 7 = 8$

31. $\underline{\ ?\ } \div 2 = 5$

32. $64 \div \underline{\ ?\ } = 8$

33. $18 \div \underline{\ ?\ } = 3$

34. $9 \div \underline{\ ?\ } = 1$

35. $\underline{\ ?\ } \div 9 = 3$

36. $\underline{\ ?\ } \div 8 = 5$

37. $\underline{\ ?\ } \div 7 = 4$

38. $24 \div \underline{\ ?\ } = 3$

39. $48 \div \underline{\ ?\ } = 6$

40. $25 \div \underline{\ ?\ } = 5$

Skills to Remember

Write the digit in the tens place.

41. 432   42. 7604   43. 28   44. 35,196   45. 8172

Write the digit in the hundreds place.

46. 3584   47. 67,312   48. 192   49. 2837   50. 495

Write the place of the red digit.

51. 7183   52. 14,697   53. 13,452   54. 8306   55. 9563

Number Patterns

▶ What is the next number in this pattern?

16, 8, 4, 2, ?

- First find the rule.

Think: 16, 8, 4, 2

÷2 ÷2 ÷2

| Rule: Divide by 2. |

- Then complete the pattern.

16, 8, 4, 2, ?

16, 8, 4, 2, 1

Think: 2 ÷ 2 = 1

The next number in the pattern is 1.

▶ Here is another pattern. What is the next number?

2, 7, 6, 11, 10, ?

Think: 2, 7, 6, 11, 10

+5 −1 +5 −1

| Rule: Add 5. Subtract 1. |

2, 7, 6, 11, 10, ?

2, 7, 6, 11, 10, 15

Think: 10 + 5 = 15

The next number is 15.

Write the rule for each pattern.
Then write the next number.

1. 10, 12, 14, 16, ?

2. 30, 40, 50, 60, ?

3. 25, 30, 35, ?

4. 10, 13, 16, 19, ?

Write the rule. Complete the pattern.

5. 27, 9, 3, _?_

6. 33, 31, 29, _?_

7. 42, 38, 34, _?_

8. 4, 8, 16, 32, _?_

9. 16, 20, 18, 22, 20, _?_

10. 54, 51, 52, 49, 50, _?_

11. 4, 12, 10, 30, 28, _?_

12. 5, 10, 13, 26, 29, _?_

13. 1, 4, 4, 7, 7, 10, _?_

14. 10, 12, 6, 8, 4, 6, _?_

Write a pattern of eight numbers for each rule.

15. Rule: Add 6.

16. Rule: Subtract 3.

17. Rule: Multiply by 2.

18. Rule: Add 50.

19. Rule: Add 3. Add 1.

20. Rule: Add 10. Subtract 1.

 Finding Together

Even numbers have either 0, 2, 4, 6, or 8 as their ones digits.

Odd numbers are all the whole numbers that are not even.

Is the sum or product odd or even? Write *O* or *E*.

21. Even + Even

22. Even × Even

23. Odd + Odd

24. Odd × Odd

25. Even + Odd

26. Odd × Even

27. In your Math Journal, write two or three examples for each of exercises 21–26 to prove your answers.

Math Journal

5-4 Estimating in Division

You can estimate quotients before you divide.

Estimate: 2672 ÷ 8

- Find where the quotient begins.

  Try dividing thousands.

  $8\overline{)2672}$ 8 > 2 **Not enough thousands**

  Try dividing hundreds.

  $8\overline{)2672}$ 8 < 26 **Enough hundreds**

  So the quotient begins in the hundreds place.

  $$\overset{\text{X}}{8\overline{)2672}}$$

- Find the first digit of the quotient.

  Think: About how many 8s in 26?

  $2 \times 8 = 16$ too small

  $3 \times 8 = 24 \leftarrow$ | 26 is between

  $4 \times 8 = 32$ too large | 16 and 32.

  Try 3.

  $$\overset{3}{8\overline{)2672}}$$

- Since you are estimating, write zeros for the other digits.

  about $\overset{300}{8\overline{)2672}}$

Study these examples.

about $\overset{200}{2\overline{)523}}$ about $\overset{70}{6\overline{)425}}$ about $\overset{\$\,6.00}{5\overline{)\$32.75}}$

Write an X in the place where the quotient begins.

1. $\overset{\text{X}}{4\overline{)76}}$ 2. $6\overline{)48}$ 3. $2\overline{)451}$ 4. $7\overline{)927}$ 5. $8\overline{)745}$

6. $3\overline{)127}$ 7. $5\overline{)370}$ 8. $2\overline{)1468}$ 9. $7\overline{)4303}$

Estimate the quotient.

10. $9\overline{)95}$ **11.** $6\overline{)43}$ **12.** $2\overline{)87}$ **13.** $5\overline{)38}$ **14.** $4\overline{)92}$

15. $4\overline{)591}$ **16.** $7\overline{)862}$ **17.** $3\overline{)947}$ **18.** $2\overline{)815}$

19. $6\overline{)275}$ **20.** $9\overline{)467}$ **21.** $8\overline{)744}$ **22.** $5\overline{)342}$

23. $7\overline{)2439}$ **24.** $4\overline{)3622}$ **25.** $3\overline{)1729}$ **26.** $9\overline{)5649}$

27. $2\overline{)\$4.94}$ **28.** $3\overline{)\$6.42}$ **29.** $5\overline{)\$17.50}$ **30.** $4\overline{)\$28.58}$

Estimating with Compatible Numbers

Estimate: $53 \div 6$

> Think: What number times 6 has a product close to 53?

$9 \times 6 = 54$

So $53 \div 6$ is about 9.

Estimate: $223 \div 7$

> Think: How many tens times 7 has a product close to 22 tens?

3 tens $\times$ 7 = 21 tens

So $223 \div 7$ is about 3 tens, or 30.

Write the dividend you would use to estimate the quotient.

31. $55 \div 8$ **32.** $46 \div 6$ **33.** $362 \div 5$ **34.** $178 \div 3$

Estimate the quotient. Use compatible numbers.

35. $4\overline{)29}$ **36.** $5\overline{)33}$ **37.** $8\overline{)26}$ **38.** $3\overline{)11}$ **39.** $7\overline{)40}$

40. $3\overline{)61}$ **41.** $4\overline{)84}$ **42.** $2\overline{)63}$ **43.** $5\overline{)56}$ **44.** $3\overline{)91}$

45. $7\overline{)285}$ **46.** $5\overline{)161}$ **47.** $6\overline{)524}$ **48.** $9\overline{)472}$ **49.** $7\overline{)551}$

5-5 One-Digit Quotients

Jaime gave the same number of pencils to each of 5 friends. He had 22 pencils. How many pencils did each friend receive? How many pencils were left over?

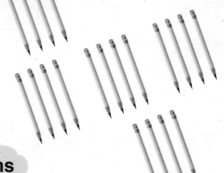

To find how many each received, divide: $22 \div 5 = \underline{\ ?\ }$

Think: $5\overline{)22}$ $5 > 2$ **Not enough tens**
 $5\overline{)22}$ $5 < 22$ **Enough ones**
The quotient begins in the ones place.

Estimate: About how many 5s in 22?

$4 \times 5 = 20$ too small

$5 \times 5 = 25$ too large

22 is between 20 and 25. Try 4.

Divide.	Multiply.	Subtract and compare.	Write the remainder.
$\begin{array}{r} 4 \\ 5\overline{)2\,2} \end{array}$	$\begin{array}{r} \times\ 4 \\ 5\overline{)2\,2} \\ 2\,0 \end{array}$	$\begin{array}{r} 4 \\ 5\overline{)2\,2} \\ -2\,0 \\ \hline 2 \end{array}$ $2 < 5$	$\begin{array}{r} 4\ \text{R 2} \\ 5\overline{)2\,2} \\ -2\,0 \\ \hline 2 \end{array}$ **remainder**

Check by multiplying and adding.

$\begin{array}{r} 4 \leftarrow \text{quotient} \\ \times\ 5 \leftarrow \text{divisor} \\ \hline 20 \\ +\ 2 \leftarrow \text{remainder} \\ \hline 22 \leftarrow \text{dividend} \end{array}$

The remainder must always be less than the divisor.

Each friend received 4 pencils. There were 2 pencils left over.

Copy and complete.

$$\begin{array}{r} 6 \\ 4\overline{)2\,4} \\ -2\,4 \\ \hline 0 \end{array}$$

1. (above)

There is no remainder.

2. $\begin{array}{r} 6 \text{ R ?} \\ 3\overline{)2\,0} \\ -1\,8 \\ \hline 2 \end{array}$

3. $\begin{array}{r} 9 \text{ R ?} \\ 5\overline{)4\,8} \\ -4\,5 \\ \hline ? \end{array}$

4. $\begin{array}{r} ? \text{ R ?} \\ 2\overline{)1\,3} \\ -?\,? \\ \hline ? \end{array}$

Divide.

5. $2\overline{)15}$ 6. $4\overline{)35}$ 7. $3\overline{)23}$ 8. $5\overline{)17}$ 9. $6\overline{)27}$

10. $6\overline{)14}$ 11. $4\overline{)26}$ 12. $5\overline{)37}$ 13. $7\overline{)50}$ 14. $4\overline{)33}$

15. $6\overline{)55}$ 16. $5\overline{)38}$ 17. $8\overline{)68}$ 18. $2\overline{)19}$ 19. $7\overline{)45}$

20. $8\overline{)23}$ 21. $3\overline{)26}$ 22. $7\overline{)35}$ 23. $7\overline{)29}$ 24. $8\overline{)38}$

25. $9\overline{)64}$ 26. $8\overline{)52}$ 27. $6\overline{)45}$ 28. $9\overline{)71}$ 29. $9\overline{)82}$

Find the quotient and the remainder.

30. $25 \div 3$ 31. $23 \div 7$ 32. $84 \div 9$

33. $50 \div 8$ 34. $38 \div 4$ 35. $57 \div 6$

PROBLEM SOLVING

36. Caryn put away 36 crayons in boxes. Each box holds 8 crayons. How many boxes could be filled? How many crayons would be left over?

37. Mika put 37 drawings in folders. She put 4 drawings in each folder. How many folders were there? How many extra drawings were there?

38. Bill placed the same number of pencils at each of 6 tables. He began with 44 pencils. At most, how many pencils could he have placed at each table? How many pencils would have been left over?

Divisibility

A number is **divisible** by another number
when the remainder is zero.

List the whole numbers from 1 to 50.

▶ Skip count to 50 by 2. Circle
the numbers you counted. Look
at the digits in the *ones* place.
What pattern do you see?

> Even numbers end in
> 0, 2, 4, 6, or 8.
> **All even numbers are
> divisible by 2.**

▶ Skip count to 50 by 5. Draw
a box around the numbers you
counted. Look at the digits
in the ones place. What pattern
do you see?

> **All numbers ending in
> 0 or 5 are divisible by 5.**

▶ Skip count to 50 by 10. Mark
an X on these numbers.
What pattern do you see?

> **All numbers ending in
> 0 are divisible by 10.**

Is the number divisible by 2? Write *yes* or *no*.

1. 28　　**2.** 75　　**3.** 700　　**4.** 144　　**5.** 807

6. 516　　**7.** 343　　**8.** 2931　　**9.** 1462　　**10.** 7749

11. 6847　　**12.** 2900　　**13.** 75,192　　**14.** 27,346　　**15.** 92,983

Is the number divisible by 5? Write *yes* or *no*.

16. 64　　**17.** 85　　**18.** 900　　**19.** 245　　**20.** 819

21. 703　　**22.** 456　　**23.** 1820　　**24.** 4795　　**25.** 9240

26. 8675　　**27.** 3299　　**28.** 10,000　　**29.** 42,685　　**30.** 74,007

Is the number divisible by 10? Write *yes* or *no*.

31. 930   **32.** 749   **33.** 6820   **34.** 5000   **35.** 8304

36. 1006   **37.** 4673   **38.** 52,651   **39.** 66,830   **40.** 90,060

Copy and complete the table.

41.

Divisible by	60	88	75	600	494	750	2313	1026	8750
2	yes	?	?	?	?	?	no	?	?
5	yes	?	yes	?	?	?	?	?	?
10	?	?	?	yes	?	?	?	?	?

Divisibility by 3

If the sum of the digits of a number is divisible by 3, that number is divisible by 3.

$27 \rightarrow 2 + 7 = 9 \rightarrow 9 \div 3 = 3$
27 is divisible by 3.

$435 \rightarrow 4 + 3 + 5 = 12 \rightarrow 12 \div 3 = 4$
435 is divisible by 3.

Is the number divisible by 3? Write *yes* or *no*.

42. 72   **43.** 54   **44.** 253   **45.** 534   **46.** 312

47. 932   **48.** 210   **49.** 842   **50.** 1065   **51.** 4906

Share Your Thinking

Discuss ✓

52. What numbers are the greatest possible remainders?

Divisors	2	3	4	5	6	7	8	9
Greatest Remainders	?	?	?	?	?	?	?	?

53. What pattern do you notice in the table?

Ian cut a 72-inch length of cloth into 2 equal strips. What was the length of each strip?

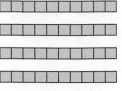

To find the length of each strip, divide: $72 \div 2 = $?

Think: $2\overline{)72}$ $2 < 7$ **Enough tens**

Estimate: About how many 2s in 7?

$3 \times 2 = 6$
$4 \times 2 = 8$ ← $\boxed{\text{7 is between 6 and 8. Try 3.}}$

Divide the tens.	Multiply.	Subtract and compare.	Bring down the ones.
$\begin{array}{r} 3 \\ 2\overline{)7\,2} \end{array}$	$\begin{array}{r} 3 \\ 2\overline{)7\,2} \\ 6 \end{array}$	$\begin{array}{r} 3 \\ 2\overline{)7\,2} \\ -6 \\ \hline 1 \end{array}$ ← $\boxed{1 < 2}$	$\begin{array}{r} 3 \\ 2\overline{)7\,2} \\ -6 \downarrow \\ \hline 1\,2 \end{array}$

Repeat the steps to divide the ones.

Estimate: About how many 2s in 12?

$6 \times 2 = 12$ ← $\boxed{\text{Try 6.}}$

Divide the ones.	Multiply.	Subtract and compare.	Check.
$\begin{array}{r} 3\,6 \\ 2\overline{)7\,2} \\ -6 \downarrow \\ \hline 1\,2 \end{array}$	$\begin{array}{r} 3\,6 \\ 2\overline{)7\,2} \\ -6 \downarrow \\ \hline 1\,2 \\ 1\,2 \end{array}$	$\begin{array}{r} 3\,6 \\ 2\overline{)7\,2} \\ -6 \downarrow \\ \hline 1\,2 \\ -1\,2 \\ \hline 0 \end{array}$	$\begin{array}{r} 3\,6 \\ \times\ \ 2 \\ \hline 7\,2 \end{array}$

$36 \times 2 = 2 \times 36$

No remainder

The length of each strip was 36 inches.

Copy and complete.

$$
\begin{array}{r}
1\ 0 \\
\textbf{1.}\ 4\overline{)4\ 0} \\
-4\ \downarrow \\
\hline
0\ 0 \\
-\ ? \\
\hline
?
\end{array}
\qquad
\begin{array}{r}
2\ ? \\
\textbf{2.}\ 4\overline{)8\ 4} \\
-8\ \downarrow \\
\hline
0\ 4 \\
-\ ? \\
\hline
?
\end{array}
\qquad
\begin{array}{r}
1\ ? \\
\textbf{3.}\ 6\overline{)7\ 8} \\
-?\ \downarrow \\
\hline
?\ ? \\
-?\ ? \\
\hline
?
\end{array}
\qquad
\begin{array}{r}
?\ ? \\
\textbf{4.}\ 2\overline{)3\ 4} \\
-?\ \downarrow \\
\hline
?\ ? \\
-?\ ? \\
\hline
?
\end{array}
$$

Estimate. Then divide.

5. $5\overline{)60}$ **6.** $6\overline{)84}$ **7.** $4\overline{)64}$ **8.** $7\overline{)91}$ **9.** $3\overline{)69}$

10. $8\overline{)96}$ **11.** $4\overline{)92}$ **12.** $6\overline{)96}$ **13.** $3\overline{)48}$ **14.** $9\overline{)99}$

Find the quotient.

15. $84 \div 3$ **16.** $80 \div 5$ **17.** $56 \div 4$

18. $45 \div 3$ **19.** $90 \div 2$ **20.** $88 \div 2$

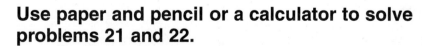

Choose a Computation Method

Read division problems carefully before you decide
whether to use paper and pencil or a calculator
when you solve them.

Use paper and pencil or a calculator to solve problems 21 and 22.

21. Reg made 80 pompons. He sewed 9 pompons on each costume. At most, how many costumes were there? How many pompons were left over?

22. Kate cut an 84-inch long ribbon into 3 equal parts. How many inches long was each part?

23. If you need to know the remainder, why is it easier to use paper and pencil?

Math Journal

179

5-8 More Two-Digit Quotients

Luz had 80 favors to separate equally into 6 party bags. At most, how many favors could she put in each bag? How many favors would be left over?

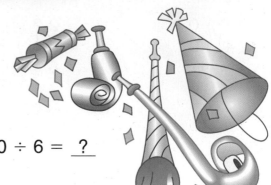

To find how many in each bag, divide: 80 ÷ 6 = __?__

Think: 6)‾80‾   6 < 8   **Enough tens**

Estimate: About how many 6s in 8?

$$1 \times 6 = 6 \longleftarrow \boxed{\text{Try 1.}}$$
$$2 \times 6 = 12$$

Divide the tens.	Multiply.	Subtract and compare.	Bring down the ones.
$\begin{array}{r} 1 \\ 6\overline{)8\,0} \end{array}$	$\begin{array}{r} \overset{\times}{}1 \\ 6\overline{)8\,0} \\ 6 \end{array}$	$\begin{array}{r} 1 \\ 6\overline{)8\,0} \\ -6 \\ \hline 2 \end{array}$ $\boxed{2 < 6}$	$\begin{array}{r} 1 \\ 6\overline{)8\,0} \\ -6\downarrow \\ \hline 2\,0 \end{array}$

Repeat the steps.

Divide the ones.	Multiply.	Subtract and compare.	Check.
$\begin{array}{r} 1\,3 \\ 6\overline{)8\,0} \\ -6\downarrow \\ \hline 2\,0 \end{array}$	$\begin{array}{r} \times1\,3 \\ 6\overline{)8\,0} \\ -6\downarrow \\ \hline 2\,0 \\ 1\,8 \end{array}$	$\begin{array}{r} 1\,3 \;\; \text{R}\,2 \\ 6\overline{)8\,0} \\ -6\downarrow \\ \hline 2\,0 \\ -1\,8 \\ \hline 2 \end{array}$ $\boxed{2 < 6}$	$\begin{array}{r} 1\,3 \\ \times\;\;\; 6 \\ \hline 7\,8 \\ +\;\;\; 2 \\ \hline 8\,0 \end{array}$

At most, she could put 13 favors into each bag.
There would be 2 favors left over.

180

Copy and complete.

1. $\overset{\displaystyle 1\ 0 \quad R\ \underline{?}}{5\overline{)5\ 4}}$
$\begin{array}{r}-5\downarrow \\ \hline 4 \\ -\ ? \\ \hline ?\end{array}$

2. $\overset{\displaystyle 2\ ? \quad R\ \underline{?}}{3\overline{)7\ 4}}$
$\begin{array}{r}-6\downarrow \\ \hline 1\ 4 \\ -?\ ? \\ \hline ?\end{array}$

3. $\overset{\displaystyle 1\ ? \quad R\ \underline{?}}{8\overline{)9\ 8}}$
$\begin{array}{r}-?\downarrow \\ \hline ?\ ? \\ -?\ ? \\ \hline ?\end{array}$

4. $\overset{\displaystyle ?\ ? \quad R\ 3}{4\overline{)9\ 9}}$
$\begin{array}{r}-8\downarrow \\ \hline 1\ ? \\ -?\ ? \\ \hline 3\end{array}$

Estimate. Then divide.

5. $4\overline{)49}$
6. $2\overline{)81}$
7. $5\overline{)92}$
8. $6\overline{)83}$
9. $3\overline{)92}$

10. $8\overline{)91}$
11. $7\overline{)87}$
12. $3\overline{)58}$
13. $4\overline{)89}$
14. $2\overline{)74}$

15. $3\overline{)37}$
16. $5\overline{)63}$
17. $7\overline{)94}$
18. $6\overline{)67}$
19. $8\overline{)89}$

20. $5\overline{)86}$
21. $2\overline{)93}$
22. $4\overline{)51}$
23. $2\overline{)47}$
24. $7\overline{)79}$

25. $6\overline{)97}$
26. $4\overline{)86}$
27. $6\overline{)99}$
28. $5\overline{)87}$
29. $9\overline{)98}$

30. $61 \div 2$
31. $47 \div 3$
32. $71 \div 4$
33. $76 \div 5$

34. $92 \div 9$
35. $84 \div 8$
36. $96 \div 7$
37. $85 \div 6$

PROBLEM SOLVING

38. There were 65 balloons at Willy's party. He tied 6 balloons to each tree in his yard and the extra balloons to his mailbox. What is the greatest number of trees that could have been in Willy's yard? How many balloons would he have tied to his mailbox?

39. Val hid 96 eggs in the yard. Each of 7 children found the same number of eggs. What is the greatest number of eggs each child could have found? How many eggs would still have remained hidden?

5-9 Three-Digit Quotients

Divide: 745 ÷ 2 = __?__

Use the division steps to find three-digit quotients.

- Divide the hundreds.

  Estimate: __?__ × 2 = 7

  3 × 2 = 6

  4 × 2 = 8

  Try 3.

  $$\begin{array}{r} 3 \\ 2\overline{)745} \\ -6\downarrow \\ \hline 1\,4 \end{array}$$

- Divide the tens.

  Estimate: __?__ × 2 = 14

  7 × 2 = 14

  Try 7.

  $$\begin{array}{r} 37 \\ 2\overline{)745} \\ -6\downarrow \\ \hline 1\,4\downarrow \\ -1\,4\downarrow \\ \hline 0\,5 \end{array}$$

- Divide the ones.

  Estimate: __?__ × 2 = 5

  2 × 2 = 4

  3 × 2 = 6

  Try 2.

  $$\begin{array}{r} 372 \quad R\,1 \\ 2\overline{)745} \\ -6\downarrow \\ \hline 1\,4 \\ -1\,4 \\ \hline 0\,5 \\ -\;4 \\ \hline 1 \end{array}$$

  This 0 need not be written.

- Check.

  $$\begin{array}{r} 372 \\ \times\quad 2 \\ \hline 744 \\ +\quad 1 \\ \hline 745 \end{array}$$

Division Steps

- Estimate.
- Divide.
- Multiply.
- Subtract.
- Compare.
- Bring down.
- Repeat the steps as necessary.
- Check.

Remember: Write the remainder in the quotient.

Copy and complete.

1.
```
      1 2 5  R 3
  5)6 2 8
   -5↓
    1 2
   -? ?↓
      2 8
     -? ?
        3
```
Check.
```
      1 2 5
  ×       5
      6 2 5
  +       ?
      6 2 8
```

2.
```
      2 4 3  R 2
  3)7 3 1
   -6↓
    1 3
   -1 2↓
      1 1
     -  ?
        2
```
Check.
```
      2 4 3
  ×       3
      7 2 9
  +       ?
      7 3 1
```

3.
```
      2 6 ?
  3)8 0 7
   -?↓
    2 0
   -? ?↓
      2 7
     -? ?
        0
```
Check.
```
      2 6 ?
  ×       3
      ? ? ?
```

4.
```
      ? 2 ?  R ?
  2)6 5 1
   -6↓
    0 5
    -?↓
      1 1
     -1 0
        ?
```
Check.
```
      ? 2 ?
  ×       2
      ? ? ?
  +       ?
      6 5 1
```

Estimate. Then divide.

5. 2)632 **6.** 4)976 **7.** 3)733 **8.** 4)762 **9.** 7)931

10. 5)568 **11.** 7)868 **12.** 4)907 **13.** 6)918 **14.** 4)872

15. 8)936 **16.** 5)860 **17.** 2)524 **18.** 7)802 **19.** 2)922

20. 3)988 **21.** 3)537 **22.** 6)714 **23.** 5)815 **24.** 3)884

PROBLEM SOLVING

25. At the supermarket 950 apples were placed in 3 piles. Each pile contained the same number of apples. At most, how many apples were there in each pile? How many apples were left over?

More Difficult Quotients

Handcraft Toys had 274 trains to ship to 8 stores. The same number of trains were shipped to each store. At most, how many trains did each store receive? How many trains were left over?

To find how many each received, divide: $274 \div 8 = \underline{\ ?\ }$

> Think: $8\overline{)274}$ $8 > 2$ **Not enough hundreds**
> $8\overline{)274}$ $8 < 27$ **Enough tens**

Estimate: $3 \times 8 = 24 \leftarrow$ | Try 3. |
           $4 \times 8 = 32$

Divide the tens.	Divide the ones.	Check.

Divide the tens.

$$
\begin{array}{r}
3 \\
8\overline{)2\,7\,4} \\
-2\,4\downarrow \\
\hline
3\,4
\end{array}
$$

Divide the ones.

$$
\begin{array}{r}
3\,4 \quad \text{R 2}\\
8\overline{)2\,7\,4} \\
-2\,4\downarrow \\
\hline
3\,4 \\
-3\,2 \\
\hline
2
\end{array}
$$

Check.

$$
\begin{array}{r}
3\,4 \\
\times\ \ 8 \\
\hline
2\,7\,2 \\
+\ \ \ 2 \\
\hline
2\,7\,4
\end{array}
$$

Each store received at most 34 trains. There were 2 trains left over.

Copy and complete.

1.
$$
\begin{array}{r}
7\,? \\
8\overline{)6\,0\,8} \\
-5\,6\downarrow \\
\hline
?\,8 \\
-?\,? \\
\hline
?
\end{array}
$$

2.
$$
\begin{array}{r}
8\,? \quad \text{R }\underline{\ ?\ }\\
5\overline{)4\,3\,3} \\
-?\,?\downarrow \\
\hline
3\,? \\
-?\,? \\
\hline
3
\end{array}
$$

3.
$$
\begin{array}{r}
?\,? \quad \text{R 4}\\
6\overline{)3\,5\,8} \\
-3\,0\downarrow \\
\hline
?\,8 \\
-?\,? \\
\hline
4
\end{array}
$$

4.
$$
\begin{array}{r}
?\,? \quad \text{R }\underline{\ ?\ }\\
9\overline{)4\,7\,2} \\
-4\,?\downarrow \\
\hline
?\,2 \\
-?\,? \\
\hline
?
\end{array}
$$

Estimate. Then find the quotient.

5. $3\overline{)105}$ **6.** $4\overline{)232}$ **7.** $6\overline{)258}$ **8.** $3\overline{)186}$ **9.** $5\overline{)130}$

10. $6\overline{)436}$ **11.** $7\overline{)201}$ **12.** $5\overline{)359}$ **13.** $4\overline{)354}$ **14.** $7\overline{)182}$

15. $9\overline{)756}$ **16.** $3\overline{)202}$ **17.** $9\overline{)337}$ **18.** $6\overline{)576}$ **19.** $8\overline{)197}$

20. $6\overline{)220}$ **21.** $4\overline{)228}$ **22.** $7\overline{)195}$ **23.** $5\overline{)295}$ **24.** $6\overline{)335}$

25. $7\overline{)308}$ **26.** $9\overline{)823}$ **27.** $8\overline{)692}$ **28.** $7\overline{)666}$ **29.** $9\overline{)717}$

Divide.

30. $657 \div 9$ **31.** $267 \div 8$ **32.** $396 \div 4$

33. $462 \div 5$ **34.** $498 \div 6$ **35.** $591 \div 7$

PROBLEM SOLVING Use a calculator or paper and pencil.

36. The dividend is 272. The quotient is 34. What is the divisor?

37. The dividend is 359. The divisor is 7. What is the remainder?

38. Peg packs 594 wooden animals into 6 boxes of the same size. At most, how many wooden animals does she pack into each box?

39. Janice has 108 wooden pogs to put into plastic containers. Each container holds one dozen pogs. How many containers does Janice need to fit all the pogs?

40. There are 147 tops at the factory store. If the same number of tops are sold on each of 5 days, what is the greatest number of tops that could be sold each day? How many tops would not be sold?

41. Brendan carves 193 figurines of people for dollhouses. There are 4 people in each dollhouse family. At most, how many families does he carve? How many figurines are left over?

5-11 | Zeros in the Quotient

 Hands-On Understanding

You Will Need: base ten blocks, paper, pencil

Divide: 317 ÷ 3 = __?__

Step 1   Model 317. Then separate the hundreds equally into 3 sets.

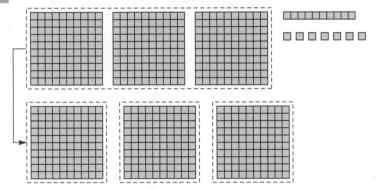

$$\begin{array}{r} 1 \\ 3\overline{)3\,1\,7} \\ -3\downarrow \\ \hline 0\,1 \end{array}$$

How many hundreds are in each equal set?
Are there any hundreds left over?

Step 2   You cannot separate 1 ten into the 3 sets.
Regroup the 1 ten as 10 ones.

$$\begin{array}{r} 1\,0 \\ 3\overline{)3\,1\,7} \\ -3\downarrow \\ \hline 0\,1 \\ -0\downarrow \\ \hline 1\,7 \end{array}$$

How many tens are there now? How many ones?

Step 3   Separate the ones equally into the 3 sets.

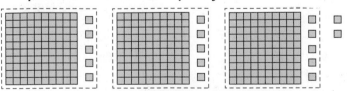

$$\begin{array}{r} 1\,0\,5\ \text{R}\,2 \\ 3\overline{)3\,1\,7} \\ -3\downarrow \\ \hline 0\,1 \\ -\ 0\downarrow \\ \hline 1\,7 \\ -1\,5 \\ \hline 2 \end{array}$$

How many ones are in each equal set? How
many ones are left over? What is the quotient
and the remainder when you divide 317 by 3?

Here is how to divide 317 by 3 using paper and pencil.

Think: $3\overline{)317}$ $3 = 3$ **Enough hundreds**

Estimate: $\underline{?} \times 3 = 3$
$\qquad\quad 1 \times 3 = 3$ ← $\boxed{\text{Try 1.}}$

Divide the hundreds.	Divide the tens.	Divide the ones.	Check.
$\begin{array}{r} 1 \\ 3\overline{)317} \\ -3\downarrow \\ \hline 0\,1 \end{array}$	$\begin{array}{r} 10 \\ 3\overline{)317} \\ -3\downarrow \\ \hline 0\,1 \\ -0\downarrow \\ \hline 17 \end{array}$	$\begin{array}{r} 105\ \text{R}\,2 \\ 3\overline{)317} \\ -3\downarrow \\ \hline 0\,1 \\ -0\downarrow \\ \hline 17 \\ -15 \\ \hline 2 \end{array}$	$\begin{array}{r} 105 \\ \times\ \ \ 3 \\ \hline 315 \\ +\ \ \ \ 2 \\ \hline 317 \end{array}$

$\boxed{\begin{array}{l}3 > 1\ \textbf{Not}\\ \textbf{enough tens}\\ \text{Write 0 in the}\\ \text{tens place.}\end{array}}$

Divide and check. You may use base ten blocks.

1. $4\overline{)800}$ **2.** $2\overline{)600}$ **3.** $3\overline{)390}$ **4.** $4\overline{)840}$ **5.** $5\overline{)550}$

6. $3\overline{)918}$ **7.** $4\overline{)824}$ **8.** $8\overline{)832}$ **9.** $9\overline{)954}$ **10.** $6\overline{)642}$

11. $6\overline{)609}$ **12.** $2\overline{)817}$ **13.** $4\overline{)842}$ **14.** $7\overline{)745}$ **15.** $5\overline{)508}$

16. $5\overline{)841}$ **17.** $9\overline{)985}$ **18.** $2\overline{)615}$ **19.** $3\overline{)902}$ **20.** $8\overline{)847}$

Communicate

21. Examine the divisors and the first two digits of the dividends in exercises 6–20. What do you notice that can help you predict *before* you divide that there will be a zero in the tens place in the quotient?

Discuss

Larger Numbers in Division

Divide: 4925 ÷ 7 = ___?___

Think: $7\overline{)4925}$ 7 > 4 **Not enough thousands**
$7\overline{)4925}$ 7 < 49 **Enough hundreds**

- Divide the hundreds.
  Estimate: ___?___ × 7 = 49
          7 × 7 = 49
          Try 7.

$$
\begin{array}{r}
7 \\
7\overline{)4925} \\
-49 \\
\hline
02
\end{array}
$$

- Divide the tens.
  Estimate: 7 > 2
          Not enough tens
          Write 0 in the
          tens place.

$$
\begin{array}{r}
70 \\
7\overline{)4925} \\
-49 \\
\hline
2 \\
-0 \\
\hline
25
\end{array}
$$

- Divide the ones.
  Estimate: ___?___ × 7 = 25
          3 × 7 = 21
          4 × 7 = 28
          Try 3.

$$
\begin{array}{r}
703 \text{ R 4} \\
7\overline{)4925} \\
-49 \\
\hline
02 \\
-0 \\
\hline
25 \\
-21 \\
\hline
4
\end{array}
$$

- Check.

$$
\begin{array}{r}
703 \\
\times\quad 7 \\
\hline
4921 \\
+\quad 4 \\
\hline
4925
\end{array}
$$

188

Copy and complete.

1.
```
     2 3 ? ?
  4)9 5 6 0
  -? ↓
    ? 5
   -? ? ↓
     ? 6
    -? ? ↓
      ? 0
```

2.
```
     5 ? ?  R ?
  4)2 2 4 2
  -? ? ↓
    2 4
   -? ? ↓
     0 2
     -? ↓
      ?
```

3.
```
     3 ? ?  R ?
  7)2 5 0 1
  -? ? ↓
    4 0
   -3 5 ↓
     5 1
    -4 9
      ?
```

Estimate. Then divide.

4. 4)7576   5. 6)1344   6. 3)8217   7. 7)2982   8. 5)7870

9. 8)4336   10. 2)5566   11. 9)1962   12. 3)4545   13. 7)4361

14. 6)2418   15. 8)7249   16. 9)4567   17. 7)5320   18. 8)3600

19. 4)6204   20. 6)5043   21. 5)9990   22. 3)8181   23. 3)1311

24. 9)2884   25. 7)3225   26. 9)2772   27. 8)2884   28. 9)3675

Find the quotient and any remainder.

29. $1332 \div 6$ 30. $2562 \div 4$ 31. $2454 \div 5$

32. $1753 \div 7$ 33. $4638 \div 6$ 34. $6834 \div 7$

PROBLEM SOLVING Use a calculator or paper and pencil.

35. Felipe has 2943 stamps. He keeps an equal number of stamps in each of 3 stamp albums. How many stamps does Felipe keep in each stamp album?

36. In 9 months Jill collected 941 stamps and Joe collected 931 stamps. The total number of stamps collected each month was the same. How many stamps did they collect each month together?

Dividing Money

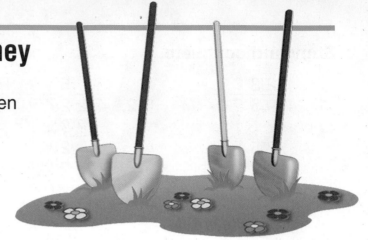

Meghan bought 4 identical garden spades for $95.92. What did each spade cost?

To find the cost of each, divide: $95.92 ÷ 4 = _?_

Write the dollar sign and decimal point in the quotient above the dollar sign and decimal point in the dividend.	Divide as usual.	Check

$$
\begin{array}{r}
\$. \\
4\overline{)\$9\,5.9\,2}
\end{array}
$$

$$
\begin{array}{r}
\$2\,3.9\,8 \\
4\overline{)\$9\,5.9\,2} \\
-8 \\
\hline
1\,5 \\
-1\,2 \\
\hline
3\,9 \\
-3\,6 \\
\hline
3\,2 \\
-3\,2 \\
\hline
\end{array}
$$

$$
\begin{array}{r}
\$2\,3.9\,8 \\
\times4 \\
\hline
\$9\,5.9\,2
\end{array}
$$

Each spade cost $23.98.

Study these examples.

$$
\begin{array}{r}
\$.0\,7 \\
7\overline{)\$.4\,9} \\
-4\,9 \\
\hline
\end{array}
$$

Check.
$$
\begin{array}{r}
\$.0\,7 \\
\times7 \\
\hline
\$.4\,9
\end{array}
$$

$$
\begin{array}{r}
\$.9\,0 \\
5\overline{)\$4.5\,0} \\
-4\,5 \\
\hline
0\,0
\end{array}
$$

Check.
$$
\begin{array}{r}
\$.9\,0 \\
\times5 \\
\hline
\$4.5\,0
\end{array}
$$

There are no dimes in the quotient. Write a zero.

Copy and complete.

$$
\begin{array}{r}
\$2.0\,1 \\
4\,\overline{)\$8.0\,4} \\
-\underline{8}\downarrow \\
0 \\
-\,?\downarrow \\
? \\
-\underline{\,?} \\
\end{array}
$$

1.

$$
\begin{array}{r}
\$\;\;5.?\,? \\
9\,\overline{)\$4\,9.9\,5} \\
-\,?\,?\downarrow \\
4\,9 \\
-4\,5\downarrow \\
?\,? \\
-\underline{?\,?} \\
\end{array}
$$

2.

$$
\begin{array}{r}
\$.1\,? \\
7\,\overline{)\$.8\,4} \\
-\,?\downarrow \\
1\,4 \\
-\underline{?\,?} \\
\end{array}
$$

3.

$$
\begin{array}{r}
\$\;\;.0\,? \\
8\,\overline{)\$0.5\,6} \\
-\underline{5\,6} \\
\end{array}
$$

4.

Estimate. Then find the quotient.

5. $5\overline{)\$1.35}$ 6. $2\overline{)\$4.94}$ 7. $4\overline{)\$2.44}$ 8. $7\overline{)\$2.31}$ 9. $2\overline{)\$8.58}$

10. $4\overline{)\$20.84}$ 11. $8\overline{)\$24.16}$ 12. $7\overline{)\$14.28}$ 13. $3\overline{)\$24.72}$ 14. $5\overline{)\$18.10}$

15. $9\overline{)\$49.77}$ 16. $6\overline{)\$14.82}$ 17. $7\overline{)\$27.93}$ 18. $8\overline{)\$20.88}$ 19. $5\overline{)\$26.00}$

20. $7\overline{)\$21.63}$ 21. $7\overline{)\$17.01}$ 22. $4\overline{)\$63.00}$ 23. $9\overline{)\$73.53}$ 24. $6\overline{)\$22.20}$

PROBLEM SOLVING

25. Help Meghan copy and complete the order form.

Amount	Description	Cost per Item	Total Cost
2 pairs	Gardening Gloves	?	$23.96
3	Lawn Chairs	?	$50.94
6	Tulip Bulbs	$.95	?
8	Daylily Plants	?	$98.80
5	Flower Pots	?	$14.95
24	Gladiola Bulbs	$1.45	?
4	Trowels	?	$31.96
2	Grass Rakes	$18.09	?
		Total	?

Order of Operations

Tim and Tom were given this problem to solve.

$$6 + 54 \div 2 - 4 \times 5 = \underline{\ ?\ }$$

Tim did this:
$$6 + 54 = 60$$
$$60 \div 2 = 30$$
$$30 - 4 = 26$$
$$26 \times 5 = 130$$

Tom did this:
$$54 \div 2 = 27$$
$$4 \times 5 = 20$$
$$6 + 27 = 33$$
$$33 - 20 = 13$$

Whose answer was correct?

Tom's answer was correct.
He used special mathematical rules
called the **order of operations.**

These are the rules for the
order of operations:

- *First* multiply or divide.
  Work in order from left to right.

- *Then* add or subtract.
  Work in order from left to right.

Solve us first.

Solve us next.

$$6 + 54 \div 2 - 4 \times 5 = \underline{\ ?\ } \qquad 6 + 27 - 20 = 13$$

| First multiply and divide from left to right. | Then add and subtract from left to right. |

Study these examples.

$$100 - 6 \times 7 \div 2 = \underline{\ ?\ }$$
$$100 - \quad 42 \ \div 2 = \underline{\ ?\ }$$
$$100 - \qquad 21 \qquad = 79$$

$$3 \times 4 \times 6 + 5 \div 5 = \underline{\ ?\ }$$
$$12 \ \times 6 + 5 \div 5 = \underline{\ ?\ }$$
$$72 \ + \ 1 \ = 73$$

Use the order of operations to solve.

1. $18 - 5 + 6$

2. $9 + 6 - 7$

3. $8 \times 6 \div 4$

4. $54 \div 6 \times 3$

5. $20 + 20 - 16$

6. $85 - 15 \times 2$

7. $10 \div 5 + 5 \times 3$

8. $8 - 4 \div 4 + 4$

9. $24 + 4 \div 4 - 5$

10. $35 - 5 + 10 \div 2$

11. $6 \times 6 + 10 \div 5 - 1$

12. $64 \div 8 \times 10 - 40 - 5$

13. $30 \div 6 \times 9 + 9 - 1$

14. $25 \times 3 - 50 \div 2 + 25$

15. $18 + 6 \div 2 - 11 + 5$

16. $20 \div 4 + 54 \div 6 + 4$

17. $7 \times 30 - 10 + 150 \div 3$

18. $45 \div 5 - 1 + 3 \times 7$

19. $44 \div 2 \times 3 - 12 + 4$

20. $20 \times 5 - 50 \times 2 + 0$

21. $30 + 20 - 25 \div 5 \times 5$

22. $200 \div 4 \times 3 - 50 + 1$

 **Connections: History**

Where did operations signs come from? The signs + and − were first used for addition and subtraction in Germany by Johann Widmann in 1498. The first use of × for multiplication is credited to William Oughtred in 1631. In 1659, the Swiss mathematician Johann Heinrich Rahn was the first to use ÷ to indicate division.

Compute.

23. $46 \times 8 + 10 - 50 \div 2 + 75 \div 3 - 100$

24. $500 - 10 \times 6 + 22 \div 2 - 48 \div 6 - 125$

5-15 Finding Averages

Aidan scored 75, 85, 90, 80, and 90 on math tests last term. What was his average test score?

To find an average:

Add the numbers.	Divide the sum by the number of addends.

```
   75
   85
   90
   80
 + 90
  420
```

Think:
5 addends

```
        84 ←—— average
    5)420
     -40↓
       20
      -20
```

Aidan's average test score was 84.

Study this example.

Find the average: $2.44, $3.68, $4.20, $1.64

```
$  2.44
   3.68
   4.20
 + 1.64
 $11.96
```

Think:
4 addends

```
        $ 2.99 ←—— average
    4)$11.96
     - 8 ↓
       3 9
      -3 6↓
         36
        -36
```

Find the average.

1. 36, 42, 72

2. 256, 498

3. 93, 126, 117

4. 500, 250

5. 49, 93, 86

6. 88, 0, 78, 90

Find the average.

7. 23, 37, 41, 19

8. 56, 18, 42, 64

9. 633, 495, 711

10. 420, 504, 297

11. $4.32, $.88, $4.00, $.76

12. 488, 128, 952, 720

13. 72, 216, 96, 108

14. $1.84, $2.76, $4.08, $2.32

15. 58, 77, 95, 49, 81

16. 93, 102, 115, 83, 42

17. 517, 423, 648, 212, 555

18. $4.25, $6.71, $3.24, $5.06, $4.94

19. $8.44, $.31, $2.97, $3.13, $.80

PROBLEM SOLVING Use the information in the grade book.

20. What was Bob's average test score?

21. What was Eric's average test score?

22. What was Carly's average test score? Was her average greater or less than Dawn's?

Students' Names	Test Scores				
	A	B	C	D	E
Bob	75	63	77	80	90
Carly	82	73	68	72	85
Dawn	75	76	83	87	94
Eric	82	68	85	85	80
Gary	86	85	92	82	70

23. How many points greater or less was Carly's average than Eric's average?

24. Did the five students have a higher average score on Test A or Test B?

25. Did the five students have the lowest average score on Test A, Test B, or Test C?

26. List the students in order from the highest average to the lowest average.

5-16 Problem Solving: Interpret the Remainder

Problem: A diner has 98 mugs. They should be stored in stacks of 8. How many more mugs are needed to make the last stack complete?

1 IMAGINE You are storing mugs in stacks of 8.

2 NAME

Facts: 98 mugs in all
8 mugs in each stack

Question: How many more mugs are needed?

3 THINK Divide because a whole is being separated into equal sets of 8.
Find the remainder.

That will tell how many mugs are in the *incomplete* stack.

number of mugs   mugs in stack
98   ÷   8   = ? R ?

4 COMPUTE

```
   1 2  R 2
8)9 8
 -8 ↓
   1 8
  -1 6
     2
```

Think: What number plus the remainder will make a stack of 8?

? + 2 = 8
6 + 2 = 8

The diner needs 6 more mugs to make complete stacks.

5 CHECK Multiply and add to check division.

12 × 8 = 96 and 96 + 2 = 98

The answer checks.

Interpret the remainder to solve each problem.

1. Jason uses 9-inch strips of plastic.
He can buy a 75-inch roll of plastic or
a 125-inch roll of plastic. Which roll
will have less wasted material?

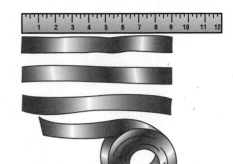

IMAGINE   Put yourself in the problem.

NAME   *Facts:*   9-inch strips of plastic
75-inch or 125-inch rolls of plastic

Question: Which roll will have less plastic
left over when it is cut into 9 inch strips?

THINK   Divide both 75 inches and 125 inches by 9.
Compare the remainders.
Look for the smaller remainder.

COMPUTE ⟶ **CHECK**

2. Each CD bin at Sound City holds 8 disks.
How many bins are needed to hold 195 disks?

3. Each treasure hunt team will have 5 people. So far 42 people
have signed up. How many more people are needed to make
every team equal? How many teams will there be?

4. Cans of juice are sold in packs of 6. The Day
Center needs 103 cans of juice. How many
packs should the center buy?

5. A soccer card club has 7 members. Together they have 1305 cards.
How many more cards do they need to share the cards equally?

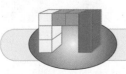

Make Up Your Own

6. Write a problem that uses a remainder. Have a classmate solve it.

Problem-Solving Applications

Solve each problem and explain the method you used.

1. Nora buys a 32-minute cartoon video. Each cartoon is 4 minutes long.
   a. How many cartoons are on the video?

   b. The video costs $6. How much does Nora spend for each cartoon?

2. Kwam watches a 1-hour cartoon special. How many 5-minute cartoons can be shown if there are no commercials? What if there are 15 minutes of commercials?

3. A cartoon channel shows 192 cartoons each day. If it shows 8 cartoons each hour, how many hours a day does the channel broadcast?

4. There are 28 characters in a film. Half of them are animals. How many characters are not animals?

5. A movie is 84 minutes long. A hopping frog appears every third minute. How many times does the frog appear?

6. There are 128 different animal videos displayed equally on 8 shelves. How many videos are on each shelf?

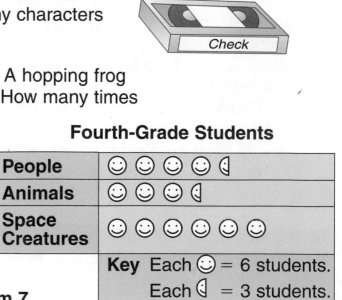

Fourth-Grade Students

People	☺ ☺ ☺ ☺ ◖
Animals	☺ ☺ ☺ ◖
Space Creatures	☺ ☺ ☺ ☺ ☺ ☺

Key Each ☺ = 6 students.
Each ◖ = 3 students.

Use the pictograph for problem 7.

7. Fourth-grade students used cartoons to illustrate their stories. How many more students used space creatures than animals?

Choose a strategy from the list or use another strategy you know to solve each problem.

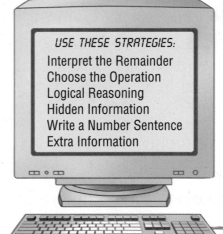

USE THESE STRATEGIES:
Interpret the Remainder
Choose the Operation
Logical Reasoning
Hidden Information
Write a Number Sentence
Extra Information

8. Mae draws 24 pictures to make 1 second of an animated cartoon. How many pictures does she draw for a 1-minute cartoon?

9. A cartoon, made up of 5760 drawings, uses the same number of drawings for each of the 4 minutes it runs. How many drawings are used per minute?

10. Chris watches a 30-minute cartoon show. If it shows as many 8-minute cartoons as possible, will there be enough time left to show a 7-minute cartoon?

11. A videotape includes 4 cartoons. They are 5 minutes, 6 minutes, 8 minutes, and 9 minutes long. What is their average length?

12. Another cartoon tape is 60 minutes long and costs $8.95. How much will 3 tapes cost?

13. Three cartoon characters are a chicken, a dog, and an octopus. Flick has more legs than Click, but fewer legs than Glick. Name each animal.

14. A video store orders 60 cartoon videotapes. Each shipping box holds 8 tapes. How many boxes will the store receive?

Make Up Your Own

15. Write a problem modeled on problem 13. Have a classmate solve it.

Chapter Review and Practice

Estimate. Then divide. *(See pp. 172–175, 178–189.)*

1. 4)32 **2.** 5)45 **3.** 3)27 **4.** 5)65 **5.** 6)72

6. 2)53 **7.** 6)93 **8.** 4)75 **9.** 6)86 **10.** 8)808

11. 3)723 **12.** 5)621 **13.** 8)337 **14.** 7)256 **15.** 4)160

16. 6)$36.36 **17.** 8)$72.64 **18.** 5)$17.55 **19.** 7)2772 **20.** 8)4074

Write the rule. Complete the pattern. *(See pp. 170–171.)*

21. 4, 7, 10, 13, _?_ , _?_ . **22.** 2, 6, 18, 54, _?_ , _?_ .

23. 8, 15, 13, 20, 18, _?_ , _?_ . **24.** 69, 64, 66, 61, _?_ , _?_ .

Answer *yes* or *no*. *(See pp. 176–177.)*

25. Is 45 divisible by 2? by 3? by 5? by 10?

26. Is 300 divisible by 2? by 3? by 5? by 10?

Find the average. *(See pp. 194–195.)*

27. 67, 36, 89, 44 **28.** 436, 219, 116

PROBLEM SOLVING   *(See pp. 166–169, 174–175, 178–181, 190–191.)*

29. Crayons were put on each of 8 tables. There were 84 crayons. How many crayons were put on each table? How many were left over?

30. What was Billy's average score for basketball if he scored the following points: 24, 30, 18, 15, 28?

31. There are 278 students in the Oak School. About how many are in each of the nine classrooms?

32. A bag of apples costs $1.62. There are 6 apples in the bag. How much does each apple cost?

(See Still More Practice, p. 465.)

FACTOR TREES

Algebra ✓

A **composite number** has more than two factors.

$$6 = 1 \times 6$$
$$= 2 \times 3$$

A **prime number** has exactly two factors, itself and 1.

$$5 = 1 \times 5$$

A **prime factor** is a prime number that is a factor of a composite number.

You can use a **factor tree** to help you find all the prime factors of a composite number.

Look at these factor trees for 12.

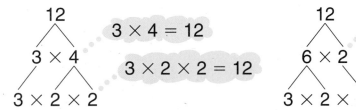

12
3 × 4
3 × 2 × 2

$3 \times 4 = 12$
$3 \times 2 \times 2 = 12$

12
6 × 2
3 × 2 × 2

$6 \times 2 = 12$
$3 \times 2 \times 2 = 12$

3 and 2 are prime numbers.
So the prime factorization of 12 is $3 \times 2 \times 2$.

Copy and complete each factor tree.

1. 30
6 × 5
? × 2 × ?

2. 40
4 × 10
? × 2 × ? × ?

3. 54
6 × 9
3 × ? × ? × ?

4. 36
9 × ?
? × ? × ? × ?

Draw a factor tree for each number.

5. 16   **6.** 10   **7.** 20   **8.** 24   **9.** 27

10. 32   **11.** 48   **12.** 35   **13.** 56   **14.** 72

Performance Assessment

What might the missing numbers be?

1. The divisor is 1. What are the dividend and quotient?
2. The dividend is 0. What are the quotient and divisor?
3. The quotient is 1. What are the divisor and dividend?

Estimate. Then divide.

4. $6)\overline{45}$ 5. $5)\overline{880}$ 6. $8)\overline{268}$ 7. $7)\overline{\$8.26}$ 8. $6)\overline{660}$

9. $2)\overline{408}$ 10. $6)\overline{804}$ 11. $5)\overline{610}$ 12. $7)\overline{700}$ 13. $3)\overline{406}$

14. $4)\overline{\$24.16}$ 15. $3)\overline{\$13.23}$ 16. $7)\overline{\$7.84}$ 17. $8)\overline{\$12.48}$ 18. $9)\overline{\$11.70}$

Write the rule. Complete the pattern.

19. 3, 7, 11, 15, ? , ? .

20. 57, 54, 51, 48, ? , ? .

21. 1, 2, 4, 8, ? , ? .

22. 3, 8, 7, 12, 11, ? , ? .

Use the order of operations to solve.

23. $6 \times 5 + 10 \div 5$

24. $32 \div 8 \times 10 - 5 - 10$

PROBLEM SOLVING   *Use a strategy you have learned.*

25. Find the cost of one eraser if Carlos paid $.54 for 6 erasers. Is the quotient divisible by 2?

26. Each shelf holds 8 dictionaries. Jean has 37 dictionaries. How many shelves does she need?

27. Marie has 1467 stamps. She gives the same number to each of her 4 nephews. At most, how many does each boy receive? How many are left over?

28. On four days Jan read 156 pages, 274 pages, 856 pages, and 306 pages. What was the average number of pages read by Jan per day?

Cumulative Review II

Choose the best answer.

1. Choose the standard form of the number.

   90,000,000 + 500,000 + 17

   a. 90,517
   b. 9,500,017
   c. 90,005,017
   d. 90,500,017

2. Round $37.59 to the nearest dollar.

   a. $37.00
   b. $37.60
   c. $38.00
   d. $40.00

3. Find the missing number.

   16 − __?__ = 7

   a. 8
   b. 9
   c. 11
   d. 23

4. $4.21 + $1.33 + $3.35

   a. $6.45
   b. $7.89
   c. $8.89
   d. not given

5.
   $$\begin{array}{r} 48,166 \\ + 57,369 \end{array}$$

   a. 90,797
   b. 95,348
   c. 105,535
   d. not given

6.
   $$\begin{array}{r} 80,000 \\ - 47,789 \end{array}$$

   a. 32,211
   b. 42,211
   c. 47,789
   d. not given

7. What is the period of the underlined digits?

   <u>56</u>,722,086

   a. billions
   b. hundreds
   c. thousands
   d. millions

8. 3 × $8.43

   a. $25.29
   b. $24.39
   c. $2.49
   d. $25.39

9. Find the missing number.

   72 = __?__ × 8

   a. 6
   b. 7
   c. 8
   d. 9

10. Find the missing number.

    $$2\overline{)?}\;\;0$$

    a. 0
    b. 1
    c. 2
    d. 4

11. 60 × 530

    a. 3180
    b. 12,800
    c. 30,900
    d. not given

12.
    $$\begin{array}{r} \$5.27 \\ \times \quad 46 \end{array}$$

    a. $168.28
    b. $224.86
    c. $242.42
    d. not given

13. $6\overline{)97}$

    a. 11 R6
    b. 12 R3
    c. 16 R1
    d. not given

14. $8\overline{)968}$

    a. 101
    b. 121
    c. 131
    d. not given

For Your Portfolio

Solve each problem. Explain the steps and the strategy or strategies you used for each. Then choose one from problems 1–4 for your Portfolio.

1. Dolores drew the diagram below to help her solve a problem. What multiplication sentence was she using?

   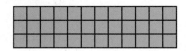

2. Dawind needs to find the average of $5.19, $.99, $1.65, and $4.53. What is the average? If Dawind increases each of the four amounts by $2.00, how will the average change?

3. José has 46 rare coins. Nel has about 8 times that number. About how many rare coins does Nel have?

4. Find the next number in each of the patterns below.
   55, 48, 41, 34, 27, ...
   4, 2, 6, 4, 12, ...

Tell about it.

5. Explain how you estimated to solve problem 3. Using your method, do you think the estimate is less than or greater than the actual product? Explain why.

6. Describe a different pattern in problem 4 using 4, 2, 6, as the first three numbers.

Communicate

For Rubric Scoring

Listen for information on how your work will be scored.

7. Fill in the boxes using the digits 1, 4, 5, 6, 8 so that the product agrees with the clues. Each digit 1, 4, 5, 6, 8 should be used only once.

   Clue #1: The product is between 30,000 and 50,000.

   Clue #2: To the nearest ten, the multiplier rounds to 60.

   Clue #3: The answer must contain the digits 0, 4, 6, 7, and 9.

   Clue #4: The tens digit in the answer is 3 more than the ones digit.

Measurement

6

In this chapter you will:

Estimate and compute metric units, with renaming

Investigate time and temperature –both Fahrenheit and Celsius

Learn about memory keys on a calculator

Solve two-step problems

Critical Thinking/Finding Together

Measure the length of various objects using the nonstandard units below.

- **cubit** (distance from elbow to fingertip)
- **span** (distance between outstretched thumb and pinky)

from

TAKE A NUMBER

Imagine a world
Without mathematics:

No rulers or scales,
No inches or feet,
No dates or numbers
On house or street,
No prices or weights,
No determining heights,
No hours running through
Days and nights.
No zero, no birthdays,
No way to subtract
All of the guesswork
Surrounding the fact.
No sizes for shoes,
Or suit or hat....
Wouldn't it be awful
To live like that?

Mary O'Neill

6-1 Measuring with Inches

You can use a ruler to measure an object
to the **nearest inch, nearest half inch,**
and **nearest quarter inch.**

When you measure length, align the object
you are measuring with the beginning of the ruler.

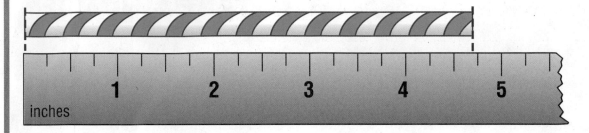

- To the nearest inch, the straw is about 5 in. long.

- To the nearest half inch, the straw is about
  $4\frac{1}{2}$ in. long.

- To the nearest quarter inch, the straw is
  about $4\frac{3}{4}$ in. long.

Think: Each inch on
the ruler is divided
into 2 half inches
and 4 quarter inches.

**Measure each to the nearest inch, nearest half inch,
and nearest quarter inch.**

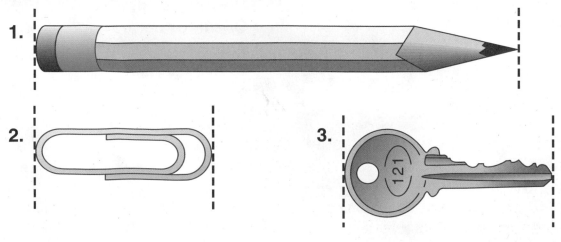

1.

2.

3.

Draw a line segment for each length.

4. 3 in.   **5.** 2 in.   **6.** $1\frac{1}{2}$ in.   **7.** $4\frac{1}{2}$ in.

8. $5\frac{1}{2}$ in.   **9.** $3\frac{1}{4}$ in.   **10.** $6\frac{1}{4}$ in.   **11.** $2\frac{1}{4}$ in.

12. $3\frac{3}{4}$ in.   **13.** $1\frac{3}{4}$ in.   **14.** $4\frac{3}{4}$ in.   **15.** $6\frac{3}{4}$ in.

Estimate the length of each to the nearest inch.
Then measure each line to check your estimates.

16. |————————————————————————|

17. |——————————————————————————————|

18. |—————————————|

19. |——————————————————————|

20. |————————————————————————————————————|

PROBLEM SOLVING

Communicate ✓

21. Suppose your ruler was broken like the ruler at the right. Could you use it to measure the line in exercise 18? How?

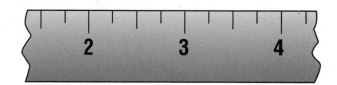

 **Project**

Discuss ✓

22. Measure your classroom and some of the objects in it. Decide what you will measure and what measuring tools you will use. Record each measurement. Discuss your results with your class.

6-2 Renaming Units of Length

Can Dontae fit a 72-in.-long shelf in a closet that is 5 ft wide?

Compare: 72 in. _?_ 5 ft

Before you can compare measurements in different units, you need to **rename** units.

▶ You can make a table to rename units.

ft	1	2	3	4	5
in.	12	24	36	48	60

Think: 1 ft = 12 in.

5 ft = 60 in.   72 > 60   So 72 in. > 5 ft.

▶ You can multiply the larger unit to rename units.

5 ft = _?_ in.
5 ft = (5 × 12) in. = 60 in.

72 > 60   So 72 in. > 5 ft.

▶ You can divide the smaller unit to rename units.
Use a calculator.

72 in. = _?_ ft
72 in. = (72 ÷ 12) ft = 6 ft

6 > 5   So 72 in. > 5 ft.

Press these keys:

Dontae cannot fit a 72-in.-long shelf in a closet that is 5 ft wide.

Copy and complete. You may make a table
or compute with paper and pencil or a calculator.

1. 2 yd = _?_ in.   **2.** 96 in. = _?_ ft   **3.** 18 ft = _?_ yd

4. 7 ft = _?_ in.   **5.** 4 yd = _?_ ft   **6.** 144 in. = _?_ yd.

Compare. Write <, =, or >. You may make a table or compute.

7. 6 yd _?_ 9 ft   **8.** 8 ft _?_ 84 in.   **9.** 36 in. _?_ 3 ft

10. 7 ft _?_ 108 in.   **11.** 20 yd _?_ 45 ft   **12.** 8 yd _?_ 18 ft

---

Mile

The **mile (mi)** is a customary
unit of length.

> 5280 feet (ft) = 1 mile (mi)
> 1760 yards (yd) = 1 mile (mi)

Miles are used to measure
long lengths called **distances**.

It takes about 25 minutes
to walk 1 mile.

Compare. Write <, =, or >. Explain which method you used.

13. 2 mi _?_ 3520 yd   **14.** 5280 yd _?_ 3 mi   **15.** 3 mi _?_ 21,120 ft

16. 10,560 ft _?_ 2 mi   **17.** 6 mi _?_ 42,240 ft   **18.** 7040 yd _?_ 4 mi

Computing Customary Units

▶ Last month a sunflower was 4 feet 7 inches tall. Then it grew 1 foot 8 inches taller. How tall is the sunflower now?

To find how tall it is now, add: 4 ft 7 in. + 1 ft 8 in. = __?__

Add the smaller units first. Rename units as needed.

```
  4 ft   7 in.
+ 1 ft   8 in.
  5 ft  15 in.  = 5 ft + 1 ft + 3 in. = 6 ft 3 in.
```

15 in. = 12 in. + 3 in.
     =   1 ft + 3 in.   The sunflower is 6 ft 3 in. tall now.

▶ A lilac stem was 7 feet 7 inches high. Sam pruned 2 feet 5 inches off the stem. How high was the stem after pruning?

To find how high, subtract: 7 ft 7 in. − 2 ft 5 in. = __?__

Subtract the smaller units first. Rename units as needed.

```
  7 ft 7 in.
− 2 ft 5 in.
  5 ft 2 in.
```
The stem was 5 ft 2 in. high after pruning.

Study these examples.

```
   8 yd 1 ft              3 ft = 1 yd
+  2 yd 2 ft
  10 yd 3 ft = 10 yd + 1 yd = 11 yd
```

```
  4 ft 11 in.
−      9 in.
  4 ft  2 in.
```

Add.

1. 6 ft 2 in.
   + 3 ft 5 in.

2. 7 yd 1 ft
   + 1 yd 1 ft

3. 3 yd 2 ft
   + 5 yd

4. 8 ft 6 in.
   +   5 in.

5. 10 ft 3 in.
   + 4 ft 10 in.

6. 5 ft 8 in.
   + 7 ft 11 in.

7. 6 ft 9 in. + 9 ft 5 in.

8. 4 yd 2 ft + 3 yd 2 ft

Subtract.

9. 4 ft 8 in.
   − 1 ft 3 in.

10. 9 ft 4 in.
    − 9 ft 2 in.

11. 7 yd 2 ft
    −   2 ft

12. 2 yd 2 ft
    − 2 yd

13. 12 ft 9 in.
    − 2 ft 4 in.

14. 8 ft 10 in.
    − 5 ft 6 in.

15. 9 ft 7 in. − 4 ft 3 in.

16. 4 ft 6 in. − 2 ft 6 in.

PROBLEM SOLVING

17. Amy's fence is 18 ft 10 in. long. She adds a 3 ft 5 in. section to the fence. How long is the fence now?

18. Joe painted 6 ft of a fence that is 20 ft 6 in. long. How much of the fence is not painted?

Challenge

Subtract. Rename when necessary.

19. ³4̸ ft ¹⁶4̸ in.
    − 1 ft 6 in.
    ───────────
    2 ft 10 in.

20. 3 yd 1 ft
    − 1 yd 2 ft

21. 8 ft 6 in.
    − 3 ft 11 in.

22. 9 yd 2 ft
    − 8 yd 4 ft

23. 5 ft 1 in.
    − 4 ft 11 in.

24. 12 ft
    − 6 ft 8 in.

6-4 | Customary Units of Capacity

Discover Together

You Will Need: measuring cup for liquid; pint, quart, and gallon containers; paper towels; paper and pencil

Record each of your answers.

Look at the markings on the measuring cup. Find the 1-cup mark. This mark also shows the number of **fluid ounces (fl oz)** that are equal to 1 cup.

1. How many fluid ounces are equal to 1 cup?

Pour water into the gallon container to the 1-gallon mark. Then fill as many quart containers as you can with the gallon of water.

2. How many quart containers did you fill? How many quarts are equal to 1 gallon?

Pour the water from one of the quart containers into as many pint containers as you can.

3. How many pint containers did you fill? How many pints are equal to 1 quart?

Pour the water from one of the pint containers to the 1-cup mark of as many measuring cups as you can.

4. How many cups of water did you pour? How many cups are equal to 1 pint?

Communicate

Discuss ✓

5. How many pints are equal to 1 gallon? Explain how you found out.

6. How many cups are equal to 1 quart? 1 gallon? Explain how you found out.

7. How many fluid ounces are equal to 1 pint? 1 quart? 1 gallon? Explain how you found out.

Algebra ✓

Copy and complete each table.

8.

gal	1	2	?	4	?
qt	4	8	?	?	20
pt	8	?	24	?	?

9.

pt	1	2	3	?	5
c	2	?	?	8	?
fl oz	16	32	?	?	?

Complete.

10. 2 pt = __?__ c

11. 8 c = __?__ pt

12. 16 qt = __?__ gal

13. 2 gal = __?__ qt

14. 10 pt = __?__ c

15. 48 pt = __?__ gal

PROBLEM SOLVING

16. Would you need 6 c, 6 pt, or 6 gal of paint to paint a room?

17. Would you drink 1 fl oz, 1 c, or 1 qt of milk at lunch?

18. Ted's pail holds 2 qt of water. He filled the pail 6 times to wash his mother's car. How many gallons of water did Ted use?

19. How many 14-fl oz cans of broth are needed for a recipe that calls for 1qt of broth?

6-5 Customary Units of Weight

The **ounce (oz)** and the **ton (T)** are customary units of weight.

16 ounces (oz) = 1 pound (lb)
2000 pounds (lb) = 1 ton (T)

A letter weighs about 1 ounce.

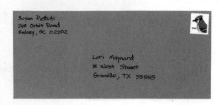

A compact car weighs about 1 ton.

Write *oz, lb,* or *T* for the unit you would use to measure the weight of each.

1. a carrot

2. an elephant

3. an electric guitar

4. a fire engine

5. a person

6. a toaster

7. a dog

8. a canary

9. a dump truck

Write the letter of the best estimate.

10. an orange
  a. 6 oz
  b. 1 lb
  c. 2 lb

11. a cat
  a. 30 lb
  b. 12 oz
  c. 12 lb

Algebra ✓

Copy and complete each table.

12.

oz	16	32	?	64	?	?
lb	1	2	3	?	5	6

13.

lb	2000	?	6000	?	?
T	1	2	?	4	?

Compare. Write <, =, or >. You may make a table
or compute.

14. 5 lb _?_ 96 oz   **15.** 6 T _?_ 12,000 lb   **16.** 4 lb _?_ 58 oz

17. 8 lb _?_ 112 oz   **18.** 144 oz _?_ 6 lb   **19.** 8500 lb _?_ 5 T

20. 48 oz _?_ 2 lb   **21.** 2 T _?_ 2000 lb   **22.** 10 lb _?_ 1600 oz

**Match. Write the letter of the tool you would use
to measure each.**

23. length of a pencil   **a.** ruler

24. water for a vase   **b.** scale

25. length of the classroom   **c.** measuring cup

26. weight of your teacher   **d.** yardstick

PROBLEM SOLVING

27. Akeem has 5 sisters. He
gives a 4-oz plum to each
sister. In all do the plums
weigh more or less than 1 lb?

28. A truck can carry 3000 lb of
cargo. Can it carry two tractors
that each weigh 1000 lb and
a 625-lb plow?

29. Can a truck that weighs 7500 lb safely cross
a bridge with a 3 T weight limit?

 **Share Your Thinking**   Discuss

30. Name two things that are
small *and* heavy.

31. Name two things that are
big *and* light.

32. Does the size of an object always give a clue to
how much it weighs? Explain your answer.

6-6 Measuring with Metric Units

The **decimeter (dm)** is a metric unit of length.

| 10 centimeters (cm) = 1 decimeter (dm) |

▶ In base ten blocks each tens rod is about 1 decimeter long.

▶ You can use a metric ruler to measure an object to the **nearest centimeter** or the **nearest decimeter**.

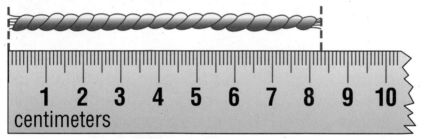

The length of the piece of yarn is between 8 cm and 9 cm. It is closer to 8 cm.
To the nearest centimeter, the piece of yarn is 8 cm long.
To the nearest decimeter, the piece of yarn is 1 dm long.

Measure each to the nearest centimeter.

1. |—————————————|

2. |————————————————————————|

3. |————————————————————|

Draw a line segment for each length.

4. 2 cm   **5.** 8 cm   **6.** 12 cm   **7.** 10 cm   **8.** 2 dm

9. 16 cm   **10.** 1 dm   **11.** 20 cm   **12.** 3 dm   **13.** 3 cm

Estimate each to the nearest centimeter and to the nearest decimeter. Then measure to check your estimates.

14. the length of your shoe

15. the length of your desk

16. the width of your hand

17. the length of a dollar bill

18. the length of this book

19. the width of this book

20. For exercises 14–19, was it easier to estimate in centimeters or in decimeters? Why?

PROBLEM SOLVING

21. Jesús and Ray measure the same wall. Jesús says it is 360 centimeters long. Ray says it is 36 decimeters long. Can they both be right? Explain your answer.

 **Connections: Language Arts**

22. Find the meanings of these prefixes commonly used in the metric system of measurement.
centi-   deci-   kilo-   milli-

6-7 Working with Metric Units

The **millimeter (mm)** and the **kilometer (km)** are other metric units of length.

| 10 millimeters (mm) = 1 centimeter (cm) |
| 100 millimeters (mm) = 1 decimeter (dm) |
| 10 decimeters (dm) = 1 meter (m) |
| 1000 meters (m) = 1 kilometer (km) |

▶ A dime is about 1 millimeter thick.

It takes about 15 minutes to walk 1 kilometer.

▶ Make a table or compute to rename units.

Compare: 3000 m __?__ 4 km

1 km = 1000 m

• Make a table:

m	1000	2000	3000
km	1	2	3

3 < 4   So 3000 m < 4 km.

• Multiply:

4 km = (4 × 1000) m
4 km = 4000 m

3000 < 4000   So 3000 m < 4 km.

• Divide:

3000 m = (3000 ÷ 1000) km
3000 m = 3 km

3 < 4   So 3000 m < 4 km.

You may use a calculator to divide when you rename units.

Compare. Write <, =, or >.
You may make a table or compute.

1. 40 km _?_ 400 m   **2.** 2000 mm _?_ 200 cm   **3.** 9 dm _?_ 900 cm

4. 6 m _?_ 80 dm   **5.** 7 dm _?_ 7000 mm   **6.** 500 cm _?_ 5 dm

7. 2000 m _?_ 5 km   **8.** 9 dm _?_ 1 m   **9.** 8000 cm _?_ 80 m

PROBLEM SOLVING Use the map below.

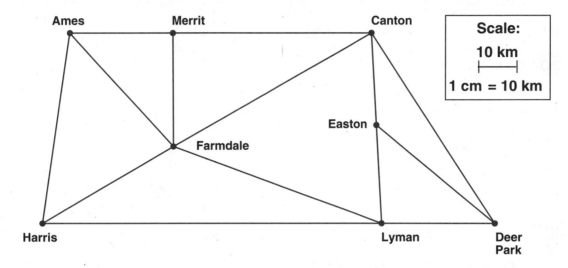

Scale:
10 km
⊢——⊣
1 cm = 10 km

10. What is the shortest route from Ames to Lyman? How many kilometers long is this route?

11. Is the route from Canton to Deer Park longer or shorter than the route from Canton to Lyman?

12. Mr. Yuan wants to travel from Harris to Deer Park to Canton. About how many kilometers will he travel?

13. Ms. Rau must travel from Lyman to Merrit. Should she go through Farmdale or through Canton? Why?

14. Is the route from Ames to Canton to Farmdale longer or shorter than the route from Easton to Deer Park to Lyman to Farmdale?

15. What is the shortest route from Deer Park to Ames? from Easton to Harris? How long is each route?

Update your skills. See page 19.

6-8 Metric Units of Capacity

The **milliliter (mL)** is a metric unit of liquid capacity.

1000 milliliters (mL) = 1 liter (L)

There are about 20 drops of water in 1 mL.

Write the letter of the best estimate.

1. bottle of liquid soap
   a. 1 mL   **b.** 10 mL   **c.** 1 L

2. gasoline for a car
   a. 48 mL   **b.** 48 L   **c.** 480 L

3. bowl of soup
   a. 5 mL   **b.** 500 mL   **c.** 5 L

4. ladle of soup
   a. 25 mL   **b.** 250 mL   **c.** 250 L

5. water in an aquarium
   a. 60 mL   **b.** 600 mL   **c.** 60 L

Write *mL* or *L* for the unit you would use to measure the capacity of each.

6. large jug of apple cider

7. tablespoon of syrup

8. glass of juice

9. bucket

10. washing machine

11. cup

Copy and complete the table.

12.

L	1	2	?	?	?	6	?	?
mL	1000	?	3000	?	?	?	7000	?

Algebra

Compare. Write <, =, or >.
You may make a table or compute.

13. 2 L _?_ 200 mL   **14.** 5 L _?_ 6000 mL   **15.** 8 L _?_ 8000 mL

16. 15 L _?_ 1500 mL   **17.** 4000 mL _?_ 3 L   **18.** 9000 mL _?_ 10 L

Write in order from the least amount to the greatest amount.

19. 4 L, 40 mL, 400 mL, 4 mL   **20.** 200 L, 20 mL, 20 L, 2 mL

21. 38 L, 380 mL, 380 L, 138 L   **22.** 24 L, 2400 mL, 240 mL, 240 L

PROBLEM SOLVING

23. Mr. Wood's van can travel 5 km on 1 L of gasoline. How much gasoline does the van use to travel 50 kilometers?

24. Mrs. Wood's water jug holds 4 L of water. It has 500 mL of water in it now. How much more water is needed to fill the jug?

25. Ellen and Allen both carry small canteens. Each canteen holds 750 mL of water. How much water do they need to fill both canteens?

26. The Woods began their trip with 75 L of gasoline in their gas tank. They used 68 L of gasoline. How much gasoline was left in the tank?

27. Ellen filled her 750-mL canteen four times in one day. How many liters of water did she use?

 **Project**

28. Fill a 1-qt container with water. Pour it into a 1-L container. How would you describe the capacity of a quart compared to the capacity of a liter?

Communicate

6-9 | Metric Units of Mass

The **gram (g)** is a metric unit of mass.

A paper clip has a mass of about 1 gram.

1000 grams (g) = 1 kilogram (kg)

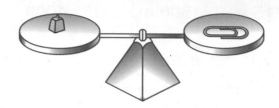

Write the letter of the best estimate.

1. an egg   **a.** 90 g   **b.** 9 kg   **c.** 90 kg

2. a shark   **a.** 100 g   **b.** 1000 g   **c.** 1000 kg

3. a worm   **a.** 14 g   **b.** 14 000 g **c.** 14 kg

4. a small dog   **a.** 880 g   **b.** 8 kg   **c.** 88 kg

5. a slice of bread   **a.** 2 g   **b.** 28 g   **c.** 28 kg

Write *g* or *kg* for the unit you would use to measure the mass of each.

6. a dinosaur   7. a mouse   8. a math book

9. a bag of oranges   10. a feather   11. a crayon

Copy and complete the table.

Algebra

12.

kg	1	?	3	?	?	?	?	8
g	1000	?	?	4000	?	?	?	?

Compare. Write <, =, or >.
You may make a table or compute.

13. 2 kg _?_ 20 g

14. 5 kg _?_ 5000 g

15. 9 kg _?_ 90 000 g

16. 8 kg _?_ 9000 g

17. 80 g _?_ 8 kg

18. 6000 g _?_ 5 kg

PROBLEM SOLVING

19. A penny has a mass of about 3 g. About what is the mass of a roll of 50 pennies? of 2 rolls of 50 pennies?

20. Pete puts 150 g of turkey into each turkey sandwich. How many kilograms of turkey does he need for 20 sandwiches?

21. Each loaf of bread that Pete uses has a mass of 500 g. He orders 10 loaves of bread. Is this more than or less than 8 kilograms?

22. A carton holds up to 30 kg. Pete has 28 kg of canned goods and 4000 g of side dishes. Can he pack them all into the carton?

23. Pete cooks two turkeys. The first has a mass of 11 000 g. The second has a mass of 17 kg. Which turkey has the greater mass? how much greater?

24. Find objects in your classroom that you think have a mass of about 50 g. Check by placing a 50-g mass on one side of a balance and each object on the other side. How close were your guesses?

Communicate ✓

Critical Thinking

Choose reasonable numbers so that each picture makes sense.

25.

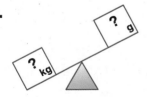

26.

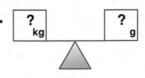

27.

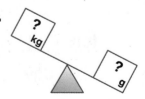

Temperature

A **thermometer** is used to measure **temperature**.

Temperature can be measured in **degrees Fahrenheit (°F)** or in **degrees Celsius (°C)**.

▶ Each line on the Fahrenheit scale stands for 2°F. Room temperature in degrees Fahrenheit is about 68°F.

Each line on the Celsius scale stands for 1°C. Room temperature in degrees Celsius is about 20°C.

▶ Use a minus sign to write temperatures below zero.

Write: ⁻5°F
Read: 5 degrees Fahrenheit below zero

Write: ⁻10°C
Read: 10 degrees Celsius below zero

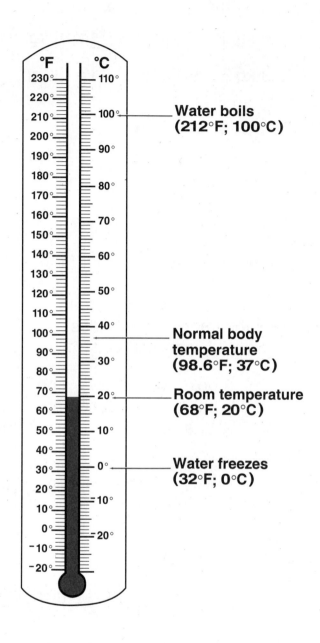

Water boils
(212°F; 100°C)

Normal body temperature
(98.6°F; 37°C)

Room temperature
(68°F; 20°C)

Water freezes
(32°F; 0°C)

Write the letter of the better estimate.

1. hot summer day   **a.** 90°C
                     b. 90°F

2. ice skating weather   **a.** ⁻10°C
                        b. 10°C

Write each temperature.

3.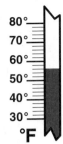

4.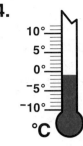

5.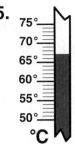

6.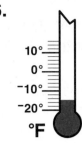

Compare. Write <, =, or >.
You may use the thermometer on page 224.

Algebra ✓

7. 30°C _?_ 120°F

8. 100°C _?_ 212°F

9. 50°F _?_ 10°C

10. 140°F _?_ 60°C

11. ⁻10°F _?_ ⁻20°C

12. 100°F _?_ 50°C

PROBLEM SOLVING Use the thermometer on page 224.

13. At 6:00 A.M. the temperature was 45°F. It rose 13°F by noon. What was the temperature at noon?

14. The temperature was 22°C at 8:00 P.M. Overnight it dropped 9°C. What was the temperature in the morning?

15. The temperature was 36°F at 7:00 P.M. It dropped 10°F by midnight. What was the temperature at midnight?

16. At 5:00 A.M. the temperature was ⁻3°C. By noon it was 6°C. By how many degrees did the temperature rise?

17. The temperature rose 11°F from 5:30 A.M. to 10:00 A.M. It was ⁻17°F at 5:30 A.M. What was the temperature at 10:00 A.M.?

 **Project**

Communicate ✓

18. Keep a record of the temperature at the same time each day for 7 days in a row. Share your temperature record with your class.

Time

You can read time after the half hour as minutes **past** the hour or as minutes **to** the next hour.

Read: 26 minutes **past** 6

Write: 6:26

Read: 43 minutes **past** 2
or
17 minutes **to** 3

Write: 2:43

Cesar is going to a movie that starts at 1:55 P.M. Will he see the movie in the morning or in the afternoon?

24 hours = 1 day

| midnight
12:00 A.M. | 6:00 A.M. | noon
12:00 P.M. | 6:00 P.M. | midnight
12:00 A.M. |

Use A.M. for times beginning at 12:00 midnight and before 12:00 noon.

Use P.M. for times beginning at 12:00 noon and before 12:00 midnight.

Cesar will see the movie in the afternoon.

Write each time.

1.

2.

3.

Give the time in minutes past the hour and in minutes to the hour.

4.

5.

6.

7.

8.

9.

Write A.M. or P.M. to make each statement reasonable.

10. Bill has breakfast at 7:15 _?_

11. School lets out at 3:00 _?_

12. Ahn goes to bed at 9:30 _?_

13. School begins at 8:00 _?_

Write the time. Use A.M. or P.M.

14. 10 minutes past 8 in the morning

15. 22 minutes to 10 at night

16. 36 minutes past 4 in the afternoon

17. 8 minutes to 9 in the morning

18. 18 minutes to noon

19. 45 minutes past midnight

 **Skills to Remember**

Skip count to find each pattern.

20. by 2 from 12 to 30

21. by 10 from 9 to 59

22. by 5 from 0 to 25

23. by 5 from 30 to 60

6-12 Elapsed Time

Jody arrived at the airport at 11:10 A.M. to meet Lisa. Lisa's plane landed at 1:24 P.M. How long did Jody wait for Lisa?

To find how much time has passed:

- Count the hours by 1s.
- Count the minutes by 5s and 1s.

Start at 11:10 A.M. Count the hours to 1:10 P.M.	11:10 ⌐ 12:10 ⌐ 1 hour 1:10 ⌐ 1 hour	2 hours

Count the minutes to 1:24 P.M.	1:10 ⌐ 5 minutes 1:15 ⌐ 5 minutes 1:20 ⌐ 1 minute 1:21 ⌐ 1 minute 1:22 ⌐ 1 minute 1:23 ⌐ 1 minute 1:24 ⌐ 1 minute	14 minutes

Jody waited 2 hours 14 minutes for Lisa.

Write how much time has passed.

1. from 8:05 A.M. to 8:30 A.M.

2. from 1:25 P.M. to 1:50 P.M.

3. from 6:30 A.M. to 6:51 A.M.

4. from 11:15 P.M. to 11:47 P.M.

5. from 11:45 P.M. to 12:04 A.M.

6. from 11:55 A.M. to 12:16 P.M.

7. from 3:25 P.M. to 4:40 P.M.

8. from 8:30 A.M. to 10:05 A.M.

Write how much time has passed.

9.

$9{:}45$ P.M. to $1{:}23$ A.M.

10. A.M. to P.M.

PROBLEM SOLVING

11. Jody and Lisa left the airport at 2:05 P.M. They drove for 47 minutes before arriving at Jody's house. What time did they arrive at Jody's house?

12. Jody and Lisa will visit their cousin, who lives 1 hour and 10 minutes away. They want to get there at 11:00 A.M. What time should they leave?

Elapsed Time on a Calendar

Lisa arrived on June 26 and left on July 8. How many days did she visit?

Count from June 26 to July 8. Count June 27 as day 1.

JUNE						
S	M	T	W	TH	F	S
			1	2	3	4
5	6	7	8	9	10	11
12	13	14	15	16	17	18
19	20	21	22	23	24	25
26	27	28	29	30		

JULY						
S	M	T	W	TH	F	S
					1	2
3	4	5	6	7	8	9
10	11	12	13	14	15	16
17	18	19	20	21	22	23
24/31	25	26	27	28	29	30

AUGUST						
S	M	T	W	TH	F	S
	1	2	3	4	5	6
7	8	9	10	11	12	13
14	15	16	17	18	19	20
21	22	23	24	25	26	27
28	29	30	31			

Lisa visited for 12 days.

PROBLEM SOLVING Use the calendar above.

13. What date is 10 days after July 22?

14. What date is 21 days before July 2?

15. What date is 4 weeks after June 15? before August 17?

TECHNOLOGY

Memory Keys

Memory keys are used to add, subtract, or recall
a value stored in a calculator's memory.

M– → Subtracts a value from the value stored in memory.

M+ → Adds a value to the value stored in memory.

MR → Recalls the value stored in memory.

▶ Use memory keys to compute: $21 \div 3 - 2 \times 3 = \underline{\ ?\ }$.

> Remember: First multiply and divide, then
> add and subtract, in order from left to right.

Press these keys:

2 1 ÷ 3 = M+ ← Adds the value 7 to memory.

Display:

2 × 3 = M– ← Subtracts 6 from the value 7 in memory.

MR ← Recalls the value 1 stored in memory.

Clear memory before computing.

So $21 \div 3 - 2 \times 3 = 1$.

▶ Use memory keys to add: $20 \text{ ft} + 72 \text{ in.} = \underline{\ ?\ }$.

$20 \text{ ft} + 72 \text{ in.} \div 12 \text{ in.} = \underline{\ ?\ }$ ← Rename 72 in. as ft. 12 in. = 1 ft

Press these keys: **Display:**

So $20 \text{ ft} + 72 \text{ in.} = 26 \text{ ft}$.

Match each expression with the correct calculator keys.

1. $16 + 5 \times 2$ **2.** $5 \times 2 - 6$ **3.** $7 + 8 - 3 \times 5$ **4.** $3 \times 5 + 7 \times 4$

a. | 5 | × | 2 | = | M+ |
| 6 | M− | MR |

b. | 3 | × | 5 | = | M+ |
| 7 | × | 4 | = | M+ | MR |

c. | 3 | × | 5 | = | M+ |
| 7 | + | 8 | = | M− | MR |

d. | 5 | × | 2 | = | M+ |
| 1 | 6 | M+ | MR |

Algebra ✓

Use the memory keys on your calculator to compute.

5. $38 + 6 \times 3$ **6.** $5 \times 90 - 37$ **7.** $108 \div 12 - 2$

8. $19 \times 8 + 9 \times 41$ **9.** $34 \div 17 + 10 \times 3$ **10.** $4 \times 6 - 24 \div 8$

11. $54 \div 2 + 48 \div 6$ **12.** $12 \times 3 + 81 \div 9$ **13.** $64 \div 4 - 4 \times 4$

14. 4 ft + 36 in. = ? ft   **15.** 96 in. + 9 ft = ? ft

16. 15 ft + 6 yd = ? yd   **17.** 12 yd + 5 ft = ? ft

18. 60 in. + 2 ft = ? in.   **19.** 180 yd − 180 ft = ? yd

20. $.45 + 3 \times \$8.01 - \$4.48$ **21.** \$7.60 + \$25.10 \div 5 - \$.05$

22. $\$36.48 \div 6 + \$.09 \times 5$ **23.** \$52.50 \times 3 - \$60.02 \div 2$

You can use a calculator to count by 5s.
Count by 5s to 30.

Press these keys:

24. Start at 25. Count by
2s to 37.

25. Start at 24 in. Count by
12 in. to 5 ft.

6-14 Problem Solving: Two-Step Problem

Problem: Maria has 3 packages to send to Hawaii in zone 8. One weighs 3 lb and the others weigh 4 lb each. How much money will she save if she uses parcel post instead of paying $13.40 for using priority mail?

Weight	Zone 8 Rates	
(lb)	Priority	Parcel Post
1	$2.90	—
2	$2.90	$2.85
3	$4.10	$4.05
4	$4.65	$4.60
5	$5.45	$5.40

1 IMAGINE Put yourself in the problem.

2 NAME *Facts:* 1—3-lb package
2—4-lb packages

Question: How much money is saved by using parcel post?

3 THINK Plan the steps to follow.

Step 1: Use the prices in the chart. Add to find the cost of sending the packages by parcel post.

3-lb cost + 4-lb cost = total cost

Step 2: Subtract to find the difference.

priority mail cost − parcel post cost = savings

4 COMPUTE Step 1: parcel post   Step 2: savings

```
        1
   $  4.0 5
      4.6 0
   +  4.6 0
   $1 3.2 5
```

```
            3 10
   $1 3.4 0
 −  1 3.2 5
   $    .1 5
```

Maria will save $.15 by using parcel post.

5 CHECK Use a calculator to check the computation in each step. Remember to press the decimal point key.

Use the Two-Step Problem strategy to solve each problem.

1. Paul sends his cousin three 28-oz fruitcakes and a 27-oz package of poppy seed muffins. What is the total weight of the package?

IMAGINE   Create a mental picture.

NAME   *Facts:*   3—28-oz fruitcakes
                       1—27-oz package of muffins

              Question: What is the total weight?

THINK   Plan the steps to follow.

              Step 1: Multiply to find the weight of the 3 fruitcakes.
                       fruitcakes: 3×28 oz = _?_

              Step 2: Add to find the total weight.

        COMPUTE ⟶ **CHECK**

2. This year, Dan's wood-carving club includes 17 children, 23 teenagers, and 46 adults. Last year, there were 54 members in all. By how much has the membership changed?

3. Dan works as a wood carver from 8:30 A.M. to 4:30 P.M. each day. How many hours does Dan work in a 5-day work week?

4. Ira sends six 2-lb parcel post packages to Hawaii. How much change will he get from $20? (*Hint:* Use the chart on page 232.)

5. Mr. Cheng bought 8 gallons of paint. Each gallon cost $12.27. He also bought a paint roller for $4.75. What was the total cost?

Solve each problem and explain the method you used.

1. The sun set at 7:52 P.M. It rose the next morning at 5:02 A.M. How much time passed from sunset to sunrise?

2. Ray caught 3 fish that were about 4 lb each. Mary caught 4 fish that were each about the same weight as Ray's. About how many pounds of fish did they catch?

3. Mrs. O'Hara packed 2 pounds of trail mix. Her family ate 7 ounces of the mix. How much was left?

4. Mr. O'Hara brought 3 rolls of fishing line. Each roll holds 525 yards of line. Did he bring more than a mile of line?

5. Mrs. O'Hara caught a 12 kg fish. How much is this in grams?

6. The O'Haras drove to Loon Lake. They left home at 8:25 A.M. Lunch at a rest stop took 45 minutes. They arrived at Loon Lake at 4:00 P.M. How long were they driving?

7. Mary's jug holds 3 L of water. It already has 500 mL in it. How much water should Mary add to fill it?

8. A hiking trail is 4 km long. There are trail markers every 8 m. How many trail markers are there?

Choose a strategy from the list or use another strategy you know to solve each problem.

USE THESE STRATEGIES:
Two-Step Problem
Extra Information
Choose the Operation
Logical Reasoning
Interpret the Remainder
Guess and Test

9. Ray heard a loon's call at 7:48 A.M. and again 13 minutes later. What time did he hear the second call?

10. Mary glues 8 pine needles onto each postcard. She has 130 pine needles. How many postcards can she make?

11. The distance across Loon Lake is 2 miles. Mary rows the boat 2640 yd across. How far away is she from the other side?

12. Six cabins are about evenly spaced along the 3 km perimeter of Moon Lake. About how far apart are the cabins?

13. There were 325 yd of line on Ray's fishing reel. He cuts off 18 feet of line. How much line is left on the reel?

14. The family leaves Loon Lake at 9:00 A.M. and arrives home at 5:30 P.M. Mr. O'Hara drives the first half of the trip and then Mrs. O'Hara drives. About what time does Mrs. O'Hara start driving?

Use the map for problems 15 and 16.

15. About how long will it take to get from Loon Lake to Moon Lake at a rate of 50 miles per hour?

Loon Lake
90 miles
80 miles
70 miles
Moon Lake
125 miles
Spoon Lake

16. From Moon Lake, Joe wants to visit both Loon and Spoon lakes. What is the distance of the shortest route? the longest route?

Write *in.*, *ft*, *yd*, or *mi* for the unit you would use to measure each.   *(See pp. 206–209.)*

1. distance across the county

2. width of a creek

3. width of a book

4. length of a pool

Add.   *(See pp. 210–211.)*

5. 3 ft 2 in.
   + 4 ft 5 in.

6. 4 yd 1 ft
   + 6 yd 1 ft

7. 6 yd 2 ft
   + 7 yd

Complete.   *(See pp. 212–219.)*

8. 4 pt = _?_ c

9. 32 oz = _?_ lb

10. 6000 lb = _?_ T

11. 6 cm = _?_ mm

12. 4000 m = _?_ km

13. 300 mm = _?_ cm

Compare. Write <, =, or >.   *(See pp. 212–223.)*

14. 6 kg _?_ 6000 g

15. 16 fl oz _?_ 2 c

16. 60 mm _?_ 6 m

17. 400 mL _?_ 4 L

18. 16 qt _?_ 1 gal

19. 5000 lb _?_ 1 T

20. 100 cm _?_ 10 m

21. 5 dm _?_ 50 m

22. 50 kg _?_ 1 g

Write the letter of the better estimate.   *(See pp. 224–225.)*

23. snow skiing weather   **a.** 25° F   **b.** 25° C

24. a day for a picnic   **a.** 32° C   **b.** 32° F

Write how much time has passed.   *(See pp. 226–229.)*

25. from 11:25 A.M. to 12:15 P.M.

26. from 11:30 P.M. to 7:15 A.M.

(See Still More Practice, p. 466.)

TIME ZONES

The clocks show the time in four different **time zones** of the United States when it is 12:00 noon Central time.

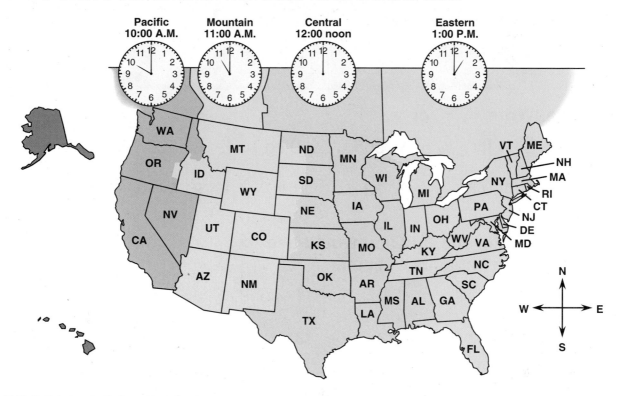

Pacific	Mountain	Central	Eastern
10:00 A.M.	11:00 A.M.	12:00 noon	1:00 P.M.

PROBLEM SOLVING Use the time-zone map.

1. What time is it in California when it is 2:00 P.M. in Maine?

2. What time is it in Georgia when it is 10:00 A.M. in Iowa?

3. Emily lives in Arizona. She will call Nat in Ohio at 6:30 P.M. Eastern time. What is that time in Arizona?

4. Xue will call Chad in Nevada at 1:45 P.M. Eastern time from New York. What time is that in Nevada?

5. It is 2:07 A.M. in Arkansas. What time is it in
   a. Texas?
   b. Oregon?
   c. Vermont?
   d. Montana?

6. A 6-hour flight to Utah leaves Delaware at 1:27 P.M. Eastern time. What is the time in Utah when the plane arrives?

237

Check Your Mastery

Performance Assessment

Complete and extend each table by 3 columns.

1.

pt	1	2
c	2	?
fl oz	?	32

2.

gal	1	?	3
qt	4	?	12
pt	8	16	?

3.

cm	2	4	?
mm	20	?	60

Compare. Write <, =, or >.

4. 36 in. _?_ 4 ft

5. 900 g _?_ 9 kg

6. 5280 yd _?_ 2 mi

7. 4 km _?_ 400 m

8. 32 oz _?_ 1 lb

9. 14 lb _?_ 208 oz

Write *mL* or *L* for the unit you would use to measure the capacity of each.

10. glass of milk

11. pond

12. cup of soda

13. gasoline tank of a car

Write *true* or *false* for each statement.

14. You can ice skate at 30° C.

15. You can wear shorts at 90° F.

16. You need a coat at 8° C.

17. 11:30 P.M. is school time.

18. Lunch is near 12:05 P.M.

19. It is usually dark at 10:30 A.M.

PROBLEM SOLVING   *Use a strategy you have learned.*

20. Joe has saved $24 a week for the last 4 weeks. He wants to buy a ukelele that costs $89.95. Does he have enough money?

21. Margaret expected 50 people to come to her party. In the first hour 17 people came in, in the second hour 18 came in, and the third hour 14 came in. Did Margaret have more or less people than she expected?

Statistics and Probability

7

In 1919, Babe Ruth hit 29 home runs, batted .322, and made $40,000.

In 1991 the average major league baseball player hit 15 home runs, batted .275, and made $840,000.

WHAT IS THE CORRECT ANSWER:
Babe Ruth < The average modern baseball player
Babe Ruth > The average modern baseball player
Babe Ruth = The average modern baseball player

From *Math Curse* by
Jon Scieszka and Lane Smith.

In this chapter you will:

Collect, organize, and interpret data
Investigate combinations
Predict probability of dependent and
 independent events
Explore tree diagrams
Solve problems by making up a question

Critical Thinking/Finding Together
A batting average of .100 means 100 hits
out of 1000 times at bat. A batting average
of .200 means 200 hits out of 1000 times at
bat. Describe Babe Ruth's batting average.

7-1 Graphing Sense

Graphs display data so that it can be easily understood. You can use graphs to compare sets of data.

▶ A **pictograph** uses pictures or symbols to represent different numbers of the same item. The **Key** tells how many each symbol stands for.

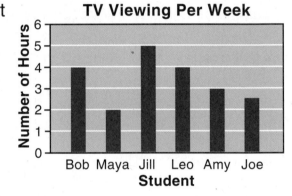

Video Sales in May	
Cartoon	▢▢▢▮
Adventure	▢▢▢▢▢
Comedy	▢▢▢▢▢
Drama	▢▢▮
Key: Each ▢ = 100 videos.	
Each ▮ = 50 videos.	

▶ A **bar graph** uses bars to represent measurements or numbers of different items. The **scale** on a bar graph tells how much or how many each bar stands for.

TV Viewing Per Week

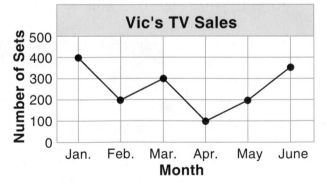

▶ A **line graph** uses points and lines on a grid to show change over a period of time. A line graph also has a scale.

Vic's TV Sales

▶ A **circle graph** uses sections of a circle to compare the parts of a whole group.

TV Favorites of Ms. Lee's Class

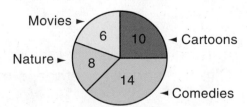

240

PROBLEM SOLVING Use the graphs on page 240.

1. What is the title of the bar graph?

2. What is the title of the circle graph?

3. Which graph has a scale of 100?

4. Which graph has a key?

5. Which kind of video had the greatest sales in May? the least? How many of each were sold?

6. How many cartoon videos were sold in May?

7. Which student watches TV the most hours per week? the fewest hours?

8. Which two students watch TV the same number of hours per week? How many hours is this?

9. Which student watches TV $2\frac{1}{2}$ hours per week?

10. In which month did Vic sell the fewest TVs? the most TVs? How many TVs did he sell in each of these months?

11. Which kind of TV show is most popular with Ms. Lee's class? least popular?

12. How many students are there in Ms. Lee's class? Explain how you found out.

 **Project**

Communicate ✓

13. Cut out different graphs from old magazines and newspapers. Paste the graphs on paper. Clip the pages together to make a graph scrapbook that you can share with your class.

7-2 | Making Pictographs

Kai made a tally of the dogs that were in each category in the dog show.

Category	Tally	Total
Sporting	ⵌ ⵌ ⵌ ⵌ ⵌ ⵌ ⵌ I	36
Terriers	ⵌ ⵌ ⵌ ⵌ	20
Working	ⵌ ⵌ ⵌ ⵌ ⵌ ⵌ ⵌ IIII	39
Hounds	ⵌ ⵌ ⵌ ⵌ ⵌ ⵌ	30
Toy	ⵌ	5
Nonsporting	ⵌ ⵌ	10

Then Kai organized his data in a pictograph.

▶ To make a pictograph:

- List each category.

- If necessary, round the data to nearby numbers.
  36 → 35   39 → 40

- Choose a picture or symbol that can represent the number in each category.

- Choose a key.
  Let each ⵦ = 10 dogs.

- Draw pictures to represent the number in each category.

- Label the pictograph. Write the title and the key.

Dogs in Dog Show				
Sporting	ⵦ	ⵦ	ⵦ	ⵧ
Terriers	ⵦ	ⵦ		
Working	ⵦ	ⵦ	ⵦ	ⵦ
Hounds	ⵦ	ⵦ	ⵦ	
Toy	ⵧ			
Nonsporting	ⵦ			

Key: Each ⵦ = 10 dogs.
Each ⵧ = 5 dogs.

About how many of the dogs in the show were sporting dogs?

▶ To find about how many, count the number of pictures for the sporting dog category. Then use the key.

There are $3\frac{1}{2}$ pictures for the sporting dog category.

Use the key: ⵦ ⵦ ⵦ ⵧ
10 + 10 + 10 + 5 = 35

Of the dogs in the show, about 35 were sporting dogs.

PROBLEM SOLVING

The pictograph at the right shows the ice cream cones Ida sold at Ida's Ice Cream on a weekend in June.

Ice Cream Cones Sold	
Vanilla	🍦 🍦 🍦 🍦 🍦 🍦
Chocolate	🍦 🍦 🍦 🍦 🍦 🍦 🍦 🍦 🍦 🍦
Strawberry	🍦 🍦 🍦 🍦 🍦
Butter Pecan	🍦 🍦 🍦 🍦 🍦 🍦 🍦
Pistachio	🍦 🍦 🍦 🍦
Cherry	🍦 🍦

Key: Each 🍦 = 50 cones.
       Each 🍦 = 25 cones.

Use the pictograph.

1. Which flavor was the most popular? How many cones of this flavor did Ida sell?

2. Ida sold 350 cones of one flavor. What flavor was this?

3. How many vanilla cones did Ida sell? how many cherry cones?

4. How many fewer cherry cones than strawberry cones did Ida sell?

5. How many ice cream cones did Ida sell altogether? List the flavors in order from greatest to least.

Use each to make a pictograph.

6.
Color of Car	Tally
Black	‖‖‖ ‖‖‖ ‖‖‖ ‖‖‖ ‖‖‖
Gray	‖‖‖ ‖‖‖ ‖‖‖ ‖‖‖ ‖‖‖ ‖‖‖ ‖‖‖
Blue	‖‖‖ ‖‖‖
Red	‖‖‖ ‖‖‖ ‖‖‖
White	‖‖‖ ‖‖‖ ‖‖‖ ‖‖‖ ‖‖‖ ‖‖‖
Green	‖‖‖

7.
Cats in the Cat Show	
Breed	Number
American Shorthair	275
Abyssinian	150
Siamese	200
Persian	250
Burmese	125
Manx	50
Rex	50
Himalayan	125

8. Write two questions for each of the pictographs you made.

Skills to Remember

Write the number that is halfway between each pair.

9. 100; 200   10. 0; 1000   11. 0; 500   12. 50; 100   13. 1000; 3000

7-3 Making Bar Graphs

Heidi found some information about the tallest tree of each species in the United States.

Heidi organized the data she found in a **vertical bar graph**.

Tallest Trees	
Tree	**Height in Feet**
Apple	70
Avocado	40
Mahogany	70
Mountain Ash	50
Pawpaw	60
Cypress	55

▶ To make a vertical bar graph:

- Use the data from the table to choose an appropriate scale. Start at 0.

- Draw and label the scale on the vertical line, or **axis**. (*Vertical* means "up and down.")

- Draw and label the horizontal axis. (*Horizontal* means "across.") List the name of each item.

- Draw vertical bars to represent each number.

- Title the graph.

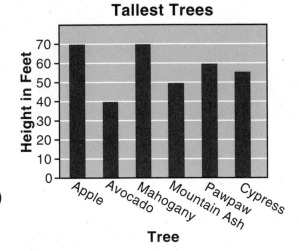

How tall is the tallest cypress tree in the United States?

▶ To find how tall, look at the bar labeled *Cypress.*

The top of the bar is *halfway* between 50 and 60.

The number that is *halfway* between 50 and 60 is 55.

So the tallest cypress tree in the United States is 55 feet tall.

PROBLEM SOLVING Use the bar graph on page 244.

1. Which tree is 60 feet tall? How much taller is it than the shortest tree?

2. Which two trees are the same height? How tall are they?

Copy and complete the horizontal bar graph. Use the table.

Top Speeds	
Animal	**Miles Per Hour**
Cat	30
Cheetah	70
Elephant	25
Grizzly Bear	30
Lion	50
Rabbit	35
Zebra	40

3.

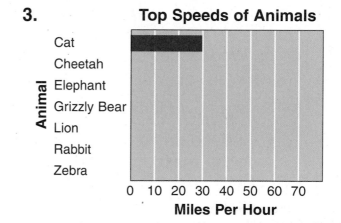

Use the completed horizontal bar graph to solve each problem.

4. Which animal has the shortest bar? the longest bar?

5. Which two animals have bars of the same length?

6. Which animals have bars that do not come to a ten?

7. Survey your classmates on their favorite animal. Collect and represent the data in a table and bar graph.

Connections: Science

Make a bar graph. Use the information below.

8.

Average Life Spans of Animals			
Baboon	– 20 years	Grizzly Bear	– 25 years
Bison	– 15 years	Pig	– 10 years
Elephant	– 40 years	Gray Squirrel	– 10 years
Hippopotamus	– 25 years	Rabbit	– 5 years

7-4 Line Graphs

Emmitt started doing sit-ups every day. The line graph shows Emmitt's progress the second week of his exercise program.

How many sit-ups did Emmitt do on Wednesday?

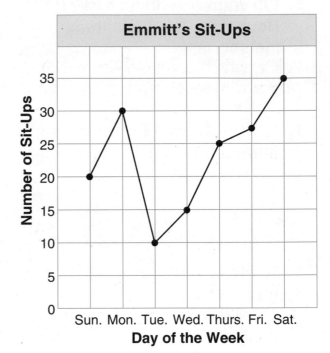

▶ To find how many:

- Find the day of the week on the **horizontal axis.**

- Move up to the point.

- Read the number on the **vertical scale** at the left.

Emmitt did 15 sit-ups on Wednesday.

About how many sit-ups did Emmitt do on Friday?

▶ To find *about* how many, move up to the point for Friday.

The point is *about* halfway between 25 and 30.

The number 27 is *about* halfway between 25 and 30.

So Emmitt did about 27 sit-ups on Friday.

Write how many sit-ups Emmitt did on the following days.

1. Monday   **2.** Thursday   **3.** Saturday   **4.** Sunday

PROBLEM SOLVING Use the line graph on page 246.

5. On which day did Emmitt do the fewest sit-ups?

6. On which day did Emmitt do the most sit-ups?

7. On which days did Emmitt do more than 25 sit-ups?

8. On which days did Emmitt do less than 20 sit-ups?

9. Between which two days was there a difference of 20 sit-ups?

10. Between which two days was there a difference of about 8 sit-ups?

Use the line graph at the right.

11. In which years were there the most 1st graders? How many 1st graders were there in these years?

12. In which years were there the fewest 1st graders? How many 1st graders was this?

13. About how many 1st graders were there in 1950? 1985?

14. In which year were there 350 first graders?

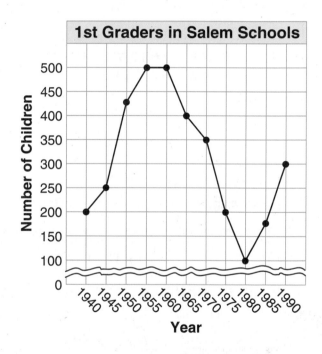

15. Is the difference in the number of 1st graders from 1985 to 1990 greater or less than that between 1975 and 1980?

16. In which 5-year period was the difference the same as the difference between 1960 and 1965?

17. In which 5-year period was the difference in the number of 1st graders the greatest?

18. In which 5-year period was there a difference of about 125 1st graders?

The town of Winterset held tryouts for a new community chorus.

The circle graph at the right shows the number of singers who were selected for the chorus.

How many singers make up the Winterset Community Chorus?

To find the number that is represented by the whole graph, add the numbers in the sections of the graph.

$$18 + 10 + 15 + 5 = 48$$

Forty-eight singers make up the Winterset Community Chorus.

Winterset Community Chorus

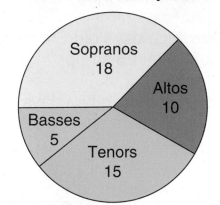

PROBLEM SOLVING Use the circle graph below.

Jeremy's Budget

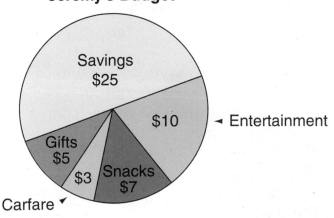

1. How much did Jeremy budget for savings? for carfare?

2. Did Jeremy budget more or less for entertainment than for gifts and snacks together? How much more or less?

3. How much money does the graph of Jeremy's budget represent in all?

PROBLEM SOLVING Use the circle graph below.

4. Which fruit is the favorite of 120 students?

5. How many students named melons as their favorite fruit?

6. Which two fruits were the favorites of the same number of students?

7. Which fruit was chosen most by students? How many students chose that fruit?

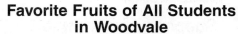

Favorite Fruits of All Students in Woodvale

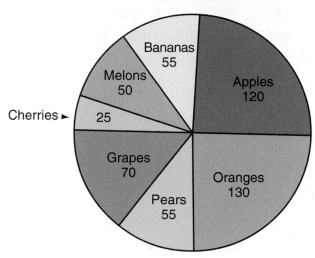

8. Which fruit was chosen as the favorite by the fewest students? Which fruit was chosen by double that number of students?

9. Were apples more or less popular than bananas and pears together? by how many votes?

10. How many students chose apples, oranges, *and* grapes? How many students in all chose apples, bananas, *and* pears?

11. How many students are there in Woodvale?

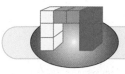

Make Up Your Own

Communicate ✓

12. Work with a partner. Decide on a survey topic and make up several questions to ask your classmates about the topic. After you survey your class display the data in a graph. Write 2–3 sentences describing the data. Which type of graph did you use? Explain why.

249

7-6 | Combinations

Suppose you went to Didi's Diner. How many different ways could you order the Early Bird Special?

Didi's Diner

Early Bird Special $7.95

Choose one from each column.

Main Dish	Side Dish	Vegetable
Roast Chicken	Potato	Broccoli
Grilled Fish	Rice	Spinach
		Carrots

▶ To find how many different ways, draw a **tree diagram.** Then count the **combinations.**

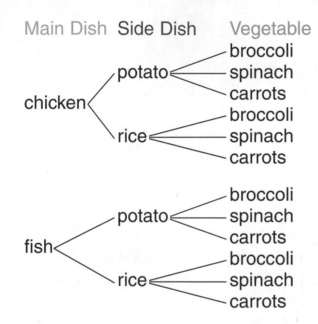

Main Dish	Side Dish	Vegetable	Combination
chicken	potato	broccoli	chicken, potato, broccoli
		spinach	chicken, potato, spinach
		carrots	chicken, potato, carrots
	rice	broccoli	chicken, rice, broccoli
		spinach	chicken, rice, spinach
		carrots	chicken, rice, carrots
fish	potato	broccoli	fish, potato, broccoli
		spinach	fish, potato, spinach
		carrots	fish, potato, carrots
	rice	broccoli	fish, rice, broccoli
		spinach	fish, rice, spinach
		carrots	fish, rice, carrots

You could order the Early Bird Special 12 different ways.

▶ You can also find the number of combinations by multiplying.

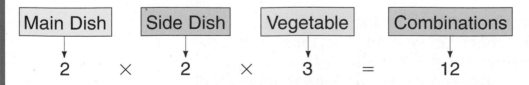

Main Dish		Side Dish		Vegetable		Combinations
2	×	2	×	3	=	12

There are 12 combinations.

250

PROBLEM SOLVING
Draw a tree diagram to solve.

1. One night, the Early Bird Special offered a choice of either ravioli or macaroni and cheese, with either string beans, peas, or cole slaw. How many different ways could you order?

2. Suppose Didi ran out of spinach. How many different ways could you order the Early Bird Special from the menu on page 250?

Draw a tree diagram or multiply to solve.

3. If you order Didi's Breakfast Special, you can choose either scrambled or poached eggs; orange, apple, or grapefruit juice; and whole wheat or white toast. How many combinations of eggs, juice, and toast could you order?

4. Mr. Gorme has breakfast at Didi's every day. He always orders either pancakes or waffles; orange, apple, grapefruit, or tomato juice; and bacon, ham, or sausage. For how many days in a row can he have breakfast without repeating an order?

5. Mr. Gorme drives a delivery van. He must wear either a white, blue, or gray shirt with either black, blue, or gray pants. How many combinations of shirt and pants can he wear?

6. At work Didi wears either a red, white, or blue blouse; a red, white, or black skirt; and a flower-print, striped, or white apron. Can she wear a different outfit every day for four weeks without repeating a combination of blouse, skirt, and apron? Explain your answer.

Communicate

251

Predicting Probability

When you pick an item from a set of items without looking, spin a spinner, or roll number cubes, you do not know beforehand what the result, or **outcome**, will be. So you make the pick, spin, or roll **at random**.

Before you pick without looking, spin, or roll, you can figure out your chances of getting a particular result. This is called finding the **probability** of the event.

▶ What is the probability of the spinner landing on red? on blue? on white?

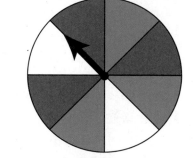

The spinner has 8 *equal* sections. Of the equal sections, 3 are red, 3 are blue, and 2 are white.

The probability of the spinner landing on

- red is 3 out of 8.

- blue is 3 out of 8.

- white is 2 out of 8.

▶ Is it equally likely that the spinner will land on

red or white?	white or blue?	blue or red?
3 > 2	2 < 3	3 = 3

So it is **more likely** that the spinner will land on red than on white.

So it is **less likely** that the spinner will land on white than on blue.

So it is **equally likely** that the spinner will land on blue or red.

PROBLEM SOLVING Use the spinner at the right.

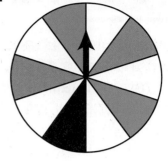

1. What is the probability of the spinner
   landing on
   a. blue?   **b.** white?
   c. black?   **d.** orange?

2. Is it *more likely*, *less likely*, or
   equally likely that the spinner
   will land on
   a. orange than white?   **b.** blue or orange?
   c. black than blue?   **d.** white than black?

Use the set of marbles.

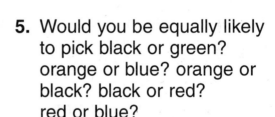

3. At random, what is the
   probability that you would pick
   a. green?   **b.** red?
   c. orange?   **d.** blue?
   e. black?   **f.** yellow?

4. Would you be more or less
   likely to pick yellow than
   green? black than yellow?
   red than blue? yellow
   than orange?

5. Would you be equally likely
   to pick black or green?
   orange or blue? orange or
   black? black or red?
   red or blue?

Challenge

6. For each spinner, describe the probability
   of landing on each color.

   a.   **b.** 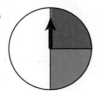   **c.**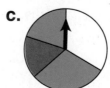

Events and Outcomes

Rae wrote A, B, and C on 3 slips of paper and put them into a bag. Then she picked a letter at random, tallied the outcome, and put the letter back into the bag.

A   B   C

There are 3 possible outcomes: A, B, and C.

Can Rae be sure of picking B on the 11th try?

▶ Rae began with 3 letters. The probability of picking B on the 1st try was 1 out of 3.

After each try, Rae put the letter back into the bag. So for each try, the probability of picking B was *still* 1 out of 3.

Rae has *the same chance* of picking B each time. She *cannot* be sure of picking B on the 11th try.

Outcomes after 10 Tries

Letter	Tally
A	IIII
B	
C	︲Hr I

Ben wrote each of the even digits on slips of paper and put them into a bag. Then he picked a digit at random and put it in his pocket. He did this for each try.

What is the probability of Ben picking 0 on the 3rd try?

Ben's Picks

Try	Digit
1st	8
2nd	6
3rd	

▶ **1st Try:** 5 digits
Probability of picking 0: 1 out of 5

2nd Try: 4 digits
Probability of picking 0: 1 out of 4

3rd Try: 3 digits
Probability of picking 0: 1 out of 3

The probability changes as the number of possible outcomes changes.

The probability of Ben picking 0 on the 3rd try is 1 out of 3.

PROBLEM SOLVING Use the information given on page 254.

1. Suppose Ben picks 4 on the 3rd try. What is the probability of his picking 0 on the 4th try?

2. If Ben picks 4 on the 3rd try, is it equally likely that he would pick 2 or 0 on the 4th try?

3. Suppose Rae started with A, B, C, D, E, and F. What would be the probability of her picking A on the 1st try? B on the 10th try? E on the 25th try? D on the 100th try?

Suppose you flip a quarter. List the possible outcomes.

4. What is the probability of it landing heads up? tails up?

5. If it landed tails up four times in a row, could you tell whether it would land heads up or tails up the 5th time? Explain your answer.

Communicate

Suppose there are 2 red marbles and 2 black marbles in a bag.

6. What is the probability of picking red? black?

7. On the 1st try you pick a red marble and put it in your pocket. On the 2nd try, what is the probability of picking red? of picking black?

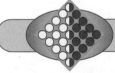

Finding Together

Discuss

Make two spinners like the ones at the right. Decide which player is EVEN and which is ODD. Spin both spinners at the same time and find the sum. If the sum is odd, ODD scores 1 point. If the sum is even, EVEN scores 1 point. The winner is the first player to score 10 points. Switch roles and play again.

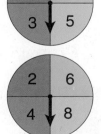

8. List all possible outcomes. Is this game fair or unfair? Explain your answer.

7-9 | Problem Solving: Make Up a Question

Problem: Penn School asked its students about school dress codes. This bar graph shows the results of the poll. What questions can you answer?

Penn School Opinion Poll

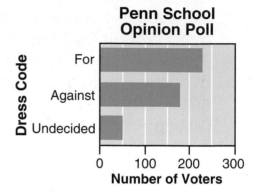

1 IMAGINE Put yourself in the problem.

2 NAME *Facts:* Penn School held an opinion poll. Results are shown in the graph.

Question: What questions can you answer from this graph?

3 THINK Study the graph and then make up several questions about it, such as:

How many people were for the dress code?
How many people were against the dress code?
How many people were polled in all?

4 COMPUTE Use the graph to find the answers.

About 225 people were for the dress code.
About 175 people were against the dress code.

Add to find the number of voters.

$$
\begin{array}{r}
{\scriptstyle 1\,1} \\
225 \\
175 \\
+\ \ 50 \\
\hline
450
\end{array}
$$

About 450 people were polled in all.

5 CHECK Use a calculator to check the sum.

Use the Make Up a Question strategy to solve each problem.

1. Each earring Jenny makes has a blue bead, a white bead, and a purple bead. The beads cost 47¢ each. What questions can you answer using this information?

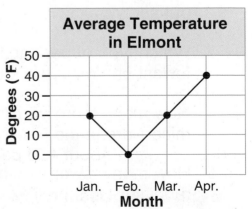

IMAGINE   Create a mental picture.

NAME   *Facts:*   Each earring has 3 beads.
The beads are blue, white, and purple.
The beads cost 47¢ each.

Question: What questions can you answer?

THINK   You could ask about price:
- How much do the beads for a pair of earrings cost?

You could ask about combinations:
- How many different bead patterns can Jenny make?

COMPUTE ——→ **CHECK**

2. Michael exercises for 12 minutes every day. He does 25 sit-ups and 10 push-ups. What questions can you answer about Michael's exercise schedule?

Exercise Schedule	
sit-ups	卌 卌 卌 卌 卌
push-ups	卌 卌

3. A puppy is 2 pounds at birth and gains about 2 pounds every week for the first month. What questions can you answer about the puppy?

4. This graph shows the average daily temperature in Elmont. What questions can you answer about the graph?

Average Temperature in Elmont

Degrees (°F): 50, 40, 30, 20, 10, 0

Month: Jan. Feb. Mar. Apr.

Problem-Solving Applications

Solve each problem and explain the method you used.

1. There were 175 dogs at the Rosedale Pet Show. There were 50 small dogs and 85 medium-size dogs. The rest were large dogs. How many large dogs were in the show? Make a pictograph about the dogs in the pet show.

Imagine 1st
Name 2nd
Think 3rd
Compute 4th
Check 5th

2. Based on the numbers in problem 1, was it more likely that a small, a medium-size, or a large dog would have won the Rosedale show?

Use the circle graph for problems 3–5.

3. Were more than half the pets dogs?

4. How many more dogs than cats were entered?

5. How many pets were entered in the Rosedale Pet Show?

Rosedale Pet Show Pets

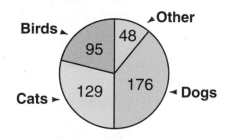

Birds 95
Other 48
Cats 129
Dogs 176

Use the bar graph for problems 6–9.

6. How many turtles were in the pet show?

7. Which type of pet had the fewest entries in the show?

8. How many more gerbils than mice were in the pet show?

9. How many fewer fish than rabbits were there?

Other Pets in the Rosedale Pet Show

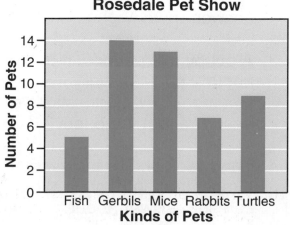

Number of Pets

Fish Gerbils Mice Rabbits Turtles
Kinds of Pets

Choose a strategy from the list or use another strategy you know to solve each problem.

10. Eight cats were finalists for best cat, and twice as many were semifinalists. There were twice as many quarterfinalists as semifinalists. How many cats were quarterfinalists?

11. Admission to the show was $3.75 for adults and $2.00 for children. Alana spent $13.25 for tickets. What tickets did she buy?

12. A dog-food supplier gave away 560 pounds of dog food. The food was bundled in 4-ounce packages. How many packages were given away?

13. A collie, a turtle, and a canary won the top three prizes. The disappointed collie buried the winner's ribbon. The first- and third-place pets both had four feet. Who won first prize?

USE THESE STRATEGIES:
Make Up a Question
Two-Step Problem
Choose the Operation
Guess and Test
Logical Reasoning
Hidden Information

Use the line graph for problems 14 and 15.

14. About how many people in all attended the pet show?

15. Did attendance increase more from Thursday to Friday or from Friday to Saturday?

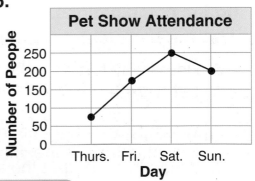

Pet Show Attendance

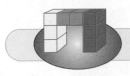

Make Up Your Own

16. Write a problem using one of the graphs from pages 258 or 259. Have a classmate solve it.

Use the list below.   *(See pp. 244–245.)*

The list shows the favorite frozen yogurt flavors of a group of students.

Inez – vanilla	Nahn – vanilla	Luis – chocolate
Mark – chocolate	Tovah – vanilla	Debi – strawberry
Saul – mint	Maura – chocolate	Shannon – mint
Cathy – strawberry	Larry – strawberry	Juan – chocolate
Shirelle – chocolate	Kyle – vanilla	Lacey – chocolate

1. Make a bar graph from the data on the list.

Use the line graph below to solve problems 2–4.   *(See pp. 246–247.)*

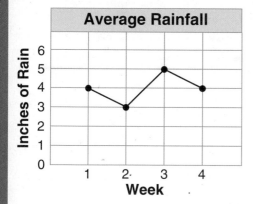

Average Rainfall

2. In which week was there the most rain?

3. How many inches of rain fell in week 2? week 3?

4. In which two weeks did the same amount of rain fall?

Find the combinations.   *(See pp. 250–251.)*

5. For lunch, Cham can buy either a tuna fish, chicken, or ham sandwich on either rye, whole-grain, or wheat bread. How many ways can he choose to buy his sandwich?

Use the spinner.   *(See pp. 252–253.)*

6. Is it equally likely that the spinner will land on

   a. red or yellow?   **b.** blue or red?

   c. yellow or white?   **d.** white or blue?

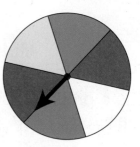

(See Still More Practice, p. 467.)

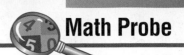

LINE PLOT

These are the math test scores of the students
in Mr. Fox's fourth-grade class.

Abby – 75	Dawn – 70	Nina – 85	Jaclyn – 65	Paul – 80
Joe – 70	Yukio – 85	Nora – 85	Lionel – 80	Lonette – 100
Nilsa – 90	Andy – 65	Sada – 75	Seiji – 85	Erika – 75
Tony – 85	Isaiah – 80	Zhou – 75	Kiri – 95	Nick – 85
Rob – 80	Carlo – 95	Greg – 90	Kate – 85	Chita – 80

Mr. Fox made a **line plot** to show the test
results. He wrote an X for each student
above the appropriate score.

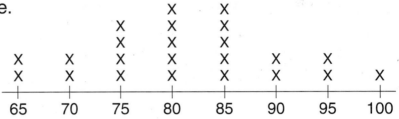

```
                                      X
                                      X
                             X        X
                    X        X        X
                    X        X        X
          X    X    X    X    X    X    X
          X    X    X    X    X    X    X    X
        --+----+----+----+----+----+----+----+--
score:   65   70   75   80   85   90   95  100
```

The **range** is the difference between
the greatest and least scores.

$$\begin{array}{r} 100 \\ -\ 65 \\ \hline \end{array}$$

range $\rightarrow$ 35

The **mode** is the score
that appears most often.

mode $\rightarrow$ 85   7 students

Math Test Scores, Ms. Anton's Class

Alison – 60	Rick – 65	Julie – 80	Kim – 70	Ruth – 80
Wendell – 70	Joann – 80	Oscar – 70	Benazar – 65	Gabriel – 70
Joyce – 85	Tate – 70	Neil – 85	Dove – 75	Gary – 90
Bao – 75	Kathy – 90	Luz – 70	Hitoshi – 90	Seve – 60

PROBLEM SOLVING

1. Use the scores above to make
a line plot.

2. What is the range of the scores?

3. What is the mode of the scores?

261

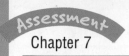
Check Your Mastery

Performance Assessment

Draw and color a spinner on which

1. it would be equally likely to land on red or yellow.

2. the probability of landing on red is 1 out of 4.

Use the pictograph below.

Magazines Sold					
Theresa	□	□	□	□	
Lyle	□	□	□	□	◻
Colette	□	□	◻		
Everly	□	□	□	□	

Key: Each □ = 10 magazines.
Each ◻ = 5 magazines.

3. Who sold the most magazines?

4. How many more magazines did Lyle sell than Colette?

5. How many magazines were sold in all?

Use the circle graph at the right.

Number of Students

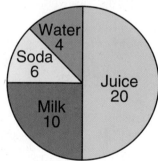

6. How many students drank
   a. milk?   **b.** juice?   **c.** soda?

7. How many more students drank milk than water? juice than soda?

8. Did fewer students drink milk than water and soda? Explain.

Use the spinner.

9. What is the probability of the spinner landing on
   a. white?   **b.** red?   **c.** blue?

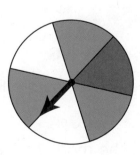

Cumulative Test I

Choose the best answer.

1. What is the value of the underlined digit in 6̲8,325,784?

    a. 6 ten millions   **b.** 6 billions
    c. 6 thousands   **d.** 6 millions

2. Round 3762 to the nearest thousand.

    a. 3000   **b.** 3800
    c. 4000   **d.** 5000

3. Estimate by rounding.

    $50.24
      3.69
   $\underline{+12.28}$

    a. $54.00
    b. $55.00
    c. $66.00
    d. $70.00

4. Subtract.

    $5.98
  $\underline{- 0.54}$

    a. $5.34
    b. $5.42
    c. $5.44
    d. not given

5. 7846 + 685

    a. 7531
    b. 8521
    c. 14,696
    d. not given

6. 4000 − 3951

    a. 49
    b. 149
    c. 1049
    d. not given

7. The product is 8. One factor is 2. What is the other factor?

    a. 4
    b. 6
    c. 10
    d. 16

8. Estimate the product.

    9×94

    a. 100
    b. 700
    c. 810
    d. 1500

9. Find the missing number.

    $\dfrac{8}{?\overline{)56}}$

    a. 6
    b. 7
    c. 8
    d. 9

10. What is the next number in the pattern?

    27, 9, 3, ?̲

    a. 0
    b. 1
    c. 3
    d. 81

11. Find the length to the nearest half inch.

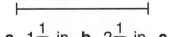

    a. $1\frac{1}{2}$ in. **b.** $2\frac{1}{2}$ in. **c.** 3 in. **d.** $3\frac{1}{2}$ in.

12. Find the length to the nearest centimeter.

    a. 7 cm   **b.** 8 cm   **c.** 67 cm   **d.** 80 cm

13. What is the probability of the spinner landing on blue?

    a. 1 out of 5
    b. 2 out of 5
    c. 1 out of 2
    d. 3 out of 5

Find the product or quotient.

14. 4×36 **15.** $7 \times \$.49$ **16.** 5×3456 **17.** $63 \times \$7.45$

18. $\begin{array}{r} 269 \\ \times\ 29 \\ \hline \end{array}$ **19.** $\begin{array}{r} 6321 \\ \times\ \ \ 8 \\ \hline \end{array}$ **20.** $6\overline{)1464}$ **21.** $5\overline{)354}$

22. $8\overline{)\$24.00}$ **23.** $7\overline{)6377}$ **24.** $800 \div 5$ **25.** $\$1.29 \div 3$

Compare. Write <, =, or >.

26. 9 ft ? 3 yd **27.** 6 lb ? 80 oz **28.** 36 in. ? 2 yd

29. 3 m ? 6000 mm **30.** 250 mL ? 25 L **31.** 2 kg ? 200 g

PROBLEM SOLVING
Use the bar graph for problems 32–35.

32. Who had the most money in pledges?

33. How much did Steve and Lita have?

34. How much more than Lita did Paco have?

35. What was the total amount in pledges for all five students?

36. Keewhan works in a restaurant. He must wear either a white, blue, or green shirt with either black, blue, or gray pants. How many different outfits can he wear? Explain how you found your answer.

Walk-a-thon Pledges

For Rubric Scoring

Listen for information about how your work will be scored.

37. Lou, Inez, and Gil are going to dance rehearsal. Lou is going a half hour before Inez. Inez is going one hour after Gil. Gil's appointment is at 12:00 P.M. What time is Lou's appointment?

Fraction Concepts

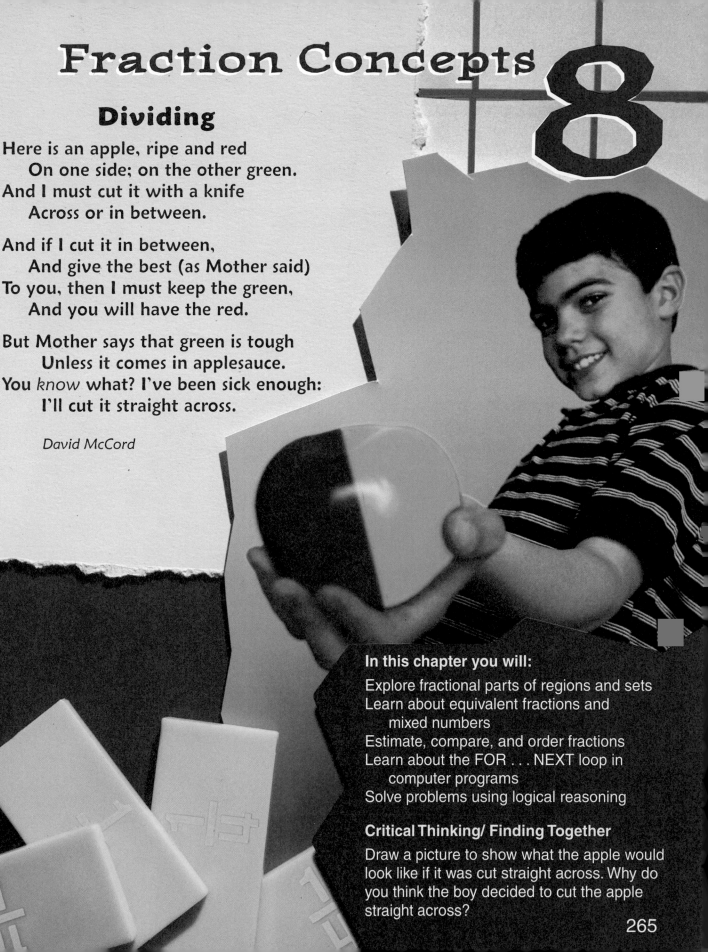

Dividing

Here is an apple, ripe and red
    On one side; on the other green.
And I must cut it with a knife
    Across or in between.

And if I cut it in between,
    And give the best (as Mother said)
To you, then I must keep the green,
    And you will have the red.

But Mother says that green is tough
    Unless it comes in applesauce.
You *know* what? I've been sick enough:
    I'll cut it straight across.

David McCord

In this chapter you will:

Explore fractional parts of regions and sets
Learn about equivalent fractions and
    mixed numbers
Estimate, compare, and order fractions
Learn about the FOR . . . NEXT loop in
    computer programs
Solve problems using logical reasoning

Critical Thinking/ Finding Together

Draw a picture to show what the apple would
look like if it was cut straight across. Why do
you think the boy decided to cut the apple
straight across?

8-1 Fraction Sense

These figures show fractional parts.

These figures do *not* show fractional parts.

 ## Discover Together

You Will Need: pattern blocks, paper, pencil

Use one layer of pattern blocks to exactly cover the figure at the right as many different ways as you can. Record all the different ways.

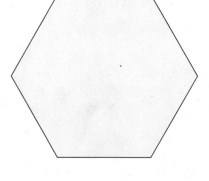

1. How many different ways could you exactly cover the figure?

2. How many of each color block did you use to cover the figure?

3. Which fractional parts did the blocks show?

4. What fractional part of the whole is 1 of each of these colors equal to?

Now exactly cover this figure as many different ways as you can. Record all the different ways.

5. How many different ways could you exactly cover the figure?

6. How many of each color block did you use to cover the figure? What fractional part of the whole is 1 of each of these colors equal to?

Arrange blue pattern blocks as shown at the right.

7. How many blocks are in the set of blue blocks? What fractional part of the set of blocks is 1 blue block?

Communicate

Discuss ✓

8. Think about your answer to question 2. Did the fractional parts get larger or smaller as you were covering the figure with more and more blocks?

9. Does a fraction give any clue about the size of the whole? Explain your reasoning.

10. Compare the fractions you wrote for question 4 to the fractions you wrote for question 7. Can the same fraction stand for both part of a whole and part of a set? Use other arrangements of pattern blocks to show other examples.

Project

11. Use what you have learned about fractions to make a book about fractional parts. Draw pictures to show how the same fraction can stand for different things.

8-2 Writing Fractions

What fractional part of the whole is yellow? What fractional part of the set is yellow?

Both the whole and the set have 6 equal parts. One of the equal parts of the whole is yellow. One of the equal parts of the set is yellow.

$\frac{1}{6}$ of the whole is yellow. $\frac{1}{6}$ of the set is yellow.

| The **numerator** names the number of equal parts. | → $\frac{1}{6}$ ← | The **denominator** names the total number of equal parts in the whole or the set. |

Read: one sixth   Write: $\frac{1}{6}$

What fractional part of the whole or of the set is purple?

$\frac{5}{6}$ of the whole is purple. $\frac{5}{6}$ of the set is purple.

numerator ⟶ 5 ← number of equal parts that are purple
denominator ⟶ 6 ← total number of equal parts in the whole or the set

Read: five sixths   Write: $\frac{5}{6}$

Write *numerator* or *denominator* for the red digit or word.

1. one half
2. six tenths
3. $\frac{5}{8}$
4. $\frac{1}{5}$

Write each as a fraction. Then circle the denominator.

5. one fourth

6. two tenths

7. one half

8. four fifths

9. three fourths

10. five eighths

11. five sixths

12. three sevenths

13. one twelfth

Write each as a fraction. Then circle the numerator.

14. seven tenths

15. three fifths

16. one eighth

17. one third

18. two sixths

19. nine twelfths

20. seven eighths

21. one ninth

22. four hundredths

Write each fraction in words. Then draw a picture to show each.

23. $\frac{1}{10}$ **24.** $\frac{2}{5}$ **25.** $\frac{1}{6}$ **26.** $\frac{3}{8}$ **27.** $\frac{2}{7}$ **28.** $\frac{5}{12}$

PROBLEM SOLVING

29. Michelle designed a banner that was $\frac{7}{8}$ purple. Write this fraction in words.

30. Louis trimmed three tenths of a group of posters in red. Write this as a fraction.

Share Your Thinking

Communicate ✓

Color fraction strips to show each fraction. Then tell a classmate how you decided which strips to use.

31. $\frac{2}{5}$ **32.** $\frac{1}{2}$ **33.** $\frac{1}{10}$

34. $\frac{3}{8}$ **35.** $\frac{5}{6}$ **36.** $\frac{11}{12}$

8-3 Estimating Fractions

▶ You can use $\frac{1}{2}$ to estimate a fraction of a region.

About what fraction of each region is blue?

about $\frac{1}{2}$ blue

more than $\frac{1}{2}$

about $\frac{3}{4}$ blue

less than $\frac{1}{2}$

about $\frac{1}{3}$ blue

▶ You can use a number line to tell whether a fraction is closer to 0, closer to $\frac{1}{2}$, or closer to 1.

Is each of these fractions closer to 0, to $\frac{1}{2}$, or to 1?

 $\frac{8}{10}$

 $\frac{2}{10}$

 $\frac{6}{10}$

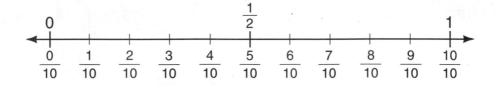

$\frac{1}{2}$ is halfway between 0 and 1.

$\frac{2}{10}$ is between 0 and $\frac{1}{2}$. It is closer to 0.

$\frac{6}{10}$ is between $\frac{1}{2}$ and 1. It is closer to $\frac{1}{2}$.

$\frac{8}{10}$ is between $\frac{1}{2}$ and 1. It is closer to 1.

Write *more than half* or *less than half* to tell about what fraction of each region is shaded.

1.    2.    3.   4.

5.    6.   7.   8.

Use the number lines. Write whether each fraction is *closer to 0*, *closer to $\frac{1}{2}$*, or *closer to 1*.

9. $\frac{3}{8}$ 10. $\frac{7}{8}$ 11. $\frac{1}{8}$ 12. $\frac{5}{8}$

13. $\frac{10}{12}$ 14. $\frac{5}{12}$ 15. $\frac{11}{12}$ 16. $\frac{7}{12}$

17. $\frac{2}{9}$ 18. $\frac{4}{9}$ 19. $\frac{7}{9}$ 20. $\frac{1}{9}$

Write whether each fraction is *closer to 0*, *closer to $\frac{1}{2}$*, or *closer to 1*. You may use number lines.

21. $\frac{3}{10}$ 22. $\frac{1}{8}$ 23. $\frac{4}{5}$ 24. $\frac{3}{7}$ 25. $\frac{1}{12}$ 26. $\frac{2}{3}$

 Finding Together

About where on each number line is the arrow pointing?

27. 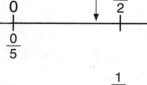   28.

29. 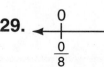   30.

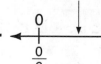

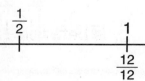

271

Equivalent Fractions

Equivalent fractions name the *same part* of a region or a set.

Equivalent Fraction Table

1				1 whole

$1 = \dfrac{2}{2} = \dfrac{3}{3} = \dfrac{4}{4} = \dfrac{5}{5} = \dfrac{6}{6} = \dfrac{7}{7} = \dfrac{8}{8} = \dfrac{9}{9} = \dfrac{10}{10} = \dfrac{12}{12}$

Notice that $\dfrac{3}{4} = \dfrac{6}{8} = \dfrac{9}{12}$.

$\dfrac{3}{4}$, $\dfrac{6}{8}$, and $\dfrac{9}{12}$ are equivalent fractions.

They all name the same part.

Write the equivalent fraction. Use the equivalent fraction table on page 272.

1. $\frac{1}{2} = \frac{?}{6}$

2. $\frac{1}{4} = \frac{?}{8}$

3. $\frac{2}{5} = \frac{?}{10}$

4. $\frac{4}{8} = \frac{?}{4}$

5. $\frac{2}{3} = \frac{?}{12}$

6. $\frac{5}{10} = \frac{?}{2}$

7. $\frac{3}{12} = \frac{?}{4}$

8. $\frac{2}{3} = \frac{?}{9}$

9. $\frac{1}{3} = \frac{?}{6}$

10. $\frac{2}{4} = \frac{?}{8}$

11. $\frac{2}{3} = \frac{?}{6}$

12. $\frac{1}{5} = \frac{?}{10}$

13. $\frac{1}{3} = \frac{?}{12}$

14. $\frac{2}{6} = \frac{?}{12}$

15. $\frac{3}{4} = \frac{?}{8}$

16. $\frac{3}{5} = \frac{?}{10}$

17. $\frac{1}{2} = \frac{?}{10}$

18. $\frac{3}{4} = \frac{?}{12}$

19. $\frac{2}{2} = \frac{?}{8}$

20. $\frac{1}{3} = \frac{?}{9}$

Does each pair show equivalent fractions? Write *yes* or *no*. Then write the equivalent fractions.

21.

22.

23.

24.

PROBLEM SOLVING Use fraction strips to show your work.

25. How many fifths are equal to four tenths?

26. How many twelfths are equal to five sixths?

 **Critical Thinking**   Algebra ✓

27. Use the equivalent fraction table. Write all the fractions:
    a. that are equal to $\frac{1}{2}$.
    b. that are equal to 1.

28. Look at fifths and tenths. Then look at sixths and twelfths. Name a fraction that is equivalent to $\frac{2}{7}$.

8-5 Writing Equivalent Fractions

Suppose you did not have an equivalent fraction chart. How would you find equivalent fractions?

$\frac{1}{4}$	$\frac{1}{4}$	$\frac{1}{4}$	$\frac{1}{4}$				
$\frac{1}{8}$	$\frac{1}{8}$	$\frac{1}{8}$	$\frac{1}{8}$	$\frac{1}{8}$	$\frac{1}{8}$	$\frac{1}{8}$	$\frac{1}{8}$

To find equivalent fractions, multiply the numerator and the denominator by the same number.

$\frac{3}{4} = \frac{?}{8}$ $4 \times \underline{?} = 8$ $4 \times 2 = 8$

$\frac{3 \times 2}{4 \times 2} = \frac{6}{8}$

So $\frac{3}{4} = \frac{6}{8}$. ← These are equivalent fractions.

Study these examples.

$\frac{1}{3} = \frac{?}{9}$ $3 \times \underline{?} = 9$
                              $3 \times 3 = 9$

$\frac{1 \times 3}{3 \times 3} = \frac{3}{9}$

So $\frac{1}{3} = \frac{3}{9}$.

$\frac{3}{5} = \frac{12}{?}$ $3 \times \underline{?} = 12$
                              $3 \times 4 = 12$

$\frac{3 \times 4}{5 \times 4} = \frac{12}{20}$

So $\frac{3}{5} = \frac{12}{20}$.

Write the equivalent fraction.

1. $\frac{1 \times 2}{3 \times 2} = \frac{?}{?}$

2. $\frac{5 \times 3}{6 \times 3} = \frac{?}{?}$

3. $\frac{2 \times 2}{5 \times 2} = \frac{?}{?}$

4. $\frac{3 \times 4}{4 \times 4} = \frac{?}{?}$

5. $\frac{1 \times 3}{8 \times 3} = \frac{?}{?}$

6. $\frac{3 \times 2}{10 \times 2} = \frac{?}{?}$

7. $\frac{1 \times 3}{7 \times ?} = \frac{3}{?}$

8. $\frac{3 \times ?}{8 \times 2} = \frac{?}{?}$

9. $\frac{2 \times 4}{3 \times ?} = \frac{?}{?}$

10. $\frac{1 \times 5}{4 \times ?} = \frac{?}{?}$

11. $\frac{5 \times ?}{7 \times 2} = \frac{?}{?}$

12. $\frac{2 \times ?}{9 \times 2} = \frac{?}{?}$

Copy and complete.

13. $\dfrac{1 \times \text{?}}{6 \times \text{?}} = \dfrac{\text{?}}{12}$

14. $\dfrac{5 \times \text{?}}{6 \times \text{?}} = \dfrac{\text{?}}{12}$

15. $\dfrac{4 \times \text{?}}{9 \times \text{?}} = \dfrac{\text{?}}{27}$

16. $\dfrac{4 \times \text{?}}{5 \times \text{?}} = \dfrac{\text{?}}{20}$

17. $\dfrac{1 \times \text{?}}{2 \times \text{?}} = \dfrac{\text{?}}{10}$

18. $\dfrac{3 \times \text{?}}{8 \times \text{?}} = \dfrac{\text{?}}{32}$

19. $\dfrac{7 \times \text{?}}{10 \times \text{?}} = \dfrac{\text{?}}{30}$

20. $\dfrac{2 \times \text{?}}{3 \times \text{?}} = \dfrac{\text{?}}{18}$

21. $\dfrac{4 \times \text{?}}{7 \times \text{?}} = \dfrac{\text{?}}{21}$

22. $\dfrac{3}{4} = \dfrac{\text{?}}{12}$

23. $\dfrac{4}{5} = \dfrac{\text{?}}{10}$

24. $\dfrac{1}{12} = \dfrac{\text{?}}{36}$

25. $\dfrac{1}{2} = \dfrac{\text{?}}{10}$

26. $\dfrac{5}{6} = \dfrac{\text{?}}{12}$

27. $\dfrac{3}{8} = \dfrac{\text{?}}{24}$

28. $\dfrac{5}{9} = \dfrac{\text{?}}{27}$

29. $\dfrac{1}{4} = \dfrac{\text{?}}{16}$

30. $\dfrac{3}{7} = \dfrac{\text{?}}{14}$

31. $\dfrac{2}{5} = \dfrac{\text{?}}{25}$

32. $\dfrac{2}{3} = \dfrac{\text{?}}{18}$

33. $\dfrac{1}{6} = \dfrac{\text{?}}{30}$

34. $\dfrac{5}{8} = \dfrac{\text{?}}{40}$

35. $\dfrac{6}{10} = \dfrac{\text{?}}{20}$

36. $\dfrac{2}{4} = \dfrac{\text{?}}{12}$

37. $\dfrac{6}{8} = \dfrac{18}{\text{?}}$

38. $\dfrac{2}{7} = \dfrac{4}{\text{?}}$

39. $\dfrac{2}{3} = \dfrac{\text{?}}{12}$

40. $\dfrac{3}{10} = \dfrac{6}{\text{?}}$

41. $\dfrac{1}{5} = \dfrac{\text{?}}{15}$

42. $\dfrac{4}{9} = \dfrac{\text{?}}{18}$

Skills to Remember

Find the product or the missing factor.

43. $\begin{array}{r} \text{?} \\ \times\ 1 \\ \hline 48 \end{array}$

44. $\begin{array}{r} 24 \\ \times\ 2 \\ \hline \text{?} \end{array}$

45. $\begin{array}{r} 16 \\ \times\ \text{?} \\ \hline 48 \end{array}$

46. $\begin{array}{r} \text{?} \\ \times\ 4 \\ \hline 48 \end{array}$

47. $\begin{array}{r} 8 \\ \times 6 \\ \hline \text{?} \end{array}$

48. $6 \times \underline{\ \text{?}\ } = 24$

49. $\underline{\ \text{?}\ } \times 3 = 24$

50. $2 \times \underline{\ \text{?}\ } = 24$

8-6 Factors

Any whole number can be represented by one or more rectangles.

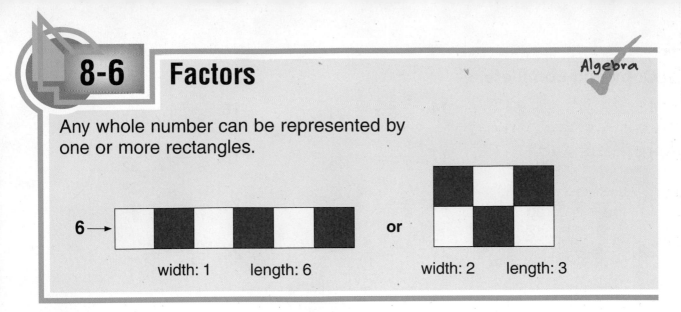

6 → width: 1   length: 6

or

width: 2   length: 3

Discover Together

You Will Need: tiles, paper, pencil

Use tiles to find as many different rectangles as you can for 24. Record each width and length.

Width		Length
1.	1	24
2.	2	?

1. How many different rectangles did you find?

The widths and lengths stand for the factors of 24.

2. What are all the factors of 24?

Now find as many different rectangles as you can for 18. Record each width and length.

3. How many different rectangles did you find?

4. What are all the factors of 18?

5. Did 18 and 24 have any rectangles and factors that were the same? Which ones?

Common factors are numbers that are factors of two or more products.

6. What are all the common factors of 24 and 18?

The **greatest common factor (GCF)** of two or more products is the greatest number that is a factor of those products.

7. What is the greatest common factor (GCF) of 24 and 18?

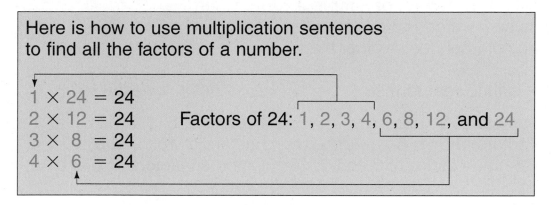

Here is how to use multiplication sentences to find all the factors of a number.

1 × 24 = 24
2 × 12 = 24   Factors of 24: 1, 2, 3, 4, 6, 8, 12, and 24
3 × 8 = 24
4 × 6 = 24

8. How would you use multiplication sentences to find all the common factors of two or more numbers?

Communicate

List all the common factors of each set of numbers. Then circle the GCF.

9. 8 and 12   **10.** 9 and 15   **11.** 6 and 15   **12.** 9 and 21

13. 10 and 30   **14.** 12 and 16   **15.** 12 and 18   **16.** 18 and 30

17. 24 and 36   **18.** 25 and 35   **19.** 36 and 42   **20.** 36 and 48

21. 8, 20, and 40   **22.** 12, 30, and 42   **23.** 10, 25, and 45

24. 18, 48, and 54   **25.** 15, 40, and 30   **26.** 20, 50, and 100

Communicate   Discuss

27. Look at the set of numbers at the right. Can the GCF be greater than 12? Explain why or why not. Then find the GCF.

12		42
	30	
	18	48
24		36

8-7 Fractions: Lowest Terms

Algebra

▶ The **terms** of a fraction are its numerator and its denominator. A fraction is in **lowest terms**, or **simplest form**, when its numerator and denominator have no common factor other than 1.

$\frac{2}{5}$ is in lowest terms.

Factors of 2: 1, 2
Factors of 5: 1, 5
Common factor of 2 and 5: 1

$\frac{6}{10}$ is *not* in lowest terms.

Factors of 6: 1, 2, 3, 6
Factors of 10: 1, 2, 5, 10
Common factors of 6 and 10: 1, 2

▶ To rename a fraction as an equivalent fraction in lowest terms or simplest form, divide the numerator and the denominator by their greatest common factor.

Write $\frac{6}{10}$ in lowest terms.

Factors of 6: 1, 2, 3, 6
Factors of 10: 1, 2, 5, 10

The GCF of 6 and 10 is 2.

$\frac{6 \div 2}{10 \div 2} = \frac{3}{5}$

Factors of 3: 1, 6
Factors of 5: 1, 5

So $\frac{6}{10}$ in lowest terms is $\frac{3}{5}$.

Copy and complete.

1. $\frac{4 \div 4}{8 \div 4} = \frac{?}{?}$

2. $\frac{3 \div 3}{9 \div 3} = \frac{?}{?}$

3. $\frac{6 \div 2}{8 \div 2} = \frac{?}{?}$

4. $\frac{8 \div 2}{10 \div 2} = \frac{?}{?}$

5. $\frac{9 \div ?}{12 \div 3} = \frac{?}{?}$

6. $\frac{14 \div 7}{21 \div ?} = \frac{?}{?}$

7. $\frac{10 \div ?}{25 \div ?} = \frac{?}{5}$

8. $\frac{12 \div ?}{42 \div ?} = \frac{?}{7}$

9. $\frac{15 \div ?}{45 \div ?} = \frac{3}{?}$

278

Is each fraction in lowest terms? Write *yes* or *no*.

10. $\frac{4}{7}$ **11.** $\frac{6}{9}$ **12.** $\frac{11}{12}$ **13.** $\frac{7}{10}$ **14.** $\frac{2}{10}$ **15.** $\frac{8}{12}$

Write each fraction in lowest terms.

16. $\frac{2}{6}$ **17.** $\frac{4}{24}$ **18.** $\frac{9}{18}$ **19.** $\frac{3}{12}$ **20.** $\frac{2}{4}$ **21.** $\frac{12}{20}$

22. $\frac{6}{18}$ **23.** $\frac{10}{20}$ **24.** $\frac{8}{24}$ **25.** $\frac{9}{15}$ **26.** $\frac{15}{20}$ **27.** $\frac{4}{10}$

28. $\frac{6}{24}$ **29.** $\frac{8}{14}$ **30.** $\frac{6}{15}$ **31.** $\frac{10}{12}$ **32.** $\frac{7}{21}$ **33.** $\frac{5}{15}$

34. $\frac{8}{18}$ **35.** $\frac{9}{27}$ **36.** $\frac{15}{18}$ **37.** $\frac{10}{15}$ **38.** $\frac{9}{15}$ **39.** $\frac{12}{18}$

PROBLEM SOLVING Express each answer in simplest form.

40. The chorus sang 12 songs at open house. Four of the songs were folk songs. What fractional part of the songs were folk songs?

41. Of 35 paintings on display in the school lobby, 7 were done in watercolors. What fractional part of the paintings were watercolors?

42. Jamie's parents looked at his notebook. Ten of the 40 pages were filled with math problems. Write this as a fraction in simplest form.

43. Glenda cut out the 26 letters of the alphabet to decorate the classroom. She cut 13 letters from green paper. What fractional part of the letters were green?

44. Writing awards were presented to 30 students. Of the awards, 6 were for poetry and 10 were for essays. What fractional part of the awards were for poetry? for essays?

45. There were 80 fourth graders in Hadley School. Of these, 35 were boys. What fractional part of the fourth graders were girls?

8-8 Mixed Numbers

Darryl is baking bread. His recipe calls for three and two thirds cups of whole-wheat flour.

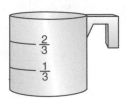  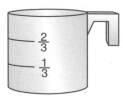 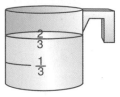

Think: $1 + 1 + 1 + \frac{2}{3}$, or $3 + \frac{2}{3}$

Read: three and two thirds   Write: $3\frac{2}{3}$

$3\frac{2}{3}$ is a mixed number.

A **mixed number** is made up of a whole number and a fraction.

whole number ⟶ $3\frac{2}{3}$ ⟵ fraction

Study these examples.

$1 + 1 + \frac{1}{4}$, or $2 + \frac{1}{4}$

Read: two and one fourth
Write: $2\frac{1}{4}$

$1 + \frac{5}{6}$

Read: one and five sixths
Write. $1\frac{5}{6}$

Write as a mixed number. Then model each.

1. four and three tenths
2. seven and two fifths
3. ten and one ninth
4. eight and five twelfths
5. two and three eighths
6. six and one half

Write a mixed number for each.

7.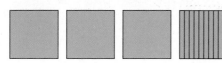

8.

9.

10.

Fractions as Whole Numbers

Some fractions can be renamed as whole numbers.

Numerator and denominator are the same.	Denominator is 1.
$\frac{4}{4} = 1$	$\frac{4}{1} = 4$
$\frac{8}{8} = 1$	$\frac{8}{1} = 8$
$\frac{75}{75} = 1$	$\frac{75}{1} = 75$
$\frac{1000}{1000} = 1$	$\frac{1000}{1} = 1000$

Rename each as a whole number.

11. $\frac{3}{3}$　　　**12.** $\frac{10}{1}$　　　**13.** $\frac{12}{1}$　　　**14.** $\frac{12}{12}$　　　**15.** $\frac{9}{1}$

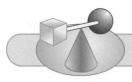 **Challenge**

To what mixed number is the arrow pointing?

16.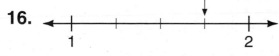

17.

18.

19.

8-9 Comparing Fractions

Compare: $\frac{5}{8}$ _?_ $\frac{3}{8}$

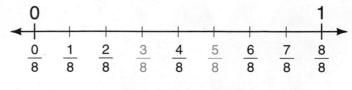

The denominators are the same.

▶ To compare fractions with the same denominators, compare the numerators.

$\frac{5}{8}$

$\frac{3}{8}$

$5 > 3 \longrightarrow \frac{5}{8} > \frac{3}{8}$

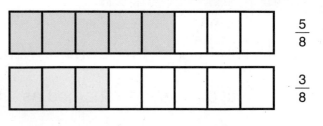

$\frac{5}{8}$

$\frac{3}{8}$

Compare: $\frac{2}{3}$ _?_ $\frac{5}{6}$

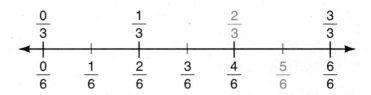

The denominators are different.

▶ To compare fractions with different denominators, first rename as equivalent fractions with the same denominators. Then compare the numerators.

$\frac{2}{3} = \frac{2 \times 2}{3 \times 2} = \frac{4}{6}$

$\frac{5}{6} \qquad\quad = \frac{5}{6}$

$4 < 5 \longrightarrow \frac{4}{6} < \frac{5}{6} \qquad$ So $\frac{2}{3} < \frac{5}{6}$.

Compare: $1\frac{2}{5}$ _?_ $1\frac{4}{5}$

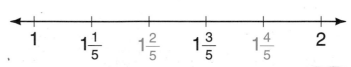

▶ To compare mixed numbers, first compare the whole numbers. Then compare the fractions.

$1\frac{2}{5}$

$1\frac{4}{5}$

$2 < 4 \longrightarrow \frac{2}{5} < \frac{4}{5} \qquad$ So $1\frac{2}{5} < 1\frac{4}{5}$.

$\boxed{1 = 1}$

Compare. Write <, =, or >. Use models to help.

1. $\frac{3}{4}$ __?__ $\frac{1}{4}$

2. $\frac{5}{8}$ __?__ $\frac{7}{8}$

3. $\frac{2}{7}$ __?__ $\frac{4}{7}$

4. $\frac{7}{9}$ __?__ $\frac{5}{9}$

5. $\frac{1}{6}$ __?__ $\frac{5}{6}$

6. $\frac{4}{5}$ __?__ $\frac{4}{5}$

7. $\frac{7}{10}$ __?__ $\frac{3}{10}$

8. $\frac{11}{12}$ __?__ $\frac{5}{12}$

9. $\frac{8}{12}$ __?__ $\frac{3}{4}$

10. $\frac{2}{3}$ __?__ $\frac{6}{9}$

11. $\frac{1}{2}$ __?__ $\frac{4}{6}$

12. $\frac{1}{4}$ __?__ $\frac{2}{8}$

13. $\frac{1}{3}$ __?__ $\frac{1}{6}$

14. $\frac{3}{5}$ __?__ $\frac{3}{10}$

15. $\frac{7}{8}$ __?__ $\frac{2}{4}$

16. $\frac{7}{12}$ __?__ $\frac{5}{6}$

17. $\frac{6}{10}$ __?__ $\frac{3}{5}$

18. $\frac{1}{2}$ __?__ $\frac{4}{8}$

19. $\frac{3}{4}$ __?__ $\frac{10}{12}$

20. $\frac{3}{10}$ __?__ $\frac{1}{2}$

21. $4\frac{3}{4}$ __?__ $4\frac{1}{4}$

22. $1\frac{2}{3}$ __?__ $2\frac{1}{3}$

23. $5\frac{1}{9}$ __?__ $2\frac{1}{9}$

24. $6\frac{2}{5}$ __?__ $6\frac{4}{5}$

25. $3\frac{3}{10}$ __?__ $3\frac{7}{10}$

26. $8\frac{5}{8}$ __?__ $8\frac{3}{8}$

27. $2\frac{4}{9}$ __?__ $4\frac{2}{9}$

28. $1\frac{3}{6}$ __?__ $1\frac{3}{6}$

PROBLEM SOLVING

29. Of the fir trees in the park, $\frac{3}{10}$ were pines and $\frac{1}{10}$ were spruce. Were there more pines or more spruce in the park?

30. At the feeding station, $\frac{1}{3}$ of the birds were sparrows and $\frac{3}{12}$ were finches. Were there more sparrows or finches at the feeding station?

31. The northern sector of the park had $3\frac{3}{4}$ mi of trails. The eastern sector had $3\frac{1}{4}$ mi of trails. Which sector had more miles of trails?

32. On Monday, $\frac{3}{4}$ of the park's visitors were schoolchildren. On Tuesday $\frac{5}{8}$ of the visitors were schoolchildren. Did more schoolchildren visit the park on Monday or on Tuesday?

Challenge

Copy and complete.

33. $1\frac{2}{3} > 1\frac{?}{?}$

34. $5\frac{3}{8} < 5\frac{?}{?}$

35. $2\frac{3}{5} < 2\frac{?}{?}$

36. $4\frac{3}{4} > 4\frac{?}{?}$

Order from least to greatest: $\frac{1}{2}$, $\frac{7}{10}$, $\frac{3}{10}$

To order fractions with *different denominators*:

- Rename as equivalent fractions with the same denominator.

$$\frac{1}{2} = \frac{1 \times 5}{2 \times 5} = \frac{5}{10}$$

$$\frac{7}{10} = \frac{7}{10}$$

$$\frac{3}{10} = \frac{3}{10}$$

- Compare the fractions by comparing the numerators.

$$3 < 5 \qquad \frac{3}{10} < \frac{5}{10}$$

$$5 < 7 \qquad \frac{5}{10} < \frac{7}{10}$$

- Arrange in order from least to greatest.

$$\frac{3}{10}, \frac{5}{10}, \frac{7}{10}$$

The order from least to greatest: $\frac{3}{10}$, $\frac{1}{2}$, $\frac{7}{10}$

Study this example.

Order from greatest to least: $\frac{3}{8}$, $\frac{1}{8}$, $\frac{7}{8}$

To order fractions with *like denominators*:

Compare the fractions by comparing the numerators.	Arrange in order from greatest to least.
$7 > 3 \quad \frac{7}{8} > \frac{3}{8}$ $3 > 1 \quad \frac{3}{8} > \frac{1}{8}$	$\frac{7}{8}, \frac{3}{8}, \frac{1}{8}$

The order from greatest to least: $\frac{7}{8}$, $\frac{3}{8}$, $\frac{1}{8}$

Write in order from least to greatest. Use models to help.

1. $\dfrac{4}{6}, \dfrac{2}{6}, \dfrac{3}{6}$

2. $\dfrac{1}{5}, \dfrac{4}{5}, \dfrac{2}{5}$

3. $\dfrac{5}{12}, \dfrac{9}{12}, \dfrac{1}{12}$

4. $\dfrac{1}{8}, \dfrac{6}{8}, \dfrac{4}{8}$

5. $\dfrac{8}{9}, \dfrac{5}{9}, \dfrac{7}{9}$

6. $\dfrac{3}{7}, \dfrac{5}{7}, \dfrac{2}{7}$

7. $\dfrac{8}{10}, \dfrac{2}{10}, \dfrac{6}{10}$

8. $\dfrac{2}{4}, \dfrac{1}{4}, \dfrac{3}{4}$

9. $\dfrac{1}{2}, \dfrac{1}{4}, \dfrac{3}{4}$

10. $\dfrac{5}{6}, \dfrac{2}{3}, \dfrac{2}{6}$

11. $\dfrac{3}{8}, \dfrac{5}{8}, \dfrac{1}{4}$

12. $\dfrac{5}{12}, \dfrac{1}{6}, \dfrac{3}{12}$

13. $\dfrac{3}{10}, \dfrac{9}{10}, \dfrac{2}{5}$

14. $\dfrac{1}{2}, \dfrac{1}{8}, \dfrac{6}{8}$

15. $\dfrac{2}{3}, \dfrac{5}{12}, \dfrac{11}{12}$

16. $\dfrac{7}{9}, \dfrac{1}{3}, \dfrac{4}{9}$

Write in order from greatest to least. Use models to help.

17. $\dfrac{1}{7}, \dfrac{6}{7}, \dfrac{4}{7}$

18. $\dfrac{4}{9}, \dfrac{8}{9}, \dfrac{2}{9}$

19. $\dfrac{1}{10}, \dfrac{7}{10}, \dfrac{8}{10}$

20. $\dfrac{5}{8}, \dfrac{2}{8}, \dfrac{7}{8}$

21. $\dfrac{9}{12}, \dfrac{3}{12}, \dfrac{6}{12}$

22. $\dfrac{3}{6}, \dfrac{5}{6}, \dfrac{1}{6}$

23. $\dfrac{3}{5}, \dfrac{1}{5}, \dfrac{4}{5}$

24. $\dfrac{3}{10}, \dfrac{9}{10}, \dfrac{2}{10}$

25. $\dfrac{1}{6}, \dfrac{1}{2}, \dfrac{2}{6}$

26. $\dfrac{5}{12}, \dfrac{9}{12}, \dfrac{1}{2}$

27. $\dfrac{2}{3}, \dfrac{2}{9}, \dfrac{5}{9}$

28. $\dfrac{3}{12}, \dfrac{3}{4}, \dfrac{7}{12}$

29. $\dfrac{6}{10}, \dfrac{9}{10}, \dfrac{1}{2}$

30. $\dfrac{2}{12}, \dfrac{1}{12}, \dfrac{2}{3}$

31. $\dfrac{1}{8}, \dfrac{3}{4}, \dfrac{5}{8}$

32. $\dfrac{3}{6}, \dfrac{7}{12}, \dfrac{2}{12}$

PROBLEM SOLVING

33. Marie cut three lengths of ribbon. They were $\dfrac{1}{2}$ yd, $\dfrac{3}{8}$ yd, and $\dfrac{5}{8}$ yd long. Which was the longest length? Which was the shortest?

34. Brad lives $\dfrac{3}{4}$ mi from school. Donna lives $\dfrac{1}{4}$ mi from school, and Chris lives $\dfrac{1}{2}$ mi from school. Who lives closest to school?

35. Jack bought $\dfrac{5}{8}$ lb turkey, $\dfrac{1}{2}$ lb ham, $\dfrac{3}{8}$ lb roast beef, and $\dfrac{1}{8}$ lb salami. Write these fractions in order from least to greatest. Then write them in order from greatest to least.

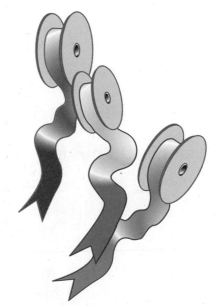

TECHNOLOGY

Probability Programs

If you tossed a coin 50 times, about how many times would you expect the coin to land on tails?

To find how many times, toss a coin or use a computer program.

The computer program below simulates a coin toss. It lets the numbers 1 and 2 stand for heads and tails. The program uses a **FOR...NEXT loop** to repeat the coin toss a given number of times.

```
10 PRINT "How many times do you want to toss a coin?"
20 INPUT T
30 FOR X = 1 to T
40 LET N = INT(RND(1)*2) + 1
50 PRINT N;
60 NEXT X
70 END
```

Waits for you to enter a number.

Tells the computer to store a number in X. The first number will be 1. The last will be what you input in line 20.

Assigns the numbers 1 or 2 to N.

Tells the computer to loop back to line 30 and store the next number in X.

Sample Output for T = 50

```
1 2 2 2 2 1 1 1 1 2 2 2 2 1 2 2 2 2 2 2 2 2 1 1 1 2
2 1 1 2 2 2 2 2 1 1 1 1 1 1 2 2 1 2 2 1 2 2 2 1
```

PROBLEM SOLVING Use the computer program above.

1. If T = 5, how many times will the program run lines 40 and 50?

2. Suppose line 30 was FOR X = 3 to 9. What value would be stored in X after lines 40 and 50 are repeated twice?

286

PROBLEM SOLVING Use the program on page 286.

3. Toss a coin 10 times. Write the output. How many 2s were printed?

4. Toss a coin 20 times. Then 30 times. Write the output for each. How many 2s were printed for each?

5. Predict the number of times a coin will land on tails if you toss it 40 times. 50 times. Run the program to check your predictions.

Write the output for each program.

6. 10 FOR X = 1 to 5
    20 PRINT X
    30 NEXT X
    40 END

7. 10 FOR I = 2 to 10
    20 PRINT 40 − I
    30 NEXT I
    40 END

8. 10 FOR H = 3 to 12
    20 PRINT H + H
    30 NEXT H
    40 END

The program below will find the **average**, or **mean**, of a set of data.

```
 5 LET Total = 0
10 PRINT "How many numbers are in the set of data?"
20 INPUT N
30 FOR X = 1 to N
40 PRINT "Enter a number from the set of data."
50 INPUT D
60 LET Total = Total + D
70 NEXT X
80 PRINT "The mean of this set of data is " Total/N
90 END
```

Find the average for each set of data.
Use the program above.

9. 12, 15, 18, 16, 21, 14

10. 124, 202, 96, 106

8-12 Problem Solving: Logic and Analogies *Algebra* ✓

Problem: Gwen, Maraya, and Sonia each buys a bracelet. One is $6\frac{3}{4}$ in., one is $6\frac{1}{2}$ in., and the third is $6\frac{4}{8}$ in.

Gwen's bracelet is longer than Sonia's. How long is Maraya's bracelet?

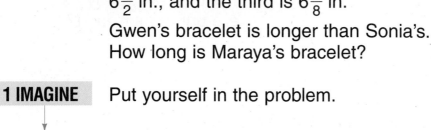

1 IMAGINE Put yourself in the problem.

2 NAME *Facts:* Bracelets are $6\frac{3}{4}$ in., $6\frac{1}{2}$ in., and $6\frac{4}{8}$ in.

Gwen's bracelet is longer than Sonia's.

Question: How long is Maraya's bracelet?

3 THINK To compare mixed numbers:

First, compare whole number parts. $6 = 6 = 6$

Next, compare fraction parts. $\frac{3}{4} \underline{\ ?\ } \frac{1}{2} \quad \frac{1}{2} \underline{\ ?\ } \frac{4}{8}$

4 COMPUTE Use fraction strips to find equivalent fractions.

$\frac{1}{4}$	$\frac{1}{4}$	$\frac{1}{4}$

$\frac{1}{8}$	$\frac{1}{8}$	$\frac{1}{8}$	$\frac{1}{8}$	$\frac{1}{8}$	$\frac{1}{8}$

$\frac{3}{4} = \frac{6}{8}$ So $6\frac{3}{4} = 6\frac{6}{8}$.

equivalent

$\frac{1}{2}$

$\frac{1}{8}$	$\frac{1}{8}$	$\frac{1}{8}$	$\frac{1}{8}$

$\frac{1}{2} = \frac{4}{8}$ So $6\frac{1}{2} = 6\frac{4}{8}$.

$\frac{6}{8} > \frac{4}{8}$ So $6\frac{3}{4} > 6\frac{1}{2}$ and $6\frac{3}{4} > 6\frac{4}{8}$.

Two bracelets are the same length.
Gwen's is longer than Sonia's, so
Gwen's bracelet is $6\frac{6}{8}$, or $6\frac{3}{4}$ in. long.
Sonia's and Maraya's must be equal in length. $6\frac{1}{2}$ in. $= 6\frac{4}{8}$ in.

5 CHECK Draw 3 lines: $6\frac{3}{4}$ in., $6\frac{1}{2}$ in., $6\frac{4}{8}$ in. Then compare.

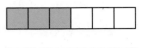

Use logical reasoning or an analogy to solve each problem.

1. Can you complete the analogy?
   $\frac{1}{2}$ is to $\frac{2}{4}$ as $\frac{3}{6}$ is to ?

IMAGINE	Draw and label the fractions.

NAME	*Facts:* $\frac{1}{2}$ and $\frac{2}{4}$ are related.

Question: What fraction is related to $\frac{3}{6}$ in the same way?

THINK	To solve an analogy, first read it aloud.

One half is to two fourths as
three sixths is to what?

Think about how $\frac{1}{2}$ and $\frac{2}{4}$ are related.

COMPUTE ⟶ **CHECK**

2. $\frac{5}{10}$ is to $\frac{10}{10}$ as $\frac{4}{8}$ is to ?

3. **ABAB** is to **CDCD** as **ABBA** is to ?

4. 8 is to $\frac{16}{24}$ as 6 is to ?

5. One worm is $4\frac{1}{4}$ in. long, another is $4\frac{3}{8}$ in.
   long, and a third is $4\frac{5}{8}$ in. long. The longest
   worm is in the garden and the shortest
   worm is on a leaf. Which worm is on a leaf?

6. A certain fraction has a numerator that is 3 less
   than its denominator. It is equivalent to $\frac{9}{18}$.
   What is the fraction?

7. $1\frac{1}{2}$ is to 1 as $3\frac{1}{2}$ is to ?

8. 6 is to $\frac{12}{18}$ as 4 is to ?

Problem-Solving Applications

Solve each problem and explain the method you used.

1. A bouquet of a dozen flowers has 4 roses. The rest are carnations. What fractional part of the bouquet is roses? is carnations?

2. Pete plants $\frac{3}{8}$ of the garden with tomatoes, $\frac{1}{2}$ with green beans, and $\frac{1}{8}$ with cucumbers. Order the sections from largest to smallest.

3. Delia's garden is $\frac{9}{12}$ flowers and $\frac{4}{16}$ herbs. Write these fractions in lowest terms. Are there more flowers or herbs in her garden?

4. A garden has $\frac{1}{10}$ red, $\frac{1}{5}$ white, $\frac{2}{10}$ yellow, and $\frac{2}{5}$ pink roses. Of which color are there the most roses? the least? Which colors share an equal number?

5. One plant is $7\frac{10}{16}$ in. tall. Another is $7\frac{3}{4}$ in. tall. The herb is the shorter plant. How tall is it?

> Remember: Express fractions in simplest form.

6. Marci has 12 sections in her flower garden and 16 sections in her herb garden. If both gardens are equal in size, which has smaller sections?

Use the circle graph for problems 7 and 8.

7. About what fractional part of Greta's garden is tulips? is red tulips?

8. What flowers make up equal parts of Greta's garden?

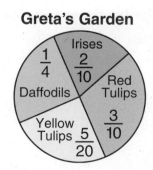

Greta's Garden

Irises $\frac{2}{10}$

$\frac{1}{4}$ Daffodils

Red Tulips

$\frac{3}{10}$

Yellow Tulips $\frac{5}{20}$

Choose a strategy from the list or use another strategy you know to solve each problem.

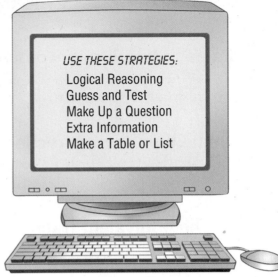

USE THESE STRATEGIES:
Logical Reasoning
Guess and Test
Make Up a Question
Extra Information
Make a Table or List

9. For every sunflower seed Deven plants, he also plants 4 zinnia seeds. Deven plants 45 seeds in all. How many zinnia seeds does he plant?

10. Diego has 10 pots. He puts marigolds in $\frac{2}{5}$ of the pots and daisies in $\frac{1}{2}$ of the pots. Are there more pots with daisies or marigolds?

11. A fraction has a denominator that is 8 greater than its numerator. It is equivalent to $\frac{1}{3}$. What is the fraction?

12. Kito plants a garden in $8\frac{1}{2}$ hours. When he is finished, $\frac{1}{4}$ of the rows are corn, $\frac{1}{3}$ are lettuce, $\frac{3}{12}$ are squash, and $\frac{1}{6}$ are zucchini. What questions can you answer about Kito's garden?

13. Ms. Tallchief plants 2 red and 2 pink geraniums in a row in her window box. How many different arrangements can she make?

14. Two fractions are equivalent. The denominator of one is the same as the numerator of the other. What are some possibilities for the two fractions?

15. Lila plants 16 bulbs. Three eighths are tulip bulbs, one eighth are daffodil bulbs, and one half are lily bulbs. Order the types of bulbs from least to greatest.

Write each as a fraction.   (See pp. 268–269.)

1. one half   **2.** three eighths   **3.** five sevenths

Write each fraction in words.

4. $\frac{5}{6}$ **5.** $\frac{3}{10}$ **6.** $\frac{7}{12}$ **7.** $\frac{3}{4}$

Estimate about what fraction of each region is shaded.   (See pp. 270–271.)

8.   **9.**   **10.**

Write the equivalent fraction.   (See pp. 272–275.)

11. $\frac{1}{9} = \frac{5}{?}$ **12.** $\frac{7}{10} = \frac{14}{?}$ **13.** $\frac{6}{9} = \frac{?}{54}$ **14.** $\frac{3}{4} = \frac{18}{?}$

Find the common factors for each set of numbers. Then find the greatest common factor.   (See pp. 276–277.)

15. 8, 12   **16.** 6, 16   **17.** 12, 20

Write each fraction in lowest terms.   (See pp. 278–279.)

18. $\frac{10}{15}$ **19.** $\frac{6}{12}$ **20.** $\frac{8}{24}$ **21.** $\frac{4}{20}$ **22.** $\frac{15}{40}$

Write in order from least to greatest.   (See pp. 282–285.)

23. $\frac{4}{10}, \frac{7}{10}, \frac{4}{5}$ **24.** $\frac{11}{12}, \frac{3}{4}, \frac{2}{12}$

Write in order from greatest to least.

25. $\frac{6}{9}, \frac{1}{3}, \frac{4}{9}$ **26.** $\frac{4}{5}, \frac{3}{10}, \frac{7}{10}$

Compare. Write <, =, or >.   (See pp. 280–283.)

27. $2\frac{7}{14}$ _?_ $3\frac{4}{14}$ **28.** $6\frac{10}{12}$ _?_ $7\frac{1}{12}$ **29.** $1\frac{2}{3}$ _?_ $2\frac{1}{3}$

          (See *Still More Practice*, p. 468.)

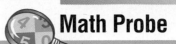

RATIO AND PERCENT

You can use a **ratio** to compare the number of violins to the number of trombones.

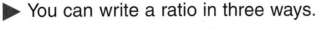

The ratio of violins to trombones is 6 to 2.

▶ You can write a ratio in three ways.

| violins | ▶ 6 to 2 ◀ | trombones |

| violins | ▶ 6 : 2 ◀ | trombones |

| violins | ▶ $\frac{6}{2}$ ◀ | trombones |

> When you write a ratio, be sure to write the numbers in the correct order.

The ratio of violins to trombones: 6 to 2, 6:2, or $\frac{6}{2}$

The ratio of trombones to violins: 2 to 6, 2:6, or $\frac{2}{6}$

▶ If you write a ratio as a fraction with a denominator of 100, you can express that ratio as a **percent (%)**.

fraction	$\frac{8}{100}$	$\frac{10}{100}$	$\frac{85}{100}$	$\frac{100}{100}$
percent	8%	10%	85%	100%

Write each ratio three ways.

1. 4 clarinets to 7 trumpets

2. 10 oboes to 1 piano

3. 5 cellos to 8 tubas

4. 9 bassoons to 6 saxophones

Write each ratio as a percent.

5. $\frac{50}{100}$

6. $\frac{1}{100}$

7. $\frac{75}{100}$

8. 25:100

9. 99:100

Write each percent as a fraction.

10. 30%

11. 5%

12. 62%

13. 48%

14. 150%

Performance Assessment

Use the number line.

1. Extend the number line to show 1.

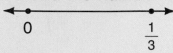

0          $\frac{1}{3}$

Draw a number line and locate each point.

2. $\frac{0}{4}$ 3. $\frac{3}{4}$ 4. $\frac{1}{2}$

Write each as a fraction.

5. five sixths   6. seven eighths   7. two tenths

Write the equivalent fraction.

8. $\frac{3}{4} = \frac{?}{12}$ 9. $\frac{1}{5} = \frac{5}{?}$ 10. $\frac{1}{3} = \frac{3}{?}$ 11. $\frac{2}{7} = \frac{?}{21}$

Find the common factors for each set.
Then find the greatest common factor.

12. 12, 24   13. 18, 36, 12

Write each in simplest form.

14. $\frac{8}{12}$ 15. $\frac{6}{10}$ 16. $\frac{12}{36}$ 17. $\frac{14}{28}$ 18. $\frac{8}{10}$

Write in order from greatest to least.

19. $\frac{14}{21}, \frac{3}{7}, \frac{3}{21}$ 20. $\frac{1}{3}, \frac{8}{9}, \frac{4}{9}$

PROBLEM SOLVING   *Use a strategy you have learned.*

21. Chet walked $\frac{5}{10}$ mile. Juanita walked $\frac{3}{5}$ mile. Toya walked $\frac{2}{5}$ mile. Who walked the longest distance? Who walked the shortest distance?

22. Mike poured this glass of juice. Is it about $\frac{1}{3}$ full or $\frac{2}{3}$ full?

Fractions: Addition and Subtraction

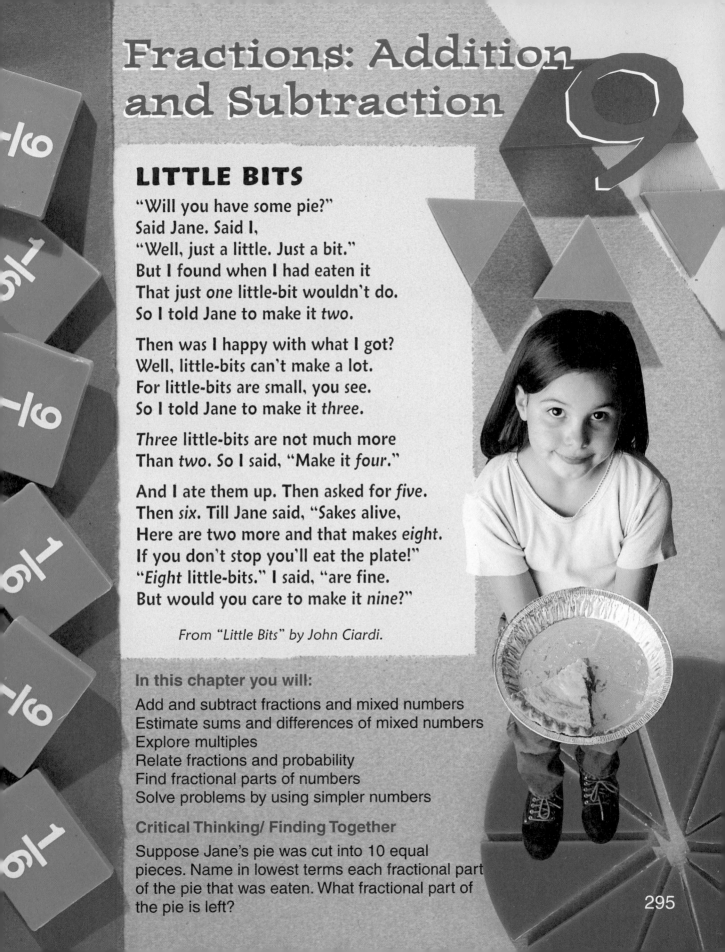

LITTLE BITS

"Will you have some pie?"
Said Jane. Said I,
"Well, just a little. Just a bit."
But I found when I had eaten it
That just *one* little-bit wouldn't do.
So I told Jane to make it *two*.

Then was I happy with what I got?
Well, little-bits can't make a lot.
For little-bits are small, you see.
So I told Jane to make it *three*.

Three little-bits are not much more
Than *two*. So I said, "Make it *four*."

And I ate them up. Then asked for *five*.
Then *six*. Till Jane said, "Sakes alive,
Here are two more and that makes *eight*.
If you don't stop you'll eat the plate!"
"*Eight* little-bits." I said, "are fine.
But would you care to make it *nine*?"

From "Little Bits" by John Ciardi.

In this chapter you will:

Add and subtract fractions and mixed numbers
Estimate sums and differences of mixed numbers
Explore multiples
Relate fractions and probability
Find fractional parts of numbers
Solve problems by using simpler numbers

Critical Thinking/ Finding Together

Suppose Jane's pie was cut into 10 equal
pieces. Name in lowest terms each fractional part
of the pie that was eaten. What fractional part of
the pie is left?

Adding: Like Denominators

Pam walked $\frac{3}{10}$ mile from her house to school. Then she walked $\frac{5}{10}$ mile from school to the bus stop. How far did Pam walk?

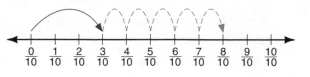

To find how far Pam walked, add: $\frac{3}{10} + \frac{5}{10} = \underline{\ ?\ }$

The denominators are the same.

To add fractions with like denominators:

- Add the numerators. $\qquad \frac{3}{10} + \frac{5}{10} = \frac{}{} \quad 8 \qquad \boxed{3 + 5 = 8}$

- Write the like denominator. $\qquad \frac{3}{10} + \frac{5}{10} = \frac{8}{10}$

- Write the sum in lowest terms. $\qquad \frac{8}{10} = \frac{8 \div 2}{10 \div 2} = \frac{4}{5}$

Factors of 8: 1, 2, 4, 8
Factors of 10: 1, 2, 5, 10   GCF: 2

Pam walked $\frac{4}{5}$ mile.

Study these examples.

$$\begin{array}{r} \frac{4}{9} \\ + \frac{5}{0} \\ \hline \frac{9}{9} = 1 \end{array} \longleftarrow \text{lowest terms} \longrightarrow \begin{array}{r} \frac{2}{7} \\ + \frac{3}{7} \\ \hline \frac{5}{7} \end{array}$$

$$\frac{1}{10} + \frac{2}{10} + \frac{4}{10} = \frac{7}{10}$$

↑ lowest terms

Add. Write the sum in lowest terms.

1. $\frac{1}{4} + \frac{2}{4}$ 　　　 2. $\frac{5}{8} + \frac{2}{8}$ 　　　 3. $\frac{1}{3} + \frac{1}{3}$ 　　　 4. $\frac{2}{7} + \frac{4}{7}$

Find the sum in lowest terms.

5. $\frac{2}{9} + \frac{1}{9}$ 6. $\frac{1}{6} + \frac{2}{6}$ 7. $\frac{2}{10} + \frac{4}{10}$ 8. $\frac{2}{5} + \frac{3}{5}$

9. $\frac{1}{8} + \frac{5}{8}$ 10. $\frac{3}{7} + \frac{4}{7}$ 11. $\frac{4}{12} + \frac{6}{12}$ 12. $\frac{2}{6} + \frac{2}{6}$

13. $\frac{3}{4} + \frac{1}{4}$ 14. $\frac{2}{9} + \frac{4}{9}$ 15. $\frac{4}{10} + \frac{4}{10}$ 16. $\frac{2}{8} + \frac{2}{8}$

17. $\frac{1}{9}$ $+ \frac{3}{9}$ 18. $\frac{3}{5}$ $+ \frac{1}{5}$ 19. $\frac{2}{10}$ $+ \frac{3}{10}$ 20. $\frac{3}{8}$ $+ \frac{5}{8}$ 21. $\frac{5}{12}$ $+ \frac{3}{12}$

22. $\frac{1}{2}$ $+ \frac{1}{2}$ 23. $\frac{5}{7}$ $+ \frac{2}{7}$ 24. $\frac{2}{8}$ $+ \frac{4}{8}$ 25. $\frac{3}{12}$ $+ \frac{3}{12}$ 26. $\frac{1}{12}$ $+ \frac{3}{12}$

27. $\frac{2}{12} + \frac{1}{12} + \frac{7}{12}$ 28. $\frac{3}{10} + \frac{2}{10} + \frac{5}{10}$ 29. $\frac{1}{8} + \frac{5}{8} + \frac{2}{8}$

PROBLEM SOLVING Write each answer in simplest form.

30. Mr. Lom rode his exercycle for $\frac{1}{4}$ hour before breakfast and $\frac{1}{4}$ hour after supper. For how much time did he ride his exercycle?

31. Jake cycled $\frac{3}{8}$ mile from his house to Rick's. Then he cycled $\frac{5}{8}$ mile from Rick's to Hal's. How far did Jake cycle?

 **Connections: Health**

For every $\frac{1}{10}$ mile that an adult walks, he or she burns about 10 calories.

32. About how many calories would a woman burn walking $\frac{2}{10}$ mile in the morning and $\frac{7}{10}$ mile in the evening?

297

9-2 Subtracting: Like Denominators

Kevin had $\frac{7}{8}$ yard of felt. He used $\frac{3}{8}$ yard to make a pirate's hat. How much felt was left?

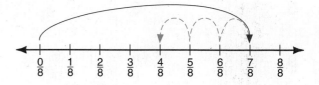

To find how much was left, subtract: $\frac{7}{8} - \frac{3}{8} = \underline{\ ?\ }$

The denominators are the same.

To subtract fractions with like denominators:

- Subtract the numerators. $\frac{7}{8} - \frac{3}{8} = \ ^4$ $7 - 3 = 4$

- Write the like denominator. $\frac{7}{8} - \frac{3}{8} = \frac{4}{8}$

- Write the difference in lowest terms. $\frac{4}{8} = \frac{4 \div 4}{8 \div 4} = \frac{1}{2}$

Factors of 4: 1, 2, 4   GCF: 4
Factors of 8: 1, 2, 4, 8

There was $\frac{1}{2}$ yard of felt left.

Study these examples.

$$\begin{array}{r} \frac{3}{4} \\ -\ \frac{3}{4} \\ \hline \frac{0}{4} = 0 \end{array}$$ ← lowest terms →

$$\begin{array}{r} \frac{7}{9} \\ -\ \frac{5}{9} \\ \hline \frac{2}{9} \end{array}$$

Find the difference in lowest terms.

1. $\frac{9}{10} - \frac{2}{10}$ **2.** $\frac{3}{5} - \frac{2}{5}$ **3.** $\frac{4}{7} - \frac{2}{7}$ **4.** $\frac{3}{4} - \frac{1}{4}$

Subtract. Write the difference in lowest terms.

5. $\frac{6}{9} - \frac{3}{9}$ **6.** $\frac{2}{3} - \frac{1}{3}$ **7.** $\frac{5}{6} - \frac{4}{6}$ **8.** $\frac{5}{8} - \frac{5}{8}$

9. $\frac{11}{12} - \frac{5}{12}$ **10.** $\frac{8}{10} - \frac{2}{10}$ **11.** $\frac{4}{5} - \frac{4}{5}$ **12.** $\frac{6}{8} - \frac{2}{8}$

13. $\frac{8}{9} - \frac{2}{9}$ **14.** $\frac{10}{12} - \frac{1}{12}$ **15.** $\frac{3}{6} - \frac{1}{6}$ **16.** $\frac{7}{10} - \frac{3}{10}$

17. $\begin{array}{r} \frac{3}{4} \\ -\frac{2}{4} \\ \hline \end{array}$ **18.** $\begin{array}{r} \frac{7}{9} \\ -\frac{1}{9} \\ \hline \end{array}$ **19.** $\begin{array}{r} \frac{11}{12} \\ -\frac{8}{12} \\ \hline \end{array}$ **20.** $\begin{array}{r} \frac{9}{10} \\ -\frac{7}{10} \\ \hline \end{array}$ **21.** $\begin{array}{r} \frac{1}{2} \\ -\frac{1}{2} \\ \hline \end{array}$

22. $\begin{array}{r} \frac{2}{7} \\ -\frac{2}{7} \\ \hline \end{array}$ **23.** $\begin{array}{r} \frac{4}{5} \\ -\frac{1}{5} \\ \hline \end{array}$ **24.** $\begin{array}{r} \frac{5}{6} \\ -\frac{1}{6} \\ \hline \end{array}$ **25.** $\begin{array}{r} \frac{10}{12} \\ -\frac{6}{12} \\ \hline \end{array}$ **26.** $\begin{array}{r} \frac{6}{10} \\ -\frac{2}{10} \\ \hline \end{array}$

PROBLEM SOLVING Write the answer in simplest form.

27. Nora bought $\frac{5}{6}$ yard of calico. Wayne bought $\frac{2}{6}$ yard of calico. How much more calico did Nora buy than Wayne?

28. Jo used $\frac{5}{8}$ yard of red linen to make a skirt. She used $\frac{1}{8}$ yard of blue linen for a scarf. Did she use more red or blue linen? How much more?

29. Ben has $\frac{1}{4}$ yard of denim. He needs $\frac{3}{4}$ yard for a school project. How much more denim does he need?

 **Skills to Remember**

Divide.

30. $6\overline{)8}$ **31.** $4\overline{)13}$ **32.** $8\overline{)43}$ **33.** $5\overline{)27}$ **34.** $2\overline{)11}$ **35.** $3\overline{)20}$

36. $48 \div 7$ **37.** $89 \div 9$ **38.** $65 \div 9$ **39.** $76 \div 8$

Improper Fractions

An **improper fraction** is a fraction greater than or equal to one. Its numerator is greater than or equal to its denominator.

$$\frac{8}{8} + \frac{8}{8} + \frac{4}{8} = \frac{20}{8}$$

$$\boxed{20 > 8}$$

Write as a mixed number in simplest form: $\frac{20}{8} = \underline{\ ?\ }$

▶ To write an improper fraction as a mixed number:

Divide the numerator by the denominator.	Write the quotient as the whole number.	Write the remainder over the divisor.

$$\frac{20}{8} = 8\overline{)20}^{\ 2\,R\ 4}$$

2

$$\frac{4}{8} = \frac{1}{2}$$

Write the fraction in lowest terms.

So $\frac{20}{8} = 2\frac{4}{8} = 2\frac{1}{2}$.   **simplest form**

▶ You can also break apart the fraction.

1   1 $\frac{1}{2}$

$$\frac{20}{8} = \frac{8}{8} + \frac{8}{8} + \frac{4}{8} = 2\frac{1}{2}$$

Study these examples.

$$\frac{12}{3} = 12 \div 3 = 4 \leftarrow \text{simplest form}$$

$$\frac{9}{7} = 7\overline{)9}^{\ 1\,R\ 2} = 1\frac{2}{7} \quad \text{simplest form}$$

Write as a whole number or mixed number in simplest form.

1.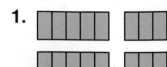

2.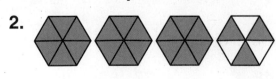

Write as a whole number or mixed number in simplest form.

3. $\frac{24}{5}$ 4. $\frac{17}{3}$ 5. $\frac{7}{2}$ 6. $\frac{9}{4}$ 7. $\frac{13}{6}$ 8. $\frac{27}{3}$

9. $\frac{30}{10}$ 10. $\frac{15}{9}$ 11. $\frac{32}{8}$ 12. $\frac{20}{6}$ 13. $\frac{10}{5}$ 14. $\frac{14}{4}$

15. $\frac{70}{12}$ 16. $\frac{58}{6}$ 17. $\frac{42}{8}$ 18. $\frac{33}{6}$ 19. $\frac{92}{8}$ 20. $\frac{26}{10}$

Add. Write each sum in simplest form.

21. $\frac{3}{5} + \frac{4}{5}$ 22. $\frac{2}{3} + \frac{2}{3}$ 23. $\frac{4}{6} + \frac{2}{6}$ 24. $\frac{2}{4} + \frac{3}{4}$

25. $\frac{4}{6} + \frac{5}{6}$ 26. $\frac{7}{8} + \frac{5}{8}$ 27. $\frac{4}{7} + \frac{5}{7}$ 28. $\frac{7}{9} + \frac{8}{9}$

29. $\frac{10}{12} + \frac{8}{12}$ 30. $\frac{3}{10} + \frac{7}{10}$ 31. $\frac{1}{2} + \frac{1}{2}$ 32. $\frac{7}{8} + \frac{3}{8}$

33. $\frac{14}{5} + \frac{16}{5}$ 34. $\frac{20}{8} + \frac{30}{8}$ 35. $\frac{25}{3} + \frac{20}{3}$ 36. $\frac{23}{9} + \frac{57}{9}$

PROBLEM SOLVING Write each answer in simplest form.

37. What is fifteen thirds in simplest form?

38. What is twenty-two eighths in simplest form?

39. Steve walked $\frac{10}{4}$ miles to the county fair. How many miles did he walk?

40. Kelvin sold $\frac{164}{8}$ gal of cider at the fair. How many gallons of cider did he sell?

41. Sue ate $\frac{5}{2}$ pies in the pie-eating contest. How many pies did she eat? Was this more or less than 3 pies?

42. Brianna has $\frac{7}{4}$ yards of material to make a skirt. Does she have more or less than 2 yards of material?

9-4 Estimating with Mixed Numbers

John brought $5\frac{1}{8}$ lb of apples, $2\frac{1}{4}$ lb of bananas, and $8\frac{1}{2}$ lb of melons to a picnic. About how many pounds of fruit did he bring to the picnic?

To find about how many pounds, estimate the sum: $5\frac{1}{8} + 2\frac{1}{4} + 8\frac{1}{2}$

▶ To estimate sums with mixed numbers, add the whole numbers.

$$
\begin{array}{r}
5\frac{1}{8} \\
2\frac{1}{4} \\
+8\frac{1}{2} \\
\hline
15
\end{array}
$$

John brought about 15 lb of fruit to the picnic.

About how many more pounds of melons than bananas did John bring?

To find about how many more, estimate the difference: $8\frac{1}{2} - 2\frac{1}{4}$

▶ To estimate differences with mixed numbers, subtract the whole numbers.

$$
\begin{array}{r}
8\frac{1}{2} \\
-2\frac{1}{4} \\
\hline
6
\end{array}
$$

John brought about 6 more pounds of melons than bananas.

Estimate the sum.

1. $6\frac{1}{5} + 9\frac{2}{10}$ 2. $8\frac{1}{4} + 8\frac{4}{12}$ 3. $3\frac{1}{2} + 7\frac{2}{6}$ 4. $1\frac{4}{9} + 4\frac{1}{6}$

5. $5\frac{2}{3} + 2\frac{4}{9}$ 6. $4\frac{3}{4} + 9\frac{3}{8}$ 7. $7\frac{2}{10} + 5\frac{1}{2}$ 8. $6\frac{3}{8} + 4\frac{1}{4}$

302

Estimate the sum.

9. $3\frac{1}{3}$
$4\frac{2}{6}$
$+\ 7\frac{1}{9}$

10. $5\frac{3}{4}$
$9\frac{1}{8}$
$+\ 6\frac{1}{4}$

11. $2\frac{1}{2}$
$8\frac{4}{10}$
$+\ 6\frac{9}{10}$

12. $9\frac{1}{4}$
$4\frac{1}{2}$
$+\ 4\frac{3}{12}$

13. $8\frac{2}{3}$
$4\frac{5}{6}$
$+\ 7\frac{1}{3}$

14. $10\frac{4}{10}$
$8\frac{1}{5}$
$+\ 10\frac{1}{10}$

15. $7\frac{3}{4}$
$12\frac{1}{4}$
$+\ 1\frac{1}{2}$

16. $14\frac{2}{3}$
$10\frac{1}{6}$
$+\ 12\frac{5}{9}$

17. $24\frac{2}{10}$
$16\frac{8}{10}$
$+\ 10\frac{3}{5}$

18. $15\frac{1}{8}$
$25\frac{3}{8}$
$+\ 6\frac{1}{4}$

Estimate the difference.

19. $18\frac{2}{3}$
$-\ 9\frac{6}{9}$

20. $9\frac{3}{4}$
$-\ 7\frac{4}{8}$

21. $13\frac{1}{5}$
$-\ 8\frac{5}{10}$

22. $7\frac{1}{2}$
$-\ 4\frac{3}{4}$

23. $11\frac{6}{9}$
$-\ 5\frac{1}{6}$

24. $15\frac{8}{12}$
$-\ 8\frac{2}{3}$

25. $6\frac{7}{8}$
$-\ 3\frac{3}{4}$

26. $14\frac{8}{10}$
$-\ 7\frac{4}{5}$

27. $13\frac{1}{2}$
$-\ 6\frac{6}{12}$

28. $12\frac{5}{6}$
$-\ 4\frac{1}{3}$

29. $22\frac{3}{9} - 12\frac{2}{3}$ 30. $48\frac{1}{2} - 30\frac{7}{10}$ 31. $19\frac{11}{12} - 11\frac{3}{4}$ 32. $25\frac{7}{12} - 15\frac{4}{6}$

PROBLEM SOLVING

33. David brought $5\frac{2}{10}$ lb of potato salad to the picnic. Sue brought $7\frac{1}{2}$ lb of potato salad. About how many pounds of potato salad were there?

34. Jerry traveled $15\frac{3}{4}$ mi to get to the picnic. Emmy traveled $6\frac{1}{2}$ mi less. About how far did Emmy have to travel?

35. Nan brought a watermelon that weighed $20\frac{1}{4}$ lb. The picnickers ate $15\frac{3}{4}$ lb of watermelon. About how many pounds of watermelon were left?

36. Sal needed 35 lb of turkey to feed the picnickers. He bought turkeys that weighed $10\frac{7}{10}$ lb, $16\frac{1}{2}$ lb, and $11\frac{3}{10}$ lb. Did Sal buy enough turkey?

9-5 Add and Subtract Mixed Numbers

Akers' Farms displays plants in trays of ten. Lucy buys $2\frac{6}{10}$ trays of plants and her brother buys $1\frac{2}{10}$ trays. How many trays of plants do they buy in all? How many more trays does Lucy buy than her brother?

▶ To find how many they buy in all, add: $2\frac{6}{10} + 1\frac{2}{10} = \underline{\ ?\ }$

Remember: Add or subtract only with like denominators.

Add the fractions.	Add the whole numbers.	Write in simplest form.
$\begin{array}{r} 2\frac{6}{10} \\ +1\frac{2}{10} \\ \hline \frac{8}{10} \end{array}$	$\begin{array}{r} 2\frac{6}{10} \\ +1\frac{2}{10} \\ \hline 3\frac{8}{10} \end{array}$	$\begin{array}{r} 2\frac{6}{10} \\ +1\frac{2}{10} \\ \hline 3\frac{8}{10} = 3\frac{4}{5} \end{array}$

They buy $3\frac{4}{5}$ trays in all.

▶ To find how many more trays Lucy buys, subtract: $2\frac{6}{10} - 1\frac{2}{10} = \underline{\ ?\ }$

Subtract the fractions.	Subtract the whole numbers.	Write in simplest form.
$\begin{array}{r} 2\frac{6}{10} \\ -1\frac{2}{10} \\ \hline \frac{4}{10} \end{array}$	$\begin{array}{r} 2\frac{6}{10} \\ -1\frac{2}{10} \\ \hline 1\frac{4}{10} \end{array}$	$\begin{array}{r} 2\frac{6}{10} \\ -1\frac{2}{10} \\ \hline 1\frac{4}{10} = 1\frac{2}{5} \end{array}$

Lucy buys $1\frac{2}{5}$ more trays than her brother.

Add. Write the sum in simplest form.

1. $6\frac{2}{6}$
   $+ 7\frac{1}{6}$

2. $8\frac{1}{4}$
   $+ 5\frac{1}{4}$

3. $4\frac{3}{5}$
   $+ 3\frac{1}{5}$

4. $2\frac{3}{8}$
   $+ 1\frac{1}{8}$

5. $9\frac{2}{7}$
   $+ 6\frac{4}{7}$

6. $5\frac{4}{12}$
   $+ 5\frac{6}{12}$

7. $7\frac{2}{9}$
   $+ 8\frac{4}{9}$

8. $26\frac{2}{4}$
   $+ 17\frac{1}{4}$

9. $36\frac{3}{10}$
   $+ 28\frac{5}{10}$

10. $47\frac{3}{8}$
    $+ 54\frac{3}{8}$

Subtract. Write the difference in simplest form.

11. $9\frac{10}{12}$
    $- 7\frac{7}{12}$

12. $5\frac{2}{3}$
    $- 1\frac{1}{3}$

13. $8\frac{3}{4}$
    $- 3\frac{1}{4}$

14. $6\frac{4}{5}$
    $- 4\frac{2}{5}$

15. $10\frac{7}{8}$
    $- 2\frac{5}{8}$

16. $57\frac{5}{6}$
    $- 48\frac{2}{6}$

17. $32\frac{6}{7}$
    $- 27\frac{1}{7}$

18. $40\frac{8}{9}$
    $- 18\frac{5}{9}$

19. $23\frac{9}{10}$
    $- 23\frac{3}{10}$

20. $12\frac{1}{2}$
    $- 7\frac{1}{2}$

Align and add or subtract. Watch the signs.

21. $18\frac{11}{12} - 9\frac{1}{12}$
22. $14\frac{7}{10} - 8\frac{3}{10}$
23. $21\frac{2}{6} + 5\frac{2}{6}$
24. $31\frac{6}{8} - 9\frac{2}{8}$

25. $1\frac{1}{8} + 19\frac{2}{8}$
26. $6\frac{2}{9} + 17\frac{1}{9}$
27. $30\frac{9}{10} - 2\frac{1}{10}$
28. $42\frac{1}{5} + 8\frac{3}{5}$

PROBLEM SOLVING

29. The fence around Lucy's garden was $7\frac{4}{12}$ ft high. She put chicken wire at the top so it is now $10\frac{7}{12}$ ft high. How many feet of wire did she add to the height of the fence?

 **Mental Math**

Add or subtract. Watch the signs.

30. $8\frac{7}{8} - 4$
31. $9 + 5\frac{2}{3}$
32. $7\frac{9}{10} + 6$
33. $10\frac{3}{4} - 9$

9-6 | Multiples

▶ The **multiples** of a number are all the products that have that number as a factor.

factors →
	0	1	2	3	4	5	6	7	8	9
	×2	×2	×2	×2	×2	×2	×2	×2	×2	×2
Multiples of 2:	0	2	4	6	8	10	12	14	16	18 ...

factors →
	0	1	2	3	4	5	6	7	8	9
	×3	×3	×3	×3	×3	×3	×3	×3	×3	×3
Multiples of 3:	0	3	6	9	12	15	18	21	24	27 ...

You can find the multiples of a number by multiplying or by skip counting.

▶ **Common multiples** are all the numbers other than 0 that are multiples of two or more numbers.

Multiples of 2: 0, 2, 4, 6, 8, 10, 12, 14, 16, 18, 20, 22, 24, ...

Multiples of 3: 0, 3, 6, 9, 12, 15, 18, 21, 24, ...

Common multiples of 2 and 3: 6, 12, 18, 24, ...

▶ The **least common multiple (LCM)** of two or more numbers is the least number that is a multiple of those numbers.

Least common multiple (LCM) of 2 and 3: 6

Is each a multiple of 2? Write *yes* or *no*.

1. 5 **2.** 40 **3.** 62 **4.** 0 **5.** 29 **6.** 88

Is each a multiple of 3? Write *yes* or *no*.

7. 33   **8.** 1   **9.** 29   **10.** 60   **11.** 48   **12.** 100

List the first eleven multiples of each.

13. 6   **14.** 4   **15.** 9   **16.** 10   **17.** 8   **18.** 5

**Write the first four common multiples for each set
of numbers. Then write the least common multiple (LCM).**

19. 2, 4   **20.** 3, 9   **21.** 4, 8   **22.** 6, 3   **23.** 5, 10

24. 2, 8   **25.** 6, 9   **26.** 8, 12   **27.** 7, 2   **28.** 2, 10

29. 4, 5   **30.** 8, 10   **31.** 4, 6   **32.** 9, 12   **33.** 3, 5

34. 2, 4, and 10   **35.** 3, 9, and 12   **36.** 2, 3, and 9

37. 6, 8, and 12   **38.** 4, 6, and 8   **39.** 5, 6, and 10

Critical Thinking

Write *true* or *false* for each statement.

40. All multiples of 3 are divisible by 3.

41. All multiples of 4 are multiples of 8.

42. No multiples of 9 are multiples of 3.

43. Some multiples of 6 are multiples of 12.

44. All multiples of 2 are even numbers.

45. No multiples of 5 are even numbers.

46. Some multiples of 3 are odd numbers.

47. All multiples of 7 are odd numbers.

Adding: Unlike Denominators

$\frac{7}{10}$ mile

Crystal Cave

$\frac{1}{2}$ mile

Cedar Lake

Camp

Mitchell hiked from camp to Crystal Cave and then to Cedar Lake. How far did Mitchell hike?

To find how far he hiked, add: $\frac{7}{10} + \frac{1}{2} = \underline{\ ?\ }$

The denominators are different.

To add fractions with unlike denominators:

Rename as fractions with like denominators.	Add the numerators. Write the like denominator.	Write the sum in lowest terms.

$$\begin{array}{r} \frac{7}{10} = \frac{7}{10} \\ + \frac{1}{2} = \frac{1 \times 5}{2 \times 5} = \frac{5}{10} \\ \hline \end{array}$$

$$\begin{array}{r} \frac{7}{10} \\ + \frac{5}{10} \\ \hline \frac{12}{10} \end{array}$$

$$\begin{array}{r} \frac{7}{10} \\ + \frac{5}{10} \\ \hline \frac{12}{10} = 1\frac{2}{10} = 1\frac{1}{5} \end{array}$$

Mitchell hiked $1\frac{1}{5}$ miles.

Study these examples.

$$\begin{array}{r} \frac{1}{4} = \frac{1 \times 2}{4 \times 2} = \frac{2}{8} \\ + \frac{3}{8} = \frac{3}{8} \\ \hline \frac{5}{8} \end{array}$$

$$\begin{array}{r} \frac{4}{6} = \frac{4}{6} \\ + \frac{1}{3} = \frac{1 \times 2}{3 \times 2} = \frac{2}{6} \\ \hline \frac{6}{6} = 1 \end{array}$$

$$\begin{array}{r} \frac{3}{4} = \frac{3 \times 3}{4 \times 3} = \frac{9}{12} \\ + \frac{1}{12} = \frac{1}{12} \\ \hline \frac{10}{12} = \frac{5}{6} \end{array}$$

Add. Write the sum in lowest terms.

1. $\begin{array}{r} \frac{1}{2} \\ + \frac{1}{4} \\ \hline \end{array}$

2. $\begin{array}{r} \frac{1}{3} \\ + \frac{1}{6} \\ \hline \end{array}$

3. $\begin{array}{r} \frac{3}{4} \\ + \frac{1}{8} \\ \hline \end{array}$

4. $\begin{array}{r} \frac{2}{3} \\ + \frac{1}{9} \\ \hline \end{array}$

5. $\begin{array}{r} \frac{3}{8} \\ + \frac{1}{2} \\ \hline \end{array}$

Find the sum in lowest terms.

6. $\dfrac{6}{8}$
$+\dfrac{1}{4}$

7. $\dfrac{1}{3}$
$+\dfrac{5}{12}$

8. $\dfrac{2}{3}$
$+\dfrac{4}{9}$

9. $\dfrac{1}{2}$
$+\dfrac{5}{8}$

10. $\dfrac{3}{5}$
$+\dfrac{3}{10}$

11. $\dfrac{7}{9}$
$+\dfrac{1}{3}$

12. $\dfrac{8}{10}$
$+\dfrac{1}{5}$

13. $\dfrac{7}{8}$
$+\dfrac{3}{4}$

14. $\dfrac{5}{12}$
$+\dfrac{1}{6}$

15. $\dfrac{2}{3}$
$+\dfrac{5}{6}$

16. $\dfrac{1}{4}$
$+\dfrac{5}{12}$

17. $\dfrac{6}{9}$
$+\dfrac{1}{3}$

18. $\dfrac{11}{12}$
$+\dfrac{3}{4}$

19. $\dfrac{1}{2}$
$+\dfrac{4}{8}$

20. $\dfrac{4}{5}$
$+\dfrac{6}{10}$

21. $\dfrac{2}{3} + \dfrac{1}{6}$

22. $\dfrac{1}{2} + \dfrac{5}{10}$

23. $\dfrac{1}{3} + \dfrac{5}{9}$

24. $\dfrac{3}{4} + \dfrac{2}{12}$

25. $\dfrac{2}{5} + \dfrac{9}{10}$

26. $\dfrac{5}{8} + \dfrac{1}{4}$

27. $\dfrac{3}{4} + \dfrac{7}{12}$

28. $\dfrac{2}{9} + \dfrac{2}{3}$

PROBLEM SOLVING Write each answer in simplest form.

29. Diego spent $\dfrac{1}{2}$ hour playing water polo and $\dfrac{3}{4}$ hour swimming. How much time did he spend on water sports?

30. Carlene hiked $\dfrac{4}{5}$ mi. Freddie hiked $\dfrac{9}{10}$ mi farther than Carlene. How far did Freddie hike?

Connections: Music

Musical notes are named by fractions.

whole note $\dfrac{1}{2}$ note $\dfrac{1}{4}$ note $\dfrac{1}{8}$ note $\dfrac{1}{16}$ note $\dfrac{1}{32}$ note

Does each set of notes equal ○? Write $<$, $=$, or $>$.

31.

32.

33.

34.

Subtracting: Unlike Denominators

Lila had $\frac{11}{12}$ ft of balsa wood.
She used $\frac{1}{4}$ ft of the wood to make a miniature chair for her dollhouse. How much wood did Lila have left?

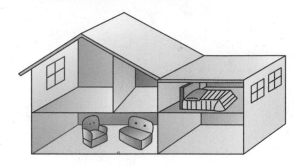

To find how much she had left,
subtract: $\frac{11}{12} - \frac{1}{4} = \frac{?}{\quad}$

The denominators are different.

To subtract fractions with unlike denominators:

Rename as fractions with like denominators.	Subtract the numerators. Write the like denominator.	Write the difference in lowest terms.
$\begin{aligned}\frac{11}{12} &= \frac{11}{12}\\ -\frac{1}{4} = \frac{1\times3}{4\times3} &= \frac{3}{12}\end{aligned}$	$\begin{aligned}&\frac{11}{12}\\ -&\frac{3}{12}\\ \hline &\frac{8}{12}\end{aligned}$	$\begin{aligned}&\frac{11}{12}\\ -&\frac{3}{12}\\ \hline &\frac{8}{12} = \frac{2}{3}\end{aligned}$

Lila had $\frac{2}{3}$ ft of wood left.

Study these examples.

$$\begin{aligned}\frac{3}{4} &= \frac{3}{4}\\ -\frac{1}{2} = \frac{1\times2}{2\times2} &= \frac{2}{4}\\ \hline &\frac{1}{4}\end{aligned}$$

$$\begin{aligned}\frac{2}{3} = \frac{2\times3}{3\times3} &= \frac{6}{9}\\ -\frac{6}{9} &= -\frac{6}{9}\\ \hline &\frac{0}{9} = 0\end{aligned}$$

Subtract. Write the difference in lowest terms.

1. $\frac{1}{4} - \frac{1}{8}$

2. $\frac{5}{8} - \frac{1}{4}$

3. $\frac{5}{6} - \frac{1}{3}$

4. $\frac{4}{5} - \frac{1}{10}$

Find the difference in lowest terms.

5. $\dfrac{2}{3}$
$-\dfrac{1}{6}$

6. $\dfrac{9}{10}$
$-\dfrac{1}{2}$

7. $\dfrac{7}{9}$
$-\dfrac{2}{3}$

8. $\dfrac{9}{12}$
$-\dfrac{3}{4}$

9. $\dfrac{7}{8}$
$-\dfrac{1}{2}$

10. $\dfrac{7}{10}$
$-\dfrac{1}{5}$

11. $\dfrac{7}{8}$
$-\dfrac{3}{4}$

12. $\dfrac{5}{6}$
$-\dfrac{2}{12}$

13. $\dfrac{2}{3}$
$-\dfrac{2}{9}$

14. $\dfrac{3}{4}$
$-\dfrac{6}{8}$

15. $\dfrac{10}{12}$
$-\dfrac{2}{6}$

16. $\dfrac{6}{8}$
$-\dfrac{1}{2}$

17. $\dfrac{1}{2}$
$-\dfrac{3}{12}$

18. $\dfrac{3}{5}$
$-\dfrac{1}{10}$

19. $\dfrac{8}{9}$
$-\dfrac{2}{3}$

20. $\dfrac{3}{4}$
$-\dfrac{8}{12}$

21. $\dfrac{9}{10}$
$-\dfrac{3}{5}$

22. $\dfrac{8}{9}$
$-\dfrac{1}{3}$

23. $\dfrac{7}{10}$
$-\dfrac{1}{2}$

24. $\dfrac{1}{3}$
$-\dfrac{2}{6}$

PROBLEM SOLVING Write each answer in simplest form.

25. Kyle worked on his model airplane for $\dfrac{3}{4}$ hour. Lief worked on his model ship for $\dfrac{1}{2}$ hour. Who worked on his model longer? how much longer?

26. Sharon decorated a valentine with pieces of ribbon. She used $\dfrac{2}{6}$ ft of red ribbon and $\dfrac{8}{12}$ ft of white ribbon. How much more white than red ribbon did Sharon use?

27. Clint had a large sheet of paper that was $\dfrac{9}{12}$ yd long. He trimmed $\dfrac{1}{3}$ yd from it. How long was the sheet of paper after trimming?

28. $\dfrac{7}{10}$ minus $\dfrac{1}{5}$ equals what number?

29. What is the sum of $\dfrac{1}{6}$ and $\dfrac{8}{12}$?

30. What is the difference between $\dfrac{8}{9}$ and $\dfrac{2}{3}$?

31. $\dfrac{3}{4}$ plus $\dfrac{3}{8}$ equals what number?

Computing Probability

There are 10 marbles in the jar:
1 is purple, 2 are white, 3 are red,
and 4 are yellow. What is the
probability that, without looking,
you would pick a marble
of each color?

The probability that you would pick:

- purple is 1 out of 10, or $\frac{1}{10}$.

- white is 2 out of 10, or $\frac{2}{10}$.

- red is 3 out of 10, or $\frac{3}{10}$.

- yellow is 4 out of 10, or $\frac{4}{10}$.

The probability of picking
any given color is the
probability of that **event**.

Probability of picking a red marble: $\frac{3}{10}$ Write: $P(\text{red}) = \frac{3}{10}$

What is the probability that you would pick
a red *or* a purple marble?

To find the probability of picking red *or* purple,
add the two probabilities by adding the fractions.

$$\frac{3}{10} + \frac{1}{10} = \frac{4}{10}$$

Probability of picking red *or* purple: $\frac{4}{10}$ Write: $P(\text{red or purple}) = \frac{4}{10}$

Find the probability of each event. Use the spinner.

1. $P(\text{green})$ **2.** $P(\text{yellow})$

3. $P(\text{red})$ **4.** $P(\text{blue})$

5. $P(\text{green or yellow})$ **6.** $P(\text{red or blue})$

7. $P(\text{green or yellow or red})$

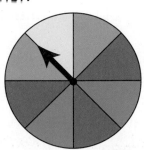

Find the probability of each event. Use the cards.

8. $P(B)$

9. $P(E)$

10. $P(C)$

11. $P(D)$

12. $P(A \text{ or } C)$

13. $P(B \text{ or } D)$

14. $P(A \text{ or } B \text{ or } E)$

15. $P(B \text{ or } C \text{ or } D)$

16. $P(\text{not } D)$

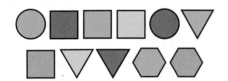

Find the probability of each event. Use the shapes.

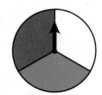

17. $P(\text{circle})$

18. $P(\text{triangle})$

19. $P(\text{square or triangle})$

Certainty and Impossibility

What is the probability of spinning red or white or blue? What is the probability of spinning green?

It is **certain** that the spinner will land on red or white or blue.

$P(\text{red or white or blue}) = \frac{3}{3} = 1 \leftarrow P(\text{certainty}) = 1$

It is **impossible** that the spinner will land on green.

$P(\text{green}) = \frac{0}{3} = 0 \leftarrow P(\text{impossible}) = 0$

Find the probability of each event. Use the marbles on page 312.

20. $P(\text{red or yellow or white})$

21. $P(\text{orange or green})$

22. $P(\text{red or purple or white or yellow})$

313

Finding Parts of Numbers

Of 12 kittens at the animal shelter,
$\frac{1}{3}$ were white and $\frac{2}{3}$ were gray. How many
of the kittens were white? How many were gray?

12 divided into
3 equal parts;
4 in each part

$12 \div 3 = 4$

$\frac{1}{3}$ of 12 = 4

12 divided into
3 equal parts;
4 in each part;
8 in 2 parts

$12 \div 3 = 4$
$2 \times 4 = 8$

$\frac{2}{3}$ of 12 = 8

To find a fractional part of a number:

- Divide the whole number by the denominator.

- Multiply the quotient by the numerator.

$\frac{1}{3}$ of 12: $12 \div 3 = 4 \longrightarrow 1 \times 4 = 4$ So $\frac{1}{3}$ of 12 = 4.

$\frac{2}{3}$ of 12: $12 \div 3 = 4 \longrightarrow 2 \times 4 = 8$ So $\frac{2}{3}$ of 12 = 8.

Of the kittens, 4 were white and 8 were gray.

Find the part of each number. You may draw a picture.

1. $\frac{1}{5}$ of 15   2. $\frac{1}{3}$ of 9   3. $\frac{1}{2}$ of 14   4. $\frac{1}{8}$ of 40

5. $\frac{1}{4}$ of 24   6. $\frac{1}{9}$ of 36   7. $\frac{1}{7}$ of 42   8. $\frac{1}{5}$ of 50

9. $\frac{1}{6}$ of 30   10. $\frac{1}{6}$ of 24   11. $\frac{1}{8}$ of 16   12. $\frac{1}{2}$ of 8

Find the missing number.

13. $\frac{2}{3}$ of 15 = _?_

14. $\frac{5}{8}$ of 16 = _?_

15. $\frac{5}{6}$ of 18 = _?_

16. $\frac{3}{7}$ of 21 = _?_

17. $\frac{3}{8}$ of 40 = _?_

18. $\frac{2}{5}$ of 25 = _?_

19. $\frac{3}{4}$ of 32 = _?_

20. $\frac{2}{9}$ of 27 = _?_

21. $\frac{5}{7}$ of 14 = _?_

22. $\frac{4}{5}$ of 45 = _?_

23. $\frac{3}{8}$ of 64 = _?_

24. $\frac{8}{9}$ of 9 = _?_

25. $\frac{3}{4}$ of 20 = _?_

26. $\frac{6}{7}$ of 28 = _?_

27. $\frac{3}{5}$ of 35 = _?_

28. $\frac{4}{6}$ of 54 = _?_

29. $\frac{2}{3}$ of 33 = _?_

30. $\frac{3}{4}$ of 40 = _?_

31. $\frac{7}{8}$ of 72 = _?_

32. $\frac{2}{5}$ of 50 = _?_

33. $\frac{4}{9}$ of 99 = _?_

PROBLEM SOLVING

34. Jamail raised 28 rabbits. Of these, $\frac{3}{4}$ were black and white. How many of the rabbits were black and white?

35. Of 30 retrievers at the kennel, $\frac{4}{5}$ were golden retrievers. How many of the retrievers were golden retrievers?

36. Farmer Green has 64 chickens. Of these, $\frac{3}{8}$ are Rhode Island Reds. How many of his chickens are Rhode Island Reds?

37. Of 150 birds that came to the feeder, $\frac{2}{3}$ were finches. How many of the birds were finches?

38. There are 32 black cows and 16 brown cows in the pasture. Three eighths of all the cows are lying down. Of these, $\frac{1}{6}$ are brown. How many brown cows are lying down?

9-11 | Problem Solving: Use Simpler Numbers

Problem: A piece of ribbon is $12\frac{7}{8}$ ft long. Tatsu cuts off two pieces that are $4\frac{3}{8}$ ft each. Does she have enough ribbon left to cut one more piece the same length?

1 IMAGINE Draw and label a picture.

2 NAME *Facts:* $12\frac{7}{8}$ ft of ribbon
Two $4\frac{3}{8}$ ft pieces cut from it

Question: Is there $4\frac{3}{8}$ ft left?

3 THINK Use simpler numbers like 12 and 4.
Use 12 for the $12\frac{7}{8}$ ft length.

Use 4 for the $4\frac{3}{8}$ ft length.

First add to find the amount of ribbon cut.
     4 ft + 4 ft = 8 ft
Then subtract to find the amount of ribbon left.
     12 ft − 8 ft = 4 ft

4 COMPUTE Now use the numbers in the problem.

$$4\frac{3}{8}\text{ ft} \qquad\qquad 12\frac{7}{8}\text{ ft}$$
$$\underline{+4\frac{3}{8}\text{ ft}} \qquad\qquad \underline{-\ 8\frac{6}{8}\text{ ft}}$$
$$8\frac{6}{8}\text{ ft cut} \qquad\qquad 4\frac{1}{8}\text{ ft left}$$

Compare: $4\frac{1}{8} < 4\frac{3}{8}$. So Tatsu does *not* have enough ribbon to cut another piece of the same length.

5 CHECK Add the lengths of the pieces. Do they equal $12\frac{7}{8}$ ft?
$$4\frac{3}{8} + 4\frac{3}{8} + 4\frac{1}{8} = 12\frac{7}{8}$$
The answer checks.

Use simpler numbers to solve each problem.

1. Frank checked his kitten's weight on the first day of each month. He kept the information on a chart. How much weight did the kitten gain between April 1 and June 1?

My Kitten's Weight

April 1	May 1	June 1
$6\frac{4}{8}$ lb	$7\frac{1}{8}$ lb	$7\frac{5}{8}$ lb

IMAGINE　　Create a mental picture.

NAME

Facts:　　April 1: $6\frac{4}{8}$ lb

　　　　　　June 1: $7\frac{5}{8}$ lb

Question:　How much weight did the kitten gain between April 1 and June 1?

THINK　　Use 6 for $6\frac{4}{8}$ and 7 for $7\frac{5}{8}$.

Subtract to find the difference.

$7 - 6 = \underline{\ ?\ }$

Then use the numbers in the problem.

COMPUTE ⟶ **CHECK**

2. One paper-clip chain is $24\frac{1}{4}$ in. long. Another is $41\frac{1}{4}$ in. long. How long will the chain be if the two chains are connected?

3. Ms. Hanley is running a $26\frac{5}{10}$ mile race. She stops for water after $7\frac{3}{10}$ miles. How much farther does she have to run?

4. A bread recipe calls for $4\frac{3}{8}$ c of white flour, $2\frac{1}{8}$ c of wheat flour, and 1 c of rye flour. How much flour does this recipe use?

 **Make Up Your Own**

5. Write a problem with fractions or mixed numbers. Use simpler numbers. Then solve it using the original numbers.

Problem-Solving Applications

**Solve each problem and explain
the method you used.**

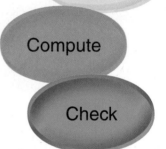

Imagine

Name

Think

Compute

Check

1. An oatmeal bar weighs $4\frac{1}{4}$ oz. How much do two oatmeal bars weigh?

2. Bags of granola weigh $6\frac{1}{8}$ oz and $12\frac{3}{8}$ oz. How much heavier is the larger bag?

3. An apple weighs $3\frac{3}{4}$ oz. A pear weighs $4\frac{1}{4}$ oz. About how much do they weigh together?

4. Todd uses 16 tablespoons of peanut butter to make sandwiches. If he spreads each sandwich with $\frac{1}{8}$ of the peanut butter, how much peanut butter will be in each sandwich?

5. Of 10 loaves of bread, $\frac{1}{5}$ have sesame seeds. How many loaves have sesame seeds?

6. Carrot bread has 100 calories per slice. Seven tenths of the calories come from carbohydrates. How many calories come from carbohydrates?

7. Nan bought $\frac{1}{8}$ lb of pecans and $\frac{1}{4}$ lb of walnuts. Did she buy more than $\frac{1}{2}$ lb of nuts?

8. A carrot is 9 in. long. Regina cuts it in thirds. How long is each piece?

Choose a strategy from the list or use another strategy you know to solve each problem.

9. Of 24 muffins for sale, $\frac{1}{2}$ are bran and $\frac{1}{4}$ are corn. How many corn muffins are for sale?

10. Jan bought half a loaf of rye bread. She gave half of her piece to Ramon. Ramon's piece weighs $\frac{1}{4}$ lb. How much did the original loaf of bread weigh?

11. A carrot has fewer calories than an apple. An oat bar has more calories than an apple. Does a carrot or an oat bar have more calories?

USE THESE STRATEGIES
Choose the Operation
Guess and Test
Use Simpler Numbers
Extra Information
Working Backwards
Logical Reasoning
Use a Graph

12. Sue's trail mix is $\frac{1}{2}$ toasted oats, $\frac{1}{4}$ raisins, and $\frac{1}{4}$ carob drops. She makes 16 oz of mix. How many ounces of raisins does she use?

13. Wes ate a snack of 150 calories. The low-fat yogurt he ate had half the calories of the oatmeal cookie. How many calories did the yogurt have?

Use the pictograph for problems 14–16.

14. What fractional part of the mini-muffins were blueberry muffins?

15. How many more apple muffins than blueberry muffins were there?

16. Liu bought $\frac{1}{2}$ of the cinnamon muffins. How many muffins did she buy?

Mini-Muffin Menu	
apple	⬡ ⬡ ⬡
blueberry	⬡ ⬡
cinnamon	⬡ ⬡ ⬡ ⬡
corn	⬡
Key: Each ⬡ = 5 mini-muffins.	

Add or subtract.
Write the answer in simplest form.   *(See pp. 296–301, 304–305, 308–311.)*

1. $\dfrac{1}{5}$
    $+\dfrac{2}{5}$

2. $\dfrac{5}{8}$
    $-\dfrac{3}{8}$

3. $\dfrac{5}{6}$
    $+\dfrac{1}{2}$

4. $\dfrac{8}{9}$
    $-\dfrac{2}{3}$

5. $\dfrac{7}{10}$
    $-\dfrac{2}{20}$

6. $4\dfrac{1}{10}$
    $+3\dfrac{1}{10}$

7. $8\dfrac{2}{6}$
    $-4\dfrac{1}{6}$

8. $5\dfrac{3}{16}$
    $-3\dfrac{3}{16}$

9. $12\dfrac{1}{8}$
    $+\ 2\dfrac{1}{8}$

10. $5\dfrac{3}{8}$
     $+4\dfrac{1}{8}$

11. $\dfrac{1}{2} + \dfrac{1}{2}$

12. $\dfrac{6}{10} - \dfrac{1}{5}$

13. $\dfrac{1}{4} + \dfrac{7}{8}$

14. $\dfrac{5}{6} - \dfrac{1}{3}$

Write as a whole number or mixed number
in simplest form.   *(See pp. 300–301.)*

15. $\dfrac{16}{8}$

16. $\dfrac{13}{4}$

17. $\dfrac{15}{6}$

18. $\dfrac{20}{5}$

19. $\dfrac{17}{3}$

Write the least common multiple of each set.   *(See pp. 306–307.)*

20. 4, 10

21. 9, 12

22. 9, 6

23. 5, 6

Estimate the sum or difference.   *(See pp. 302–303.)*

24. $11\dfrac{5}{9} - 4\dfrac{2}{3}$

25. $2\dfrac{1}{3} + 3\dfrac{1}{6} + 1\dfrac{1}{9}$

Find the part of each number.   *(See pp. 314–315.)*

26. $\dfrac{1}{9}$ of 18

27. $\dfrac{1}{8}$ of 24

28. $\dfrac{3}{4}$ of 40

29. $\dfrac{5}{8}$ of 24

PROBLEM SOLVING   *(See pp. 312–315.)*

30. Of 24 apples, $\dfrac{1}{3}$ are green.
     How many are green?

31. There are 7 marbles in a bag.
     Four are red, 2 are blue and 1 is
     green. What is the probability that
     the first one picked will be red?

(See Still More Practice, p. 469.)

LEAST COMMON DENOMINATOR

Algebra

Rafael's cookie recipe called for $\frac{1}{3}$ cup of brown sugar and $\frac{3}{4}$ cup of white sugar. How much sugar did Rafael use?

Add: $\frac{1}{3} + \frac{3}{4} = \frac{?}{\_\_}$

To add $\frac{1}{3} + \frac{3}{4}$, rename *both fractions* as fractions with the least common denominator.

The **least common denominator (LCD)** is the least common multiple of the denominators.

Multiples of 3: 0, 3, 6, 9, 12, 15, 18, 21, 24, . . .
Multiples of 4: 0, 4, 8, 12, 16, 20, 24, . . .

So the LCD of $\frac{1}{3}$ and $\frac{3}{4}$ is 12.

Rename the fractions.

$$\frac{1}{3} = \frac{1 \times 4}{3 \times 4} = \frac{4}{12}$$
$$+\ \frac{3}{4} = \frac{3 \times 3}{4 \times 3} = \frac{9}{12}$$

Add.

$$\frac{4}{12}$$
$$+\frac{9}{12}$$
$$\frac{13}{12} = 1\frac{1}{12} \quad \leftarrow \boxed{\text{simplest form}}$$

Rafael used $1\frac{1}{12}$ cups of sugar.

Write the LCD for each set of fractions.

1. $\frac{1}{2}, \frac{2}{5}$ **2.** $\frac{3}{4}, \frac{1}{6}$ **3.** $\frac{2}{3}, \frac{3}{8}$ **4.** $\frac{1}{5}, \frac{1}{6}$

5. $\frac{3}{10}, \frac{1}{4}$ **6.** $\frac{4}{5}, \frac{3}{4}$ **7.** $\frac{1}{3}, \frac{1}{5}, \frac{1}{6}$ **8.** $\frac{1}{3}, \frac{1}{4}, \frac{1}{5}$

Add or subtract. Use the LCD. Write the answer in simplest form.

9. $\frac{1}{2} + \frac{2}{7}$ **10.** $\frac{1}{4} + \frac{3}{5}$ **11.** $\frac{5}{6} - \frac{1}{9}$ **12.** $\frac{2}{3} - \frac{1}{2}$

13. $\frac{7}{8} - \frac{3}{10}$ **14.** $\frac{9}{10} + \frac{1}{6}$ **15.** $\frac{7}{9} - \frac{3}{8}$ **16.** $\frac{5}{8} + \frac{2}{3}$

Performance Assessment

Use these rule cards. Match the rule card to each pattern, then tell the next number.

Add $\frac{1}{6}$ Add $2\frac{1}{4}$ Subtract $\frac{1}{5}$

1. $\frac{9}{10}, \frac{7}{10}, \frac{5}{10}, \frac{?}{?}$ **2.** $\frac{5}{12}, \frac{7}{12}, \frac{9}{12}, \frac{?}{?}$ **3.** $1\frac{3}{4}, 4, 6\frac{1}{4}, \underline{?}$

Add or subtract. Write the answer in lowest terms.

4. $\frac{3}{10} + \frac{1}{5}$ **5.** $\frac{1}{6} + \frac{2}{6}$ **6.** $\frac{7}{8} - \frac{5}{8}$ **7.** $\frac{2}{3} - \frac{1}{6}$

8. $\frac{4}{5} - \frac{3}{10}$ **9.** $\frac{7}{8} + \frac{1}{4}$ **10.** $\frac{8}{10} - \frac{1}{2}$ **11.** $\frac{2}{3} + \frac{11}{12}$

Write as a whole number or mixed number in simplest form.

12. $\frac{7}{2}$ **13.** $\frac{16}{5}$ **14.** $\frac{21}{8}$ **15.** $\frac{30}{6}$ **16.** $\frac{38}{7}$

Write the least common multiple.

17. 2, 6   **18.** 4, 3   **19.** 4, 12   **20.** 5, 7

Find the part of each number.

21. $\frac{1}{2}$ of 26   **22.** $\frac{2}{3}$ of 21   **23.** $\frac{3}{5}$ of 25   **24.** $\frac{5}{8}$ of 64

PROBLEM SOLVING   *Use a strategy you have learned.*

25. Of 32 apples $\frac{1}{4}$ are red. How many are red?

26. One necklace is $30\frac{1}{4}$ in. long. Another is $36\frac{1}{2}$ in. long. If the two necklaces are connected, how long will the necklace be?

Cumulative Review III

Choose the best answer.

1. What is the value of the underlined digit in 6<u>8</u>,325,784?

   a. 800,000   **b.** 1,000,000
   c. 8,000,000   **d.** 80,000,000

2. How many minutes have passed from 6:45 P.M. to 7:12 P.M.?

   a. 15 min.   **b.** 27 min.
   c. 42 min.   **d.** 47 min.

3. What is the cost of 32 cassettes at $9.79 each?

   a. $19.58   **b.** $48.95
   c. $312.28   **d.** $313.28

4. Find the average of 498, 636, and 714.

   a. 507   **b.** 612
   c. 616   **d.** 1848

5. How much more than 4000 − 1967 is 1267 + 1967?

   a. 101   **b.** 201
   c. 1091   **d.** none of these

6. How many liters of water are there in 8 containers of 750 mL each?

   a. 6   **b.** 60
   c. 600   **d.** 6000

7. 465 is divisible by which number?

   a. 2   **b.** 5
   c. 10   **d.** none of these

8. Add: 4 ft 6 in. + 3 ft 8 in.

   a. 7 ft 2 in.   **b.** 7 ft 4 in.
   c. 8 ft   **d.** 8 ft 2 in.

9. What fractional part is shaded?

   a. $\frac{1}{3}$
   b. $\frac{7}{15}$
   c. $\frac{8}{15}$
   d. $\frac{3}{4}$

10. What mixed number is shown?

    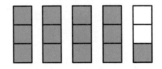

    a. $4\frac{1}{3}$
    b. $4\frac{1}{2}$
    c. $5\frac{1}{3}$
    d. $5\frac{1}{2}$

11. $\frac{7}{8}$
    $-\frac{3}{8}$

    a. $\frac{1}{3}$ **b.** $\frac{1}{2}$
    c. $\frac{3}{4}$ **d.** not given

12. $\frac{1}{3}$
    $+\frac{4}{9}$

    a. $\frac{5}{12}$ **b.** $\frac{5}{9}$
    c. $\frac{7}{9}$ **d.** not given

13. $3\frac{1}{7} + 2\frac{4}{7}$

    a. $5\frac{5}{14}$ **b.** $5\frac{4}{7}$
    c. $6\frac{5}{7}$ **d.** not given

14. $\frac{3}{4}$ of 36

    a. 9   **b.** 27
    c. 48   **d.** not given

For Your Portfolio

Solve each problem. Explain the steps and the strategy or strategies you used for each. Then choose one from problems 1–4 for your Portfolio.

1. Desmond has a canvas that is 4 ft 8 in. wide. He cuts off $2\frac{1}{2}$ ft. How much is left?

2. Sally grew $3\frac{7}{8}$ inches. Mae grew $1\frac{1}{4}$ inches. How much more did Sally grow than Mae?

3. Use the spinner below. Describe the probability of the spinner landing on each color.

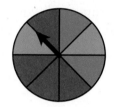

4. Mrs. Ming made this circle graph to show the grades received on tests in her Math class. How many more Bs than Cs were given?

Grades Given

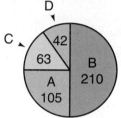

Tell about it.

5. Explain how you could change the spinner in problem 3 so that the probability of landing on green is $\frac{1}{2}$. Draw a new spinner to show this.

Communicate ✓

6. Explain how you can find the total number of all four grades given in problem 4. What is the total? What fractional part of the grades are As? Bs?

For Rubric Scoring

Listen for information on how your work will be scored.

7. Which statement below is true? Explain why.
   - Spinner A is more likely to land on 6 than Spinner B.
   - The probability of landing on 6 is the same on both spinners.

A
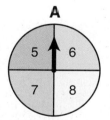

B
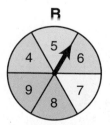

8. Draw spinners with two equal sections and four equal sections where the probability of landing on 5 is the same.

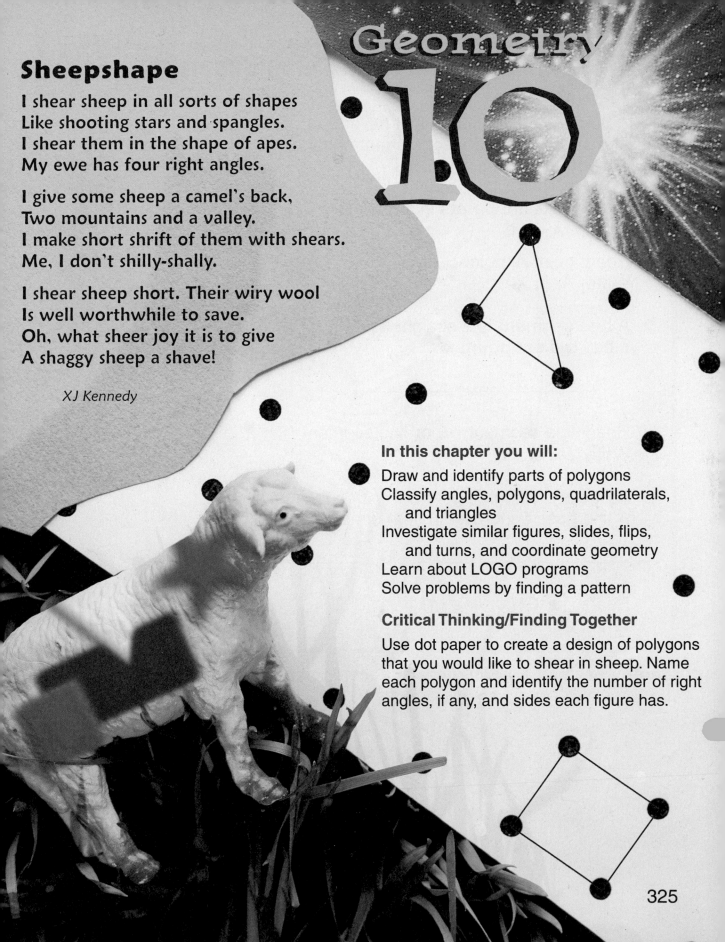

Sheepshape

I shear sheep in all sorts of shapes
Like shooting stars and spangles.
I shear them in the shape of apes.
My ewe has four right angles.

I give some sheep a camel's back,
Two mountains and a valley.
I make short shrift of them with shears.
Me, I don't shilly-shally.

I shear sheep short. Their wiry wool
Is well worthwhile to save.
Oh, what sheer joy it is to give
A shaggy sheep a shave!

XJ Kennedy

In this chapter you will:

Draw and identify parts of polygons
Classify angles, polygons, quadrilaterals,
    and triangles
Investigate similar figures, slides, flips,
    and turns, and coordinate geometry
Learn about LOGO programs
Solve problems by finding a pattern

Critical Thinking/Finding Together

Use dot paper to create a design of polygons
that you would like to shear in sheep. Name
each polygon and identify the number of right
angles, if any, and sides each figure has.

325

▶ A **plane** is a flat surface that never ends. The surface of a table or a sheet of paper are both parts of planes.

▶ *A*, *B*, and *X* are **points** in a plane.

A•   B   •X

Read: point *A*, point *B*, point *X*
Write: *A*, *B*, *X*

▶ A **line segment** is a straight figure. It has two **endpoints**.

D   E

Read: line segment *DE* **or** line segment *ED*
Write: $\overline{DE}$ **or** $\overline{ED}$

▶ A **line** is a straight figure with *no* endpoints. A line goes on forever in both directions.

G   H

> Line segment *GH* is part of line *GH*.

Read: line $\overleftrightarrow{GH}$ **or** line $\overleftrightarrow{HG}$
Write: $\overleftrightarrow{GH}$ **or** $\overleftrightarrow{HG}$

Identify each as a *point*, *line*, or *line segment*. Use symbols.

1. •H

2.

3.

4.

5.

6. L•

7.

8.

Draw and label each.

9. $\overline{TV}$ 10. K 11. $\overleftrightarrow{ST}$ 12. $\overline{FG}$

13. D 14. $\overleftrightarrow{PQ}$ 15. $\overline{LM}$ 16. Z

Write the letter(s) of each line segment.

17a.  b.   c.   d.

18a.  b.   c.   d.

Name each line two ways.

19.  20. L K 21. R S

22. W X 23. T S 24. Q R

Challenge

You can name a line by naming
any two points on the line in any order.

25. Write 6 names for this line: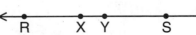
M O N

26. Write 12 names for this line:
R X Y S

327

10-2 | Rays and Angles

▶ A **ray** is a straight figure with one endpoint.
A ray goes on forever in one direction.

J   K

Read: ray *KJ*
Write: $\overrightarrow{KJ}$

Read the
endpoint first.

▶ An **angle** is formed by two rays
with the same endpoint.

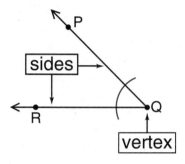

The rays form the **sides**
of the angle.

The endpoint is the **vertex**
of the angle.

Read: angle *Q* **or** angle *PQR* **or** angle *RQP*
Write: ∠*Q* **or** ∠*PQR* **or** ∠*RQP*

When you use three letters to name an angle,
the vertex is always the middle letter.

▶ When the two sides of an angle form a square corner,
the angle is called a **right angle**.

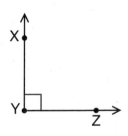

Angle *Y* is a right angle.

Draw and label each figure.

1. ∠*DEF*   **2.** $\overrightarrow{ED}$ **3.** ∠*FED*   **4.** ∠*H*   **5.** ray *EF*

Name each figure.

6.

7.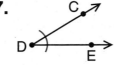

8.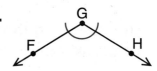

Name each angle three ways.

9.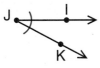

10.

11.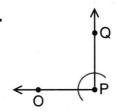

Is the angle a right angle? Write *yes* or *no*.

12.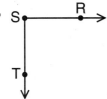

13.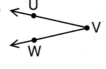

14.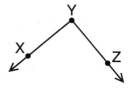

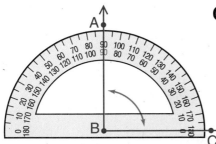

Comparing Angles

A **protractor** measures angles in degrees (°).
A right angle measures 90°.
Measure an angle by measuring the distance *between* its sides.

Acute angle
less than 90°

Obtuse angle
greater than 90°

Tell whether each angle is acute or obtuse.

Communicate

15.

16.

17.

18.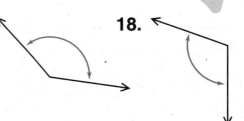

Parallel and Perpendicular Lines

▶ **Intersecting lines** are lines that meet or cross at a common point.

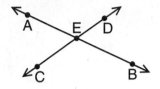

$\overleftrightarrow{AB}$ and $\overleftrightarrow{CD}$ intersect at point E.

▶ **Perpendicular lines** are intersecting lines that form right angles.

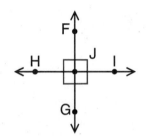

Read: line FG is perpendicular to line HI

Write: $\overleftrightarrow{FG} \perp \overleftrightarrow{HI}$

▶ **Parallel lines** are lines in the same plane that never meet.

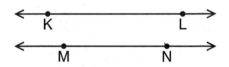

Read: line KL is parallel to line MN

Write: $\overleftrightarrow{KL} \parallel \overleftrightarrow{MN}$

Line segments can also be intersecting, perpendicular, or parallel.

Write _intersecting_ or _parallel_ to describe each pair of lines.

1.

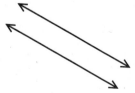

2.

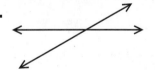

3.

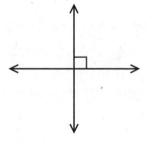

PROBLEM SOLVING
Use the figure at the right.

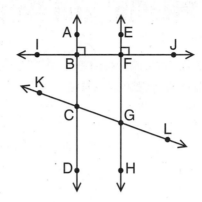

4. At what point does $\overleftrightarrow{EH}$ intersect $\overleftrightarrow{KL}$?

5. Name the pair of parallel lines.

6. What kind of angle is ∠IBA?

7. Name two pairs of perpendicular lines.

8. Is ∠FGL acute or obtuse?

Copy these lines on dot paper.

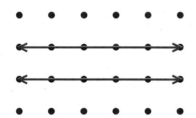

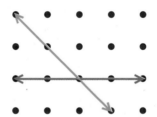

Draw a line segment that:

9. is perpendicular to both red lines.

10. is parallel to the green line and intersects the blue line.

11. intersects one red line but not the other.

12. is perpendicular to the green line.

Critical Thinking

13. Are $\overleftrightarrow{RS}$ and $\overleftrightarrow{XY}$ parallel, intersecting, or neither? Explain your answer in your Math Journal.

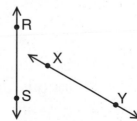

Math Journal

Circles

▶ A **circle** is a plane figure. All the points on the circle are the same distance from another point, called the **center**.

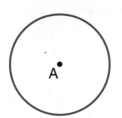

Point *A* is the center of circle *A*.

A circle is named by its center point.

▶ The parts of a circle have special names.

Any line segment with endpoints at the *center* of the circle and *on* the circle is a **radius**.

$\overline{BE}$ is a radius of circle *B*. $\overline{BC}$ and $\overline{BD}$ are also radii (plural of radius) of circle *B*.

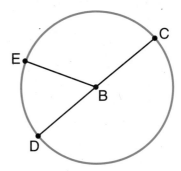

Any line segment that passes *through* the center of the circle and has *both* endpoints on the circle is a **diameter**.

The length of the diameter is always twice the length of the radius.

$\overline{CD}$ is a diameter.

PROBLEM SOLVING Use circle *F*.

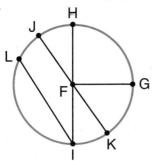

1. Name six points on the circle.

2. Name five line segments that are radii.

Communicate ✓

3. Suppose $\overline{FG}$ is 5 in. long. How long is $\overline{JF}$? How long is $\overline{HI}$? Explain how you got your answers.

332

PROBLEM SOLVING Use circle *V.*

4. Name the circle and its center.

5. How many diameters are shown? Name the diameters.

6. Is $\overline{TR}$ a radius? Explain why or why not.

 Communicate

7. Is $\overline{VX}$ a radius? Explain why or why not.

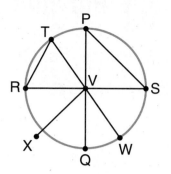

8. How many radii are shown? Name the radii.

Curves

A **simple closed curve** is a path that begins and ends at the same point and does not cross itself.

Simple Closed Curves

Not Simple   Not Closed

Write *true* or *false*. Use the picture below.

9. Some of the simple closed curves are green.

10. None of the simple closed curves are blue.

11. All circles are simple closed curves.

12. None of the simple closed curves are red.

▶ These figures are all polygons:

The **sides** of a polygon are line segments that do not cross.

> Remember:
> Polygons are *closed.*

Two sides of a polygon form an angle when they meet at a common endpoint called a **vertex** (plural: **vertices**).

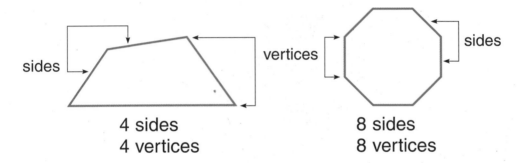

4 sides
4 vertices

8 sides
8 vertices

▶ Most polygons are named for the number of angles they have.

> *Polygon* means "many angles."

Prefix	Number of Angles	Polygon Name
tri-	3	triangle
quadri-	4	quadrilateral
penta-	5	pentagon
hexa-	6	hexagon
octa-	8	octagon

PROBLEM SOLVING
Use dot paper for problems 1–7.

1. Draw a polygon that has 8 sides and 8 vertices. What is its name?

2. Draw a polygon that has 5 sides and 5 vertices. What is its name?

3. Draw a polygon that has 3 sides. How many vertices does it have? What is its name?

4. Draw a polygon that has 6 vertices. How many sides does it have? What is its name?

5. Draw five different quadrilaterals. How many of them have at least one right angle?

6. Draw four different hexagons.

7. Draw as many octagons as you can that have all right angles. How many octagons did you draw?

8. Do you think the number of sides a polygon has is always equal to the number of its vertices? Try to draw a polygon that disproves your answer.

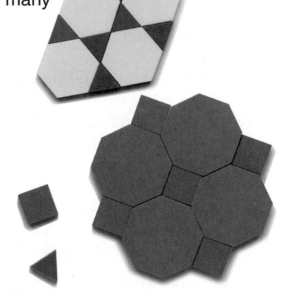

 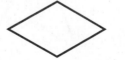 **Share Your Thinking**

Math Journal ✓

9. Copy the figures below into your Math Journal. Tell whether each is or is not a polygon.

   a. ⬭   b. ◇   c. ⬜   d. ▱

Quadrilaterals

Some quadrilaterals have special names.

▶ A **parallelogram** has opposite sides that are parallel *and* that are the same length.

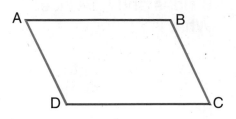

Quadrilateral *ABCD* is a parallelogram.

▶ A **rectangle** also has opposite sides that are parallel *and* that are the same length. All the angles of a rectangle are right angles.

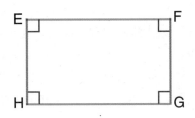

Quadrilateral *EFGH* is a rectangle.

▶ A **square** has opposite sides that are parallel. *All* its sides are the same length. All the angles of a square are right angles.

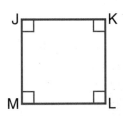

Quadrilateral *JKLM* is a square.

Is each figure a quadrilateral? Write *yes* or *no*.

1.

2.

3.

4.

5.

6.

7.

8.

PROBLEM SOLVING Use the figure below.

9. What kind of polygon is figure *DEFL*?

10. What is the special name for figure *BDLK*? figure *JLGI*? figure *ACEM*?

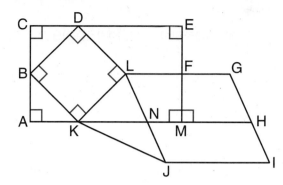

11. Name 5 quadrilaterals other than those named in questions **9** and **10**.

12. Which side is opposite $\overline{CA}$? opposite $\overline{DL}$? $\overline{BD}$? $\overline{LJ}$? $\overline{LG}$?

Use dot paper to draw a quadrilateral:

13. with 4 right angles; with 2 right angles; with 1 right angle

14. with 0 right angles and 1 pair of opposite sides that are parallel

15. whose sides are all equal in length and is *not* a square

16. with 0 right angles and 0 pairs of opposite sides that are parallel

Copy these statements into your Math Journal. Then write *true* or *false* for each statement. If a statement is false, explain why.

Math Journal

17. A quadrilateral never has 4 sides.

18. A square always has 4 right angles.

19. No triangles are quadrilaterals.

20. No quadrilateral has parallel opposite sides.

21. All rectangles are parallelograms.

22. All squares are rectangles.

23. All quadrilaterals are parallelograms.

24. Some parallelograms are also squares.

These polygons are all triangles.

 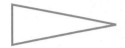 

▶ These triangles are all **right triangles**.

 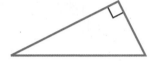 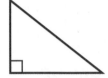

All right triangles have 1 right angle.

▶ These triangles are all **equilateral triangles**.

 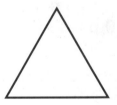   

All the sides of an equilateral triangle
are equal.

▶ These triangles are all **scalene triangles**.

None of the sides of a scalene triangle
are equal.

Classify. Write right, equilateral or scalene for each triangle.

1.

2.

3.

4.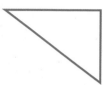

5.

6.

7.

8.

9.

10.

11.

12.

13.

14.

15.

16.

PROBLEM SOLVING

17. Suppose you wanted to draw an equilateral triangle, and you drew one side that measured 5 cm. How long would you draw each of the other sides? What would be the total length of all the sides?

Share Your Thinking

Communicate ✓

18. Explain how you decided which figures in exercises 1–16 were right triangles and which figures in exercises 9–16 were equilateral triangles.

19. Describe and name the 3 angles in a right triangle and in an equilateral triangle.

10-8 | Similar Figures

Billy used similar figures to make this pattern.

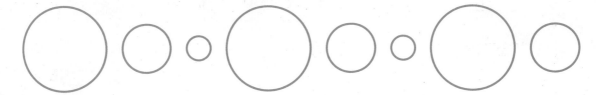

Similar figures have exactly the same shape.
They may or may not be the same size.

 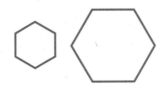 

same shape   same shape   same shape
different sizes   different sizes   same size

All **congruent** figures are also similar.

Are the figures similar? Write *yes* or *no*.

1.

2.

3.

4.

5.

6.

7.

8.

9.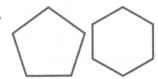

Write the letter of the figure that is similar to the first figure.

10.    a.    b.   c.

11.   a.   b.   c.

12.    a.   b.    c.

13. 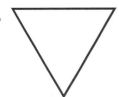   a.   b. 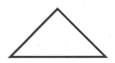   c.

Copy each figure onto dot paper. Then draw a figure that is twice as large.

14. 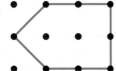   15. 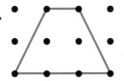   16.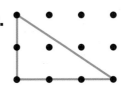

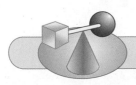

Challenge

Copy and cut out four of these triangles.

17. Fit the triangles together to form a similar triangle.

10-9 Slides and Flips

Noah and Nicole made these patterns.

Noah:

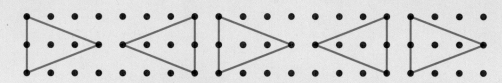

Nicole:

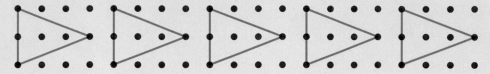

Discover Together

You Will Need: dot paper, pencil, scissors, ruler

Copy the triangle at the right onto dot paper.
Then cut it out.

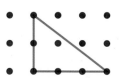

Place your triangle on another sheet of dot paper. Trace around the triangle to make a pattern in the same way that Noah made his.

1. How did you move the triangle to make the pattern?

Now place the triangle on a third sheet of dot paper. Trace around the triangle to make a pattern in the same way that Nicole made hers.

2. How did you move the triangle to make the pattern?

3. How are your two patterns alike? How are they different?

4. Explain how you know that your patterns are alike and different in the same way as Noah's and Nicole's.

342

A **slide** is a movement of a figure along a line.

A **flip** is a movement of a figure over a line so that the figure faces in the opposite direction. The lines may be imaginary.

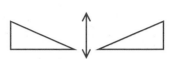

5. Is one of your patterns a slide pattern? Which one?

6. Is one of your patterns a flip pattern? Which one?

Copy the figures and movements below onto dot paper.

A.   B. 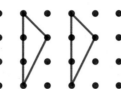 C. 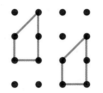 D.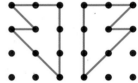

E. 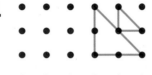 F. 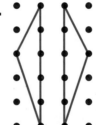 G. H.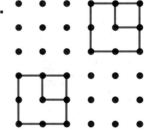

7. Which of the movements are slides? Which are flips? You may draw lines to help you decide.

Communicate

Discuss ✓

8. Can you slide a figure in any direction? Explain your answer.

9. Can you flip a figure in any direction? Explain your answer.

10. Can you always tell whether a movement is a flip or a slide? Draw a figure and its movement to prove your answer.

10-10 | Turns

▶ A **turn** is the movement of a figure around a point.

 $\frac{1}{4}$ turn   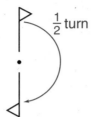 $\frac{1}{2}$ turn   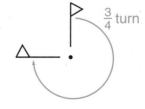 $\frac{3}{4}$ turn

▶ You can turn a figure in either direction.

Each new position is a **turn image** of the figure.

▶ If you can turn a tracing of a figure halfway around so that the tracing and the figure match exactly, the figure has **half-turn symmetry.**

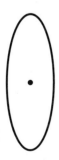   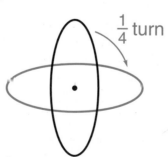 $\frac{1}{4}$ turn   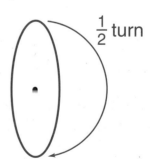 $\frac{1}{2}$ turn

This figure has half-turn symmetry.

Which figures are turn images of the first figure?

1. G   a. ◡   b. ⅁   c. ◠   d. ↺

2. F   a. ᖴ   b. Ⴈ   c. ⊥   d. Ⅎ

3. J   a. ⌐   b. ∪   c. ∩   d. ∩

4. K   a. ⋊   b. ⋉   c. ⋌   d. ⋋

5. R   a. ᴚ   b. ᴚ   c. ᴚ   d. ᴚ

Does each figure have half-turn symmetry? Write *yes* or *no*.
You may use tracing paper or dot paper and scissors.

6.

7.

8.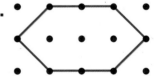

Skills to Remember

Use the graph at the right.

9. On what day was the temperature 55°F?

10. On which two days was the temperature the same?

11. What was the temperature on Tuesday?

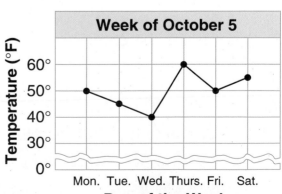

Week of October 5

Temperature (°F) vs. Day of the Week
Mon. Tue. Wed. Thurs. Fri. Sat.

345

Coordinate Geometry

You can use an **ordered pair** of numbers to locate points on a grid.

▶ What ordered pair gives the location of point *A*?

Point *A* is:
3 spaces to the right of 0. →
5 spaces up. ↑

The ordered pair (3, 5) gives the location of point *A*.

▶ What point is located at (5, 3)?

Move 5 spaces to the right of 0. →
Move 3 spaces up. ↑

Point *C* is located at (5, 3).

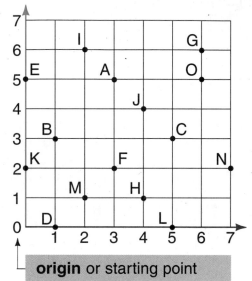

origin or starting point

In an ordered pair, the *first number* tells you to move to the right. The *second number* tells you to move up.

Use the grid to answer each question.

1. What ordered pair gives the location of point *B*?

2. What ordered pair gives the location of point *H*?

3. What point is located at (2, 6)?

4. What point is located at (4, 4)?

5. What ordered pair gives the location of point *D*?

6. What point is located at (0, 5)?

**Make and label five 10 by 10 grids on grid paper.
Draw each point. Then use line segments to connect
the points in order.**

7. (1, 1) (7, 1) (7, 7) (1, 1)   **8.** (1, 6) (5, 6) (8, 2) (3, 2) (1, 6)

9. (3, 7) (5, 7) (6, 5) (6, 3) (5, 1) (3, 1) (2, 3) (2, 5) (3, 7)

10. (4, 2) (6, 4) (5, 7) (2, 6) (1, 3) (4, 2)

11. (1, 2) (4, 3) (7, 2) (8, 8) (4, 4) (2, 6) (1, 2)

12. Name each of the polygons you have drawn.

**Make and label four more 10 by 10 grids on grid paper.
Draw each point. Then use line segments to connect
the points in order for each pair of figures.**

13. A: (1, 2) (4, 2) (1, 7) (1, 2)   **14.** A: (1, 6) (8, 6) (8, 9) (1, 6)
    B: (6, 2) (9, 2) (6, 7) (6, 2)   B: (1, 5) (8, 5) (8, 2) (1, 5)

15. A: (1, 1) (5, 1) (1, 4) (1, 1)   **16.** A: (4, 4) (4, 8) (1, 8) (4, 4)
    B: (5, 3) (9, 3) (5, 6) (5, 3)   B: (5, 3) (9, 3) (9, 6) (5, 3)

17. Name each movement of the figure in exercises 13–16.

 **Connections: Social Studies**

18. Find your town or city on a
map of your state. Tell how
finding it is like locating a
point on a grid and how
it is different.

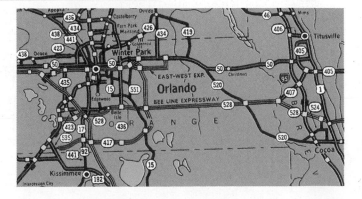

LOGO

LOGO is a computer language that can be used to draw figures. **Commands** are used to tell a small triangle called a turtle how to move around the computer screen.

LOGO turtle

Below are some commands used to move the turtle.

Command	What you enter	How the turtle moves
FORWARD	FD 20	Forward 2 steps
BACK	BK 20	Back 2 steps
RIGHT	RT 90	Makes a right angle turn (90°)
LEFT	LT 45	Makes half of a right angle turn
REPEAT	REPEAT 3[FD 20]	Repeats the command(s) in brackets 3 times
PENUP	PU	Takes turtle out of drawing mode
PENDOWN	PD	Puts turtle back in drawing mode

These commands tell the turtle to draw this figure.

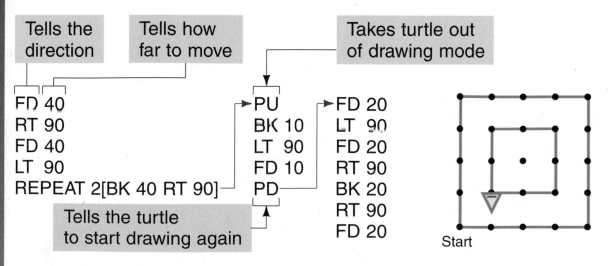

Tells the direction

Tells how far to move

Takes turtle out of drawing mode

```
FD 40          PU      FD 20
RT 90          BK 10   LT 90
FD 40          LT 90   FD 20
LT 90          FD 10   RT 90
REPEAT 2[BK 40 RT 90]  PD      BK 20
                       RT 90
                       FD 20
```

Tells the turtle to start drawing again

Start

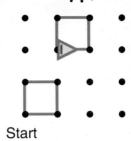

Match each movement with the correct LOGO commands.

a.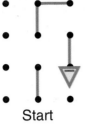
Start

b.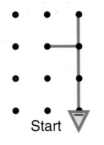
Start

c.
Start

d.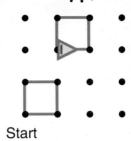
Start

1. PU
FD 20
PD
RT 90
FD 10
RT 90
BK 10
FD 30

2. REPEAT 2[FD 10 LT 90]
BK 10
PU
LT 90
FD 10
PD
LT 45
FD 14
PU → BK 14
RT 45
FD 10
PD
RT 90
FD 20
RT 90
FD 10

3. FD 10 → RT 90
PU   PU
FD 10   FD 10
PD   PD
FD 10   FD 10
RT 90
FD 10

4. REPEAT 4[FD 10 RT 90]
PU
FD 20
RT 90
FD 10
PD
REPEAT 4[FD 10 LT 90]

Write the commands to draw each.

5. a square

6. a rectangle

7. a pentagon

8. parallel lines

9. perpendicular lines

10. 2 congruent squares

11. 2 similar rectangles

12. a square within a rectangle

13.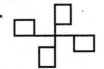

14.

15.

10-13 | Problem Solving: Find a Pattern

Algebra ✓

Problem: Tyrell draws a spiral on grid paper. He draws 4 line segments. Then he draws 5 more line segments to continue the pattern. How long is the finished spiral?

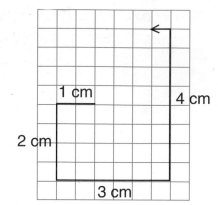

1 IMAGINE Create a mental picture of the finished spiral.

2 NAME *Facts:* The spiral has 4 line segments. The spiral will have 5 more.

Question: How long is the finished spiral?

3 THINK Measure the length of each segment. Look for sums of ten.

$$1 + 2 + 3 + 4 + 5 + 6 + 7 + 8 + 9 = \underline{\ ?\ }$$

4 COMPUTE Add to find the total length.

The sum has 4 tens:

$1 + 9$, $2 + 8$, $3 + 7$, and $4 + 6$ or 40.

$40 + 5 = 45$

The spiral is 45 cm long.

5 CHECK Cut a 45-cm string and use it to measure the spiral.
Use a calculator to check addition.

Find a pattern to solve each problem. You may use grid paper.

1. Jessie adds one more square to this drawing. It now has 2 lines of symmetry. Where does she add the square? (*Hint:* See page 23.)

Original Drawing

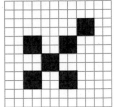

IMAGINE Put yourself in the problem.

NAME *Facts:* Jessie draws this shape. She adds one square. The finished shape has 2 lines of symmetry.

Question: Where does she add the last square?

THINK Find the line of symmetry in the original drawing. Then think of other ways to fold the shape in half.

COMPUTE ⟶ **CHECK**

2. Jacques paints these shapes in order on a belt: triangle, square, triangle, pentagon, triangle, hexagon. What are the ninth and tenth shapes?

3. Can you cut this shape into 2 congruent hexagons? 2 congruent octagons? (*Hint:* See page 22.)

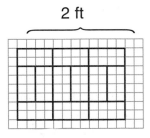

4. Troy makes a gerbil cage. The floor of the cage must have an area of 4 square units. How many different shapes can the floor be?

2 ft

5. A brick border follows this pattern. How many bricks are used in a 10-ft border?

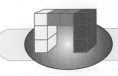

Make Up Your Own

6. Write a problem that uses a pattern. Have a classmate solve it.

Solve each problem and explain the method you used.

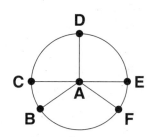

1. Sylvia's Sign Shop made an octagonal sign. How many sides does it have? how many angles?

2. The sign shop made the neon sign **HI**. Where are the parallel lines in the sign? Where are the perpendicular lines?

3. The Dilly Deli ordered a sign in the shape of a pickle outline. Is the sign a simple closed curve?

4. Roy orders a square sign from the sign shop. Does his sign have half-turn symmetry?

5. 4 is to square as 8 is to _?_ .

6. The sign for Trex Tires is a circle. Name the radii shown on the sign at the right.

7. On grid paper draw different figures using 5 squares so that the squares touch along at least one entire side. In how many different ways can the squares be arranged?

Use the table for problems 8–10.

8. How much does a hexagonal sign cost?

9. How much more expensive is a pentagonal sign than a triangular sign?

10. What is the cost of 2 rectangular signs and 1 triangular sign?

Sign Prices

Number of Sides (4 ft each)	Price per Side
3	$25
4	$20
5	$45
6	$60

Choose a strategy from the list or use another strategy you know to solve each problem.

USE THESE STRATEGIES:
Find a Pattern
Logical Reasoning
Choose the Operation
Extra Information
Make Up a Question
Draw a Picture

11. Hank, Don, and Ned are waiting in line. Don is ahead of Ned. Hank has been waiting longer than the others. What is their order in line?

12. A tool measures 16 cm. A plastic tube measures 71 cm. How much longer is the plastic tube?

13. Sylvia cuts a triangle, a square, and a pentagon out of wood. The first shape she cuts has more sides than the second but fewer sides than the third. In what order does she cut the shapes?

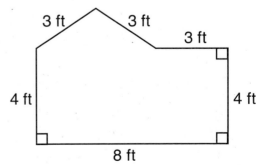

14. Marcy cut an equilateral triangle to make 4 congruent signs. Each side of the triangle is 2 ft long. How did Marcy cut the triangle?

15. Sylvia's shop has 8 rows of paint cans. There are 10 cans in the first row, 9 cans in the second row, 8 in the third, and so on. How many cans of paint are there in all?

16. How would you describe the shape of this sign for the Food Barn? How many angles does it have? What other questions can you answer about this sign?

3 ft 3 ft

3 ft

4 ft   4 ft

8 ft

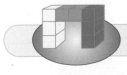

Make Up Your Own

17. Write a problem modeled on problem 12. Have a classmate solve it.

353

Identify each.   *(See pp. 326–339.)*

1. _?_ ray   a.   b.   c.

2. _?_ line segment

3. _?_ perpendicular
   lines

4. _?_ vertex   d.   e. 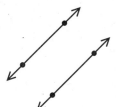   f.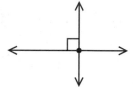

5. _?_ line

6. _?_ right triangle

7. _?_ circle   g.    h.    i.

8. _?_ parallel lines

9. _?_ point

Draw these.

10. $\overrightarrow{AX}$ 11. $\angle RST$ 12. $\overleftrightarrow{XY}$ 13. right angle *DEF*

Which figure is similar to the first figure?   *(See pp. 340–341.)*

14.   a.   b.   c.

Is each a slide? Write *yes* or *no*.   *(See pp. 342–343.)*

15.   16.   17.   18.

(See Still More Practice, pp. 469–470.)

TANGRAMS

Copy the **tangram** puzzle on grid
paper. Then cut it out.

**Flip, slide, or turn the
puzzle pieces to solve.**

1. Name the shape of each puzzle piece.

2. Which two pieces cover piece 6?
   Which two cover piece 4?

3. Which three pieces cover piece 1?
   Cover piece 1 in a different way.
   Which pieces did you use?

4. Use pieces 1 and 2. How many different
   figures can you make? Name them.

5. Make the largest possible triangle you can.
   How many pieces did you use?

6. Make two similar parallelograms. Which pieces
   did you use for the large parallelogram?
   Which did you use for the small parallelogram?

7. A **trapezoid** is a quadrilateral with only
   one pair of opposite sides that are parallel.
   Use different pairs of puzzle pieces to
   make trapezoids. How many trapezoids
   did you make? Which two pieces did you
   use for each?

8. Make trapezoids from 3 pieces, 4 pieces,
   5 pieces, and 6 pieces. Record the
   puzzle pieces you used for each.

9. Use all seven tangram pieces to make
   a rectangle and then a parallelogram.

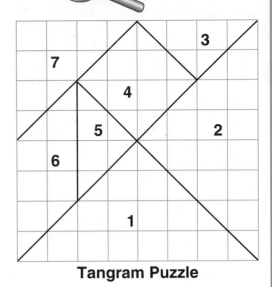

Tangram Puzzle

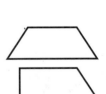

Performance Assessment

Draw these.

1. $\overleftrightarrow{LM}$
2. $\overrightarrow{BA}$
3. point R
4. ray AB

5. $\angle DEF$
6. circle A
7. $\overline{RS}$
8. line XY

Identify each.

9. _?_ parallelogram

10. _?_ perpendicular lines

11. _?_ diameter

12. _?_ radius

13. _?_ flip

14. _?_ line segment

15. _?_ ray

16. _?_ equilateral triangle

a.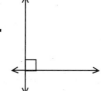

b.

c.

d.

e.

f.

g.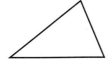

h.

i.

Does each figure have half-turn symmetry? Explain.

17.

18.

19.

Use the grid to answer each question.

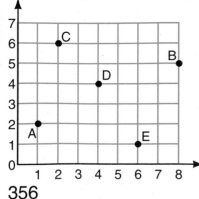

20. What ordered pair gives the location of point E?

21. What point is located at (2, 6)?

22. What ordered pair gives the location of point B?

356

Perimeter, Area, and Volume

11

POPSICLE STICKS AND GLUE

We're building a village of popsicle sticks,
Just popsicle sticks and glue:

Houses and fences, sidewalks and streets,
A school and a library, too;
Museums, churches, temples, shops,
A playground, a park, and a zoo.

Isn't it wonderful what we can do
With popsicle sticks and a new tube of glue?

Leslie D. Perkins

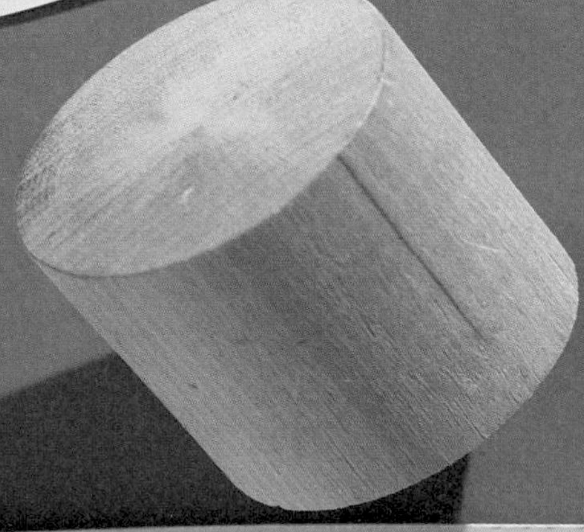

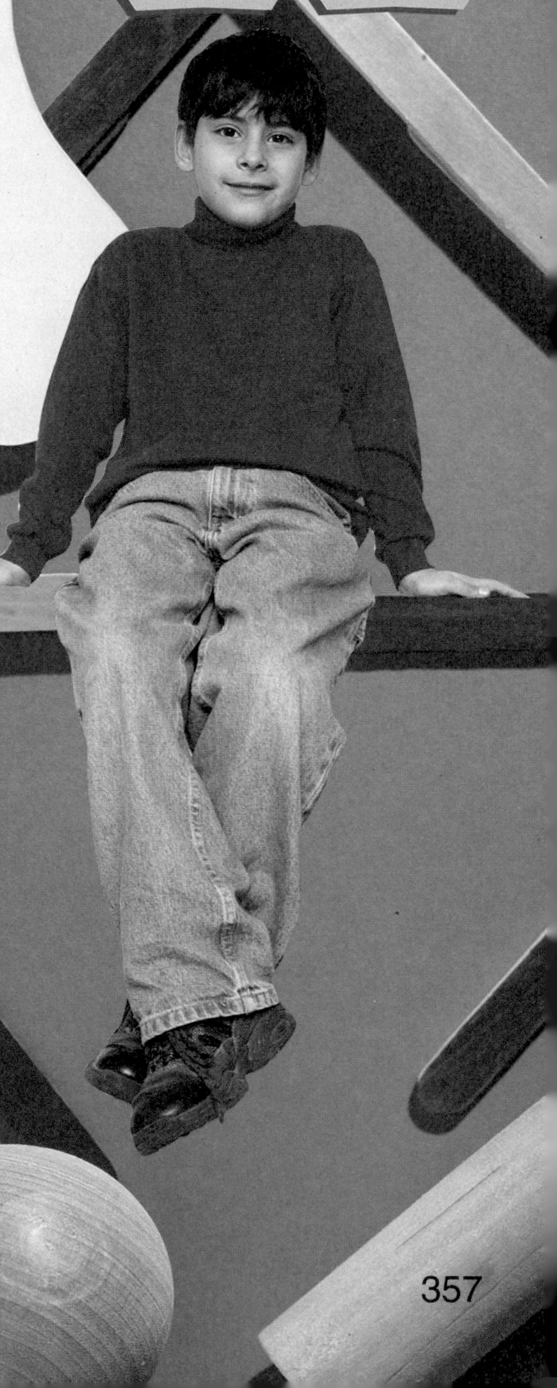

In this chapter you will:

Use models and formulas
Relate plane and space figures
Investigate spatial relationships
Solve problems using a drawing or model

Critical Thinking/Finding Together

Use popsicle sticks to build a border around your desk. How many popsicle sticks did you use? Compare your border and the number of sticks used with classmates.

11-1 Using Perimeter Formulas

Algebra

▶ You can use a **formula** to find the perimeter of a rectangle.

What is the perimeter of this rectangle?

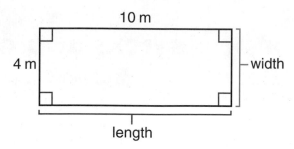

10 m

4 m

width

length

perimeter = length + width + length + width

$P \quad = \quad \ell \quad + \quad w \quad + \quad \ell \quad + \quad w$

$P = 2 \times \ell + 2 \times w$ ← formula for perimeter of a rectangle

$P = (2 \times 10) + (2 \times 4)$

$P = 20 + 8$

$P = 28$ m

The perimeter of the rectangle is 28 m.

▶ Use formulas to find the perimeter of a square and of an equilateral triangle.

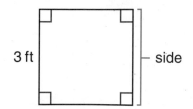

3 ft

side

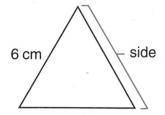

6 cm

side

$P = \text{side} + \text{side} + \text{side} + \text{side}$

$P = s + s + s + s$

$P = 4 \times s$ ← formula for perimeter of a square

$P = 4 \times 3$

$P = 12$ ft

$P = \text{side} + \text{side} + \text{side}$

$P = s + s + s$

$P = 3 \times s$ ← formula for perimeter of an equilateral triangle

$P = 3 \times 6$

$P = 18$ cm

Find the perimeter of each. Use a formula.

1.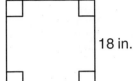
18 in.

2.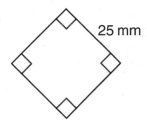
25 mm

3.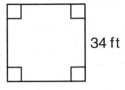
34 ft

4.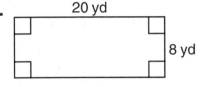
20 yd
8 yd

5.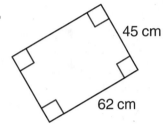
45 cm
62 cm

6.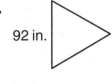
19 m
26 m

7.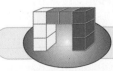
53 mm

8.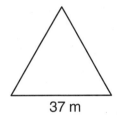
37 m

9.
92 in.

PROBLEM SOLVING Use a formula.

10. What is the perimeter of an equilateral triangle whose sides are all 17 in. long?

11. What is the perimeter of a square whose sides are all 49 m long?

12. What is the perimeter of a rectangle with a length of 72 cm and a width of 14 cm?

13. What is the perimeter of a rectangle with a width of 122 in. and a length of 15 in.?

14. A square has sides that are all 4294 mm long. What is the perimeter of the square?

Make Up Your Own

15. Make up a formula to find the perimeter of a hexagon whose sides are all the same length.

16. Make up a formula to find the perimeter of an octagon whose sides are all the same length.

11-2 Area

Area is the number of square units needed to cover a flat surface.

 1 square unit

▶ You can find the area of some figures by counting squares.

9 square units

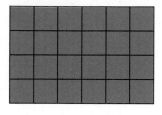

24 square units

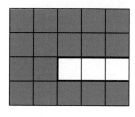

17 square units

▶ Sometimes you need to count half squares to find the area of a figure.

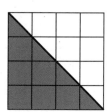

6 whole squares + 4 half squares
6 + 2 = 8
8 square units

4 half squares = 2 whole squares

▶ Sometimes you need to estimate the area of a figure.

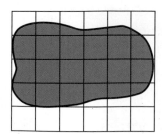

9 whole squares + 5 almost whole squares + 4 about half squares + about 1 more

9 + 5 + 2 + 1 = 17

about 17 square units

Find the area of each figure.

1.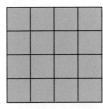

2.

3.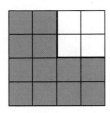

4.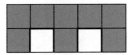

5.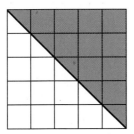

6.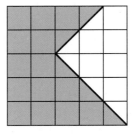

7.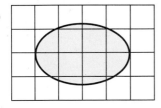

8.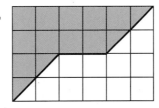

9.

Estimate the area of each figure.

10.

11.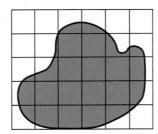

12.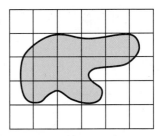

Skills to Remember

Multiply.

13.	**14.**	**15.**	**16.**	**17.**	**18.**
14	29	72	54	95	63
× 4	× 3	× 7	×48	×12	×25

Using the Area Formula Algebra

This rectangular floor is covered with tiles that are each 1 square foot. The floor is 7 ft long and 6 ft wide. What is the area of the floor?

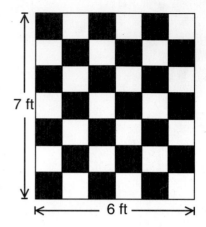

7 ft

6 ft

To find the area of the floor, you could count square feet.

or

You could use the formula for finding the area of a rectangle.

area = length × width

$A = \ell \times w$ ◄── formula for the area of a rectangle

$A = 7 \times 6$

$A = 42$ square feet (sq ft)

The area of the floor is 42 **square feet (sq ft)**.

Area is always reported in square units.

Study these examples.

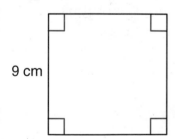

9 cm

8 yd

16 yd

$A = s \times s$ ◄── formula for the area of a square

$A = 9 \times 9$

$A = 81$ square centimeters (sq cm)

$A = \ell \times w$

$A = 16 \times 8$

$A = 128$ square yards (sq yd)

Find the area. Use the area formula.

1.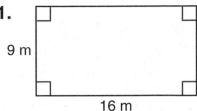

9 m

16 m

2.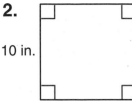

10 in.

3.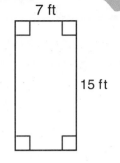

7 ft

15 ft

4.

25 cm

5.

18 m

6.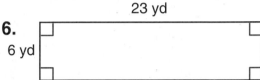

23 yd

6 yd

7.

12 in.

14 in.

8.

15 m

9.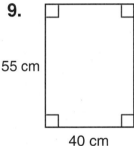

55 cm

40 cm

PROBLEM SOLVING

10. A football field is 120 yd long (including the end zones) and 55 yd wide. What is the area of a football field?

11. A baseball infield is a square that is 90 ft along each side, or base line. What is its area?

12. A tennis court is a rectangle that is 78 ft long and 27 ft wide. What is the area of a tennis court?

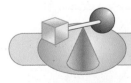

Challenge

Use grid paper.

13. Draw as many rectangles as you can that each have an area of 24 square units. Explain the patterns you notice.

▶ **Space figures** are not flat. They are sometimes called **solids**.

A **cube** is a space figure with 6 faces, 12 edges, and 8 vertices.

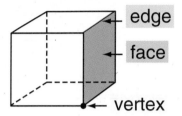

edge
face
vertex

A **face** is a flat surface surrounded by line segments.
Two faces meet at a line segment called an **edge**.
Three or more edges meet at a **vertex**.

▶ These space figures have faces, edges, and vertices.

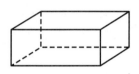 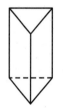 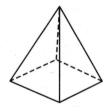

rectangular prism	**triangular prism**	**square pyramid**
6 faces	5 faces	5 faces
12 edges	9 edges	8 edges
8 vertices	6 vertices	5 vertices

▶ These space figures have 0 edges and 0 faces. Each has a curved surface.

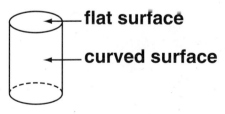  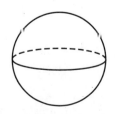

flat surface
curved surface

cylinder	**cone**	**sphere**
2 flat surfaces	1 flat surface	0 flat surfaces

Copy and complete. You need not draw the space figures.

1.	name	cube	?	?	?	?	?	?
2.	faces	?	?	?	6	?	?	?
3.	edges	12	?	?	?	?	?	?
4.	vertices	?	?	?	?	?	?	6

PROBLEM SOLVING

5. I have 2 flat surfaces, 0 edges, and 0 vertices. Which space figure am I?

6. I have 1 flat surface and a curved surface. Which space figure am I?

7. I have 5 faces and 5 vertices. How many edges do I have? Which space figure am I?

8. I am shaped like a ball. How many faces, edges, and vertices do I have? Which space figure am I?

9. I have 6 faces and 12 edges. I am not a rectangular prism. Which space figure am I?

10. I have 9 edges and 6 vertices. How many faces do I have? Which space figure am I?

 **Project**

11. Look for objects made up of combinations of space figures in the world around you. Keep a record of the figures you see. Then design a building using a combination of space figures. Tell your class what figures you used to design your building.

365

Space Figures and Polygons

▶ Each flat surface of a space figure is a plane figure.

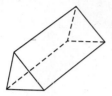

triangular
prism

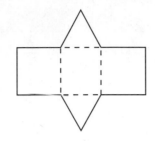

net of a triangular
prism

A **net** is the shape
made by opening
a solid figure and
laying it flat.

The net shows that a triangular prism
is made up of 5 polygons: 2 triangles
and 3 rectangles.

▶ If you could cut a space figure,
the new flat surfaces you create
would also be plane figures.

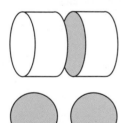

Name the shape of each shaded flat surface

1.

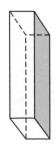

2.

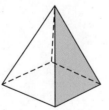

3.

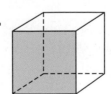

4.

**Use dot paper. Copy each net. Name each polygon.
Then cut, fold, and tape each net to make a space figure.**

5.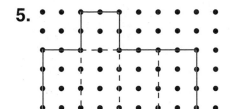

6.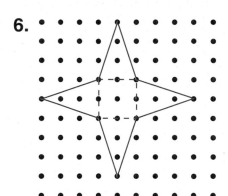

Name the shape of the new flat surface made by each cut.

7.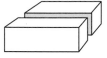

8.

9.

10.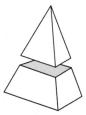

11.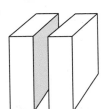

12.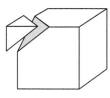

 **Project**

13. Press models of space figures into clay. Tell your teacher the names of the plane figures you have made.

14. Use dot paper. Draw a net of a cube. Cut out and fold the net. Tape the edges together.

Discover Together

You Will Need: 36 connecting cubes each, paper, pencil

Use 36 connecting cubes to build each of these rectangular prisms. Each person in your group should build a different figure.

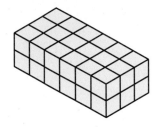 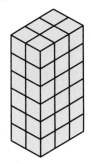 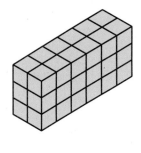

1. Compare the prisms you built. What do you notice?

Take turns. Use 36 cubes to build other rectangular prisms. Ask others in your group to build a prism just like yours. Record the length, width, and height of each prism.

2. How many different rectangular prisms did your group build?

3. How can you be sure that each prism is different from each of the other prisms?

Work together to guess how many connecting cubes you would need to build each of the space figures in exercises 4, 5, 6, and 7. Record your group's guesses. Then test the guesses by building each figure.

4.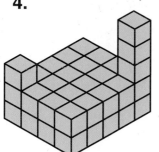

5.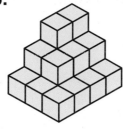

6.

7.

8. How close were your guesses to the actual number of cubes needed to build each figure?

9. Which of these figures is impossible?

a.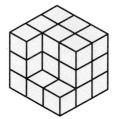

b.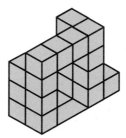

c.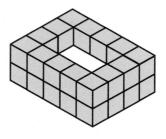

Communicate

Discuss ✓

10. How did you make your guesses for exercises 4, 5, 6, and 7?

11. How did your group decide which figure in question 9 was impossible?

Share Your Thinking

Math Journal ✓

12. Use connecting cubes to build a space figure. Draw a picture of the figure on triangle dot paper. Put your drawing in your Math Journal.

Volume

The **volume** of a space figure is the number of cubic units the figure contains.

1 cubic unit

You can find the volume of a space figure by counting the number of cubic units needed to fill it. Or you can build the figure with connecting cubes and then count the cubes.

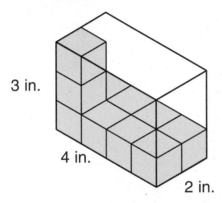

3 in.

4 in.

2 in.

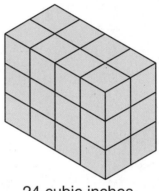

24 cubic inches

The volume of the rectangular prism is 24 cubic inches.

Find the volume of each. You may use connecting cubes.

1.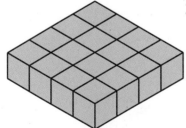

2.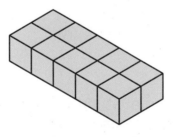

3.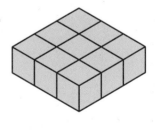

4.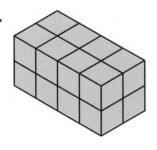

5.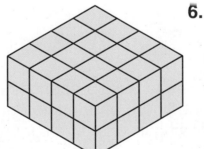

6.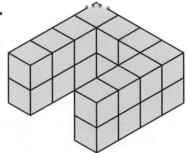

Multiplying to Find Volume

To find the volume of a cube or of a rectangular prism, multiply.

Volume = length × width × height

Volume = 6 × 2 × 4
Volume = 48 cubic meters

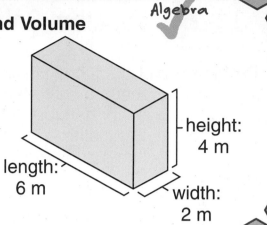

height: 4 m

length: 6 m

width: 2 m

Multiply to find the volume of each.

7.

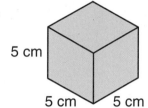

5 cm
5 cm   5 cm

8.

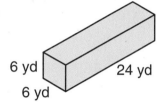

6 yd   24 yd
6 yd

9.

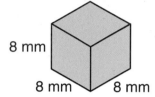

8 mm
8 mm   8 mm

10.

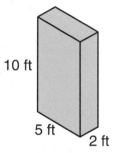

10 ft
5 ft   2 ft

11.

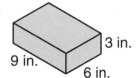

3 in.
9 in.   6 in.

12.

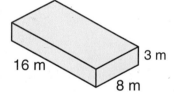

3 m
16 m   8 m

PROBLEM SOLVING

13. Trey has 18 connecting cubes. How many different rectangular prisms can he build?

 **Share Your Thinking**

Communicate

14. Could you compute to find the volume of the figure in exercise 6? Tell a classmate how you would do it.

11-8 | Problem Solving: Use a Drawing or Model

Problem: The Hobby Hut sign is a triangle. It has a light at each corner and every half foot along each side. Each side is 2 ft long. How many lights does the sign have?

1 IMAGINE Picture yourself putting light bulbs in the sign.

2 NAME *Facts:* 1 light—at each corner
1 light—every half foot along each 2-foot-long side

Question: How many lights does the sign have?

3 THINK Since each side is 2 ft long, the sign is an equilateral triangle. Draw an equilateral triangle. Use marks for each light. Multiply to find the number of lights:

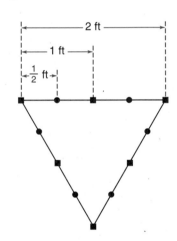

at each corner $\longrightarrow 3 \times 1 = \underline{\ ?\ }$
along each side $\longrightarrow 3 \times 3 = \underline{\ ?\ }$

Then add to find the total.

4 COMPUTE
1 light at each corner	$3 \times 1 = 3$
3 lights along each side	$3 \times 3 = 9$
The total number of lights	$3 + 9 = 12$

The sign has 12 lights.

5 CHECK Count all the lights on the sign.

Use a drawing or model to solve each problem.

1. A shape made up of tiles that are 1 cm square has an area of 9 sq cm and a perimeter of 12 cm. Describe the shape.

A = 9 sq cm

P = 18 cm

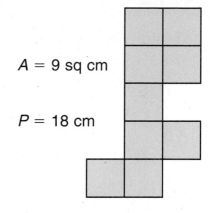

| **IMAGINE** | Create a mental picture. |

| **NAME** | *Facts:* | Area | = 9 sq cm |
| | | Perimeter | = 12 cm |

Question: What is the shape?

Second Try

| **THINK** | How should you arrange 1-sq-cm tiles to make a shape that has an area of 9 sq cm and a perimeter of 12 cm? |

A = 9 sq cm

P = 16 cm

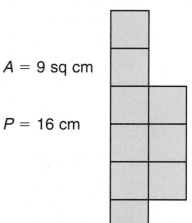

COMPUTE ⟶ **CHECK**

2. The area of a rectangle is 32 square inches. One of the sides is 4 inches long. How long are the other sides?

3. The volume of a rectangular prism is 24 cubic centimeters. The base of the prism is 2 cm by 4 cm. How tall is it?

4. How many triangles can you find in the puzzle on the right?

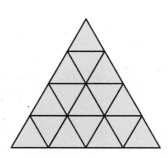

5. Bill draws a square with each side 4 cm long. He draws a dot in the center of each side and connects the dots to draw another square. About how long are the sides of the new square?

Solve each problem and explain the method you used.

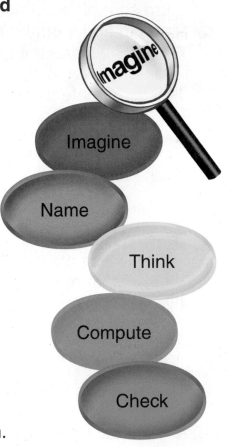

Imagine

Name

Think

Compute

Check

1. Sara babysits for Ollie and shows him how to build a tower with 27 cubes. Each cube has a volume of 1 cubic inch. What is the volume of the tower?

2. Ollie builds a tower with his 1-inch cubes. What shape is the face of each 1-inch cube in Ollie's tower?

3. Trina brings crayons. Both ends of the blue crayon are rubbed flat. What space figure does the crayon look like?

4. Benji's crib mattress is 34 in. by 30 in. What is the perimeter of the mattress? (*Hint:* See page 21.)

5. Sara draws a hexagon. Each side is 9 cm. What is the perimeter of the hexagon?

6. Trina shows Benji a shape. It has 2 circular flat surfaces. What is the shape?

7. Three children work with blocks to build a rectangular prism. It has a width of 7 in., a length of 8 in., and a height of 10 in. What is the volume of this prism?

8. Trina has a sheet of paper that is $8\frac{1}{4}$ in. by 11 in. What is the perimeter of the paper? (*Hint:* See page 21.)

Choose a strategy from the list or use another strategy you know to solve each problem.

USE THESE STRATEGIES
Use a Drawing or Model
Two-Step Problem
Logical Reasoning
Write a Number Sentence
Guess and Test
Find a Pattern

9. Ralph has a photo in his wallet. The area of the photo is 6 square inches. How long might each side be?

10. Trina made this bead pattern: 1 sphere, 2 cylinders, 3 cones, 2 spheres, 3 cylinders, 4 cones, and so on. What is the shape of the 20th figure?

11. The shortest side of a quadrilateral is 5 cm. The next side is 10 cm. The length of each succeeding side increases by 5 cm. What is the perimeter of the quadrilateral? (*Hint:* See page 21.)

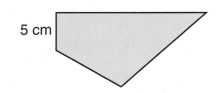

5 cm

12. Ralph's rectangular quilt has an area of 12 sq ft. One side is 3 feet long. What is the perimeter of the quilt?

13. Angie makes giant pillows. The table tells about each pillow. What space figure does Benji's pillow look like?

14. Which child's pillow is shaped like a cylinder?

	Benji's Pillow	Sara's Pillow	Trina's Pillow
faces	6	0	0
edges	12	0	0
curved surface	0	1	1
flat surface	0	2	1

15. Draw five 1-centimeter squares to make a shape so that any 2 squares touch along at least one entire side. How many different arrangements are possible?

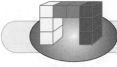

Make Up Your Own

16. Write a problem that can be solved by using a drawing or model. Have a classmate solve it.

Find the perimeter. Use a perimeter formula.   *(See pp. 358–359.)*

1.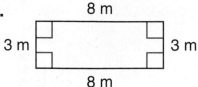
8 m
3 m   3 m
8 m

2.
36 yd

3.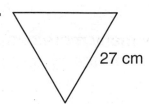
27 cm

Find the area.   *(See pp. 360–363.)*

4.

5.
15 in.

6.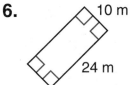
10 m
24 m

Name each space figure.   *(See pp. 364–365.)*

7.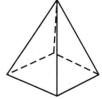

8.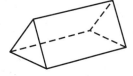

9.

Name the shape of each shaded flat surface.   *(See pp. 366–367.)*

10.

11.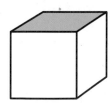

12.

Find the volume.   *(See pp. 370–371.)*

13.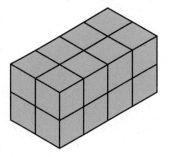

14.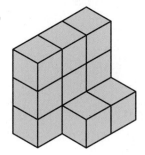

15.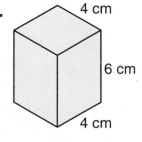
4 cm
6 cm
4 cm

   (See *Still More Practice*, p. 470.)

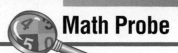
MISSING CUBIC UNITS

Maya used cubes to build this figure.
How many more cubes does she need
to finish making a rectangular prism?

Draw the rectangular prism on triangle dot paper
or use connecting cubes. Then count the
cubes that are missing.

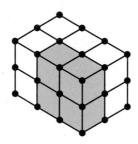

Maya needs 4 more cubes to finish making
the rectangular prism.

PROBLEM SOLVING

You may use triangle dot paper or connecting cubes.

1. How many more cubes are needed to finish building each cube?

a.

b.
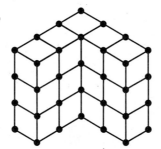

c.
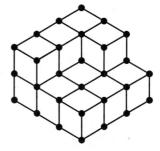

2. How many more cubes are needed to finish building
each rectangular prism?

a.
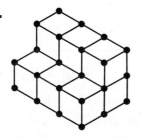

b.

c.
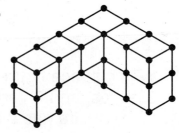

Performance Assessment

1. **a.** Jill Clark wrote her initials on a grid. Estimate the area of her initials.

   b. Write your initials on grid paper and then estimate the area they cover.

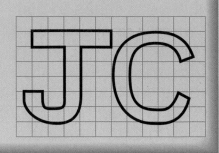

Find the perimeter. Use a perimeter formula.

2.

30 m
15 m

3.
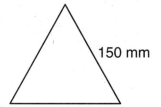
18 ft

4.
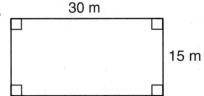
150 mm

Name the shape of the new flat surfaces made by each cut.

5.
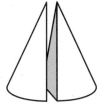

6.
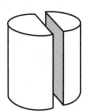

7.
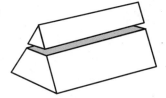

Find the volume.

8.
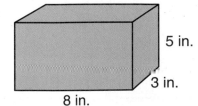
5 in.
3 in.
8 in.

9.
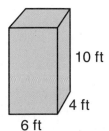
10 ft
4 ft
6 ft

10.
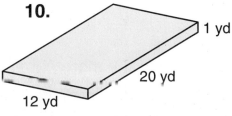
1 yd
20 yd
12 yd

PROBLEM SOLVING *Use a strategy you have learned.*

11. Which space figure has 5 faces, 9 edges, and 6 vertices? Which space figure has 2 circular flat surfaces, 0 edges, and 0 vertices?

12. The volume of a rectangular prism is 36 cubic cm. The base of the prism is 4 cm by 3 cm. How tall is it?

Cumulative Review IV

Choose the best answer.

1. Which is greatest to least?
   a. 4607; 46,070; 45,021; 46,088
   b. 46,088; 46,070; 45,021; 4607
   c. 46,070; 46,088; 45,021; 4607
   d. 46,070; 46,088; 4607; 45,021

2. 6907 + 386 + 2999
   a. 9192
   b. 10,292
   c. 13,766
   d. not given

3. 65,600 − 1,592
   a. 64,008   b. 64,018
   c. 64,192   d. not given

4. 76 × 450
   a. 5850   b. 33,900
   c. 34,206   d. not given

5. 8 + 6 ÷ 3 − 2
   a. 2 R2
   b. 8
   c. 14
   d. not given

6. The dividend is 456. The quotient is 76. What is the divisor?
   a. 4
   b. 6
   c. 8
   d. 80

7. Alana pays $11.96 for 4 identical plants. How much does each plant cost?
   a. $1.56
   b. $1.99
   c. $2.49
   d. $2.99

8. Which is normal room temperature?
   a. 20° C
   b. 37° C
   c. 50° C
   d. 68° C

9. Which is least to greatest?
   a. 4 cm; 300 mm; 40 m; 3 km
   b. 300 mm; 4 cm; 40 m; 3 km
   c. 4 cm; 300 mm; 3 km; 40 m
   d. 3 km; 4 cm; 40 m; 300 mm

10. Which graph is used to compare the parts of a whole group?
   a. circle graph   b. bar graph
   c. line graph   d. pictograph

11. $5\frac{2}{9} + 8\frac{4}{9}$

   a. $13\frac{1}{3}$ b. $13\frac{2}{9}$ c. $13\frac{2}{3}$ d. not given

12. $\frac{9}{10} - \frac{1}{2}$

   a. $\frac{1}{5}$ b. $\frac{4}{5}$ c. 1   d. not given

13. Which is a quadrilateral?

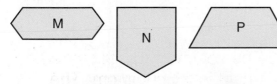

   a. M   b. N   c. P   d. all of these

14. Find the area.

6 ft
   a. 12 sq ft
   b. 24 sq ft
   c. 30 sq ft
   d. 36 sq ft

Ongoing Assessment IV

For Your Portfolio

Solve each problem. Explain the steps and the strategy or strategies you used for each. Then choose one from problems 1–4 for your Portfolio.

1. What is the perimeter of a rectangle with a length of 15 ft and a width of 9 ft?

2. Ana drew a hexagon with 6 equal sides. Each side was 7 dm long. What is the perimeter of the hexagon?

3. Juan used $\frac{1}{2}$ yd of wire for his project. Rory used $\frac{1}{4}$ yd more. How many yards of wire did both boys use?

4. Eiko buys 5 gallons of milk at $3.90 a gallon and one quart of juice for $1.99. What was the total cost?

Tell about it.

5. Explain how you could use fraction strips to solve problem 3.

6. Name a strategy that can be used to solve both problems 3 and 4. Explain.

For Rubric Scoring

Listen for information on how your work will be scored.

7. Which of these space figures have square corners? (Use models to help you do this.) How many right angles does each have?

8. How many acute angles are on the faces of a triangular prism? square pyramid? cube?

cube

triangular prism

square pyramid

Dividing by Two Digits

12

from
WHO HASN'T PLAYED GAZINTAS?

In your arithmetics
the *problem* is what sticks.
The language isn't bound
by spelling, but by sound.
So 3 gazinta 81.
The answer? 27. Done!
In *long* division, I would hint, a
lot of work gazin gazinta.

Then Tums: the sign of which is X.
Do 8 tums 1-5-6? It checks
at just one thousand two four eight.
Repeat: 1,248.

Computers work at a faster rate.

David McCord

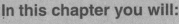

In this chapter you will:

Learn about patterns and
  estimation in division
Investigate trial quotients
  and zeros in division
Use a fraction calculator to
  add and subtract
Solve problems that have
  hidden information

Critical Thinking/
Finding Together

Explain how you can use
base ten blocks to check
that 81 ÷ 3 = 27.

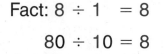

12-1 Division Patterns

Use division facts and patterns with zero to divide tens, hundreds, and thousands by multiples of 10.

Study these division patterns.

Fact: $8 \div 1 = 8$

$80 \div 10 = 8$
$800 \div 10 = 80$
$8000 \div 10 = 800$

Fact: $9 \div 3 = 3$

$90 \div 30 = 3$
$900 \div 30 = 30$
$9000 \div 30 = 300$

Fact: $28 \div 7 = 4$

$280 \div 70 = 4$
$2800 \div 70 = 40$
$28,000 \div 70 = 400$

Fact: $10 \div 2 = 5$

$100 \div 20 = 5$
$1000 \div 20 = 50$
$10,000 \div 20 = 500$

Copy and complete.

1. $9 \div 1 = \underline{\ ?\ }$
   $90 \div 10 = \underline{\ ?\ }$
   $900 \div 10 = \underline{\ ?\ }$
   $9000 \div 10 = \underline{\ ?\ }$

2. $8 \div 4 = \underline{\ ?\ }$
   $80 \div 40 = \underline{\ ?\ }$
   $800 \div 40 = \underline{\ ?\ }$
   $8000 \div 40 = \underline{\ ?\ }$

3. $6 \div 3 = \underline{\ ?\ }$
   $60 \div 30 = \underline{\ ?\ }$
   $600 \div 30 = \underline{\ ?\ }$
   $6000 \div 30 = \underline{\ ?\ }$

4. $32 \div 8 = \underline{\ ?\ }$
   $320 \div 80 = \underline{\ ?\ }$
   $3200 \div 80 = \underline{\ ?\ }$
   $32,000 \div 80 = \underline{\ ?\ }$

5. $40 \div 5 = \underline{\ ?\ }$
   $400 \div 50 = \underline{\ ?\ }$
   $4000 \div 50 = \underline{\ ?\ }$
   $40,000 \div 50 = \underline{\ ?\ }$

6. $30 \div 6 = \underline{\ ?\ }$
   $300 \div 60 = \underline{\ ?\ }$
   $3000 \div 60 = \underline{\ ?\ }$
   $30,000 \div 60 = \underline{\ ?\ }$

Divide mentally.

7. 40 ÷ 20   **8.** 20 ÷ 10   **9.** 60 ÷ 20   **10.** 70 ÷ 10

11. 360 ÷ 90   **12.** 420 ÷ 60   **13.** 560 ÷ 80   **14.** 250 ÷ 50

15. 200 ÷ 40   **16.** 300 ÷ 50   **17.** 400 ÷ 80   **18.** 100 ÷ 50

19. 8000 ÷ 20   **20.** 4000 ÷ 80   **21.** 3000 ÷ 60   **22.** 2000 ÷ 50

23. 4500 ÷ 90   **24.** 6400 ÷ 80   **25.** 1200 ÷ 30   **26.** 1800 ÷ 20

27. 30)21,000   **28.** 40)20,000   **29.** 70)35,000   **30.** 90)81,000

31. 50)40,000   **32.** 60)54,000   **33.** 30)24,000   **34.** 40)36,000

PROBLEM SOLVING

35. How many zeros are in the quotient when you divide 500 by 10?

36. How many zeros are in the quotient when you divide 4800 by 60?

37. How many zeros are in the quotient when you divide 540 by 90?

38. How many zeros are in the quotient when you divide 40,000 by 8?

Connections: Reading

You need to be able to read and understand words such as *dividend, divisor,* and *quotient* to solve some math problems.

39. The quotient is 400. The dividend is 20,000. What is the divisor?

40. The divisor is 90. The quotient is 800. What is the dividend?

41. The dividend is 50. The quotient is 5. What is the divisor?

12-2 Divisors: Multiples of Ten

Forty students share 137 marbles
equally. How many marbles does
each student get? How many
marbles are left over?

To find how many each gets,
divide: 137 ÷ 40 = _?_

Think: 40)‾1‾3‾7‾ 40 > 1   Not enough hundreds
40)‾1‾3‾7‾ 40 > 13   Not enough tens
40)‾1‾3‾7‾ 40 < 137   **Enough ones**

Estimate.   137 ÷ 40
Think: 13 ÷ 4 = _?_
Try 3.

Divide the ones.	Multiply.	Subtract and compare.	Check.

$$\begin{array}{r} 3 \\ 40)\overline{1\,3\,7} \end{array}$$

$$\begin{array}{r} \times \quad 3 \\ 40)\overline{1\,3\,7} \\ 1\,2\,0 \end{array}$$

$$\begin{array}{r} 3 \text{ R } 17 \\ 40)\overline{1\,3\,7} \\ -1\,2\,0 \\ \hline 1\,7 \end{array}$$ 17 < 40

$$\begin{array}{r} 4\,0 \\ \times \quad 3 \\ \hline 1\,2\,0 \\ +\quad 1\,7 \\ \hline 1\,3\,7 \end{array}$$

Each student gets 3 marbles.
There are 17 marbles left over.

Copy and complete.

1. $$\begin{array}{r} 4 \text{ R } \underline{?} \\ 20)\overline{85} \\ -?? \\ \hline ? \end{array}$$

2. $$\begin{array}{r} 5 \text{ R } \underline{?} \\ 50)\overline{258} \\ -??? \\ \hline ? \end{array}$$

3. $$\begin{array}{r} ? \text{ R } \underline{?} \\ 20)\overline{166} \\ -??? \\ \hline ? \end{array}$$

4. $$\begin{array}{r} ? \text{ R } 35 \\ 80)\overline{675} \\ -??? \\ \hline ?? \end{array}$$

Divide and check.

5. $40\overline{)66}$ 6. $80\overline{)98}$ 7. $20\overline{)78}$ 8. $50\overline{)85}$ 9. $30\overline{)77}$

10. $70\overline{)356}$ 11. $90\overline{)548}$ 12. $80\overline{)567}$ 13. $40\overline{)283}$ 14. $50\overline{)454}$

15. $20\overline{)175}$ 16. $50\overline{)349}$ 17. $30\overline{)199}$ 18. $70\overline{)501}$ 19. $90\overline{)317}$

20. $430 \div 70$ 21. $312 \div 50$ 22. $250 \div 40$ 23. $197 \div 60$

24. $599 \div 80$ 25. $384 \div 60$ 26. $672 \div 90$ 27. $358 \div 40$

PROBLEM SOLVING

28. Mariah will give an equal number of pencils to each of 60 students. She has 122 pencils. At most, how many pencils can she give to each student? How many pencils will she have left?

29. Brendan is sorting 150 pieces of chalk into boxes. Each box holds 20 pieces of chalk. How many boxes can he fill? How many pieces of chalk will be in the box that is not full?

30. The media center ordered 495 booklets on different health topics. Each of 80 fourth graders will read the same number of booklets. At most, how many booklets will each fourth grader read?

31. Dionne is helping Mr. Rau to stack 256 magazines. They put 30 magazines into each stack. How many stacks of 30 magazines are there? How many magazines are in the last stack?

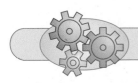 **Share Your Thinking**   Math Journal

32. In your Math Journal, tell how knowing division patterns helps you to divide by multiples of 10.

To estimate quotients with 2-digit divisors, think of nearby numbers that are compatible.

Estimate: 664 ÷ 24

664 ÷ 24

about 600   about 20

| When one number divides another evenly, the two numbers are **compatible.** |

$$\frac{30}{20)\overline{600}}$$
Think:

So 664 ÷ 24 is about 30.

Study these examples.

Estimate: 96 ÷ 31

Think: $\frac{3}{30)\overline{90}}$

So 96 ÷ 31 is about 3.

Estimate: $86.43 ÷ 38

Think: $\frac{\$\ 2.00}{40)\overline{\$80.00}}$

So $86.43 ÷ 38 is about $2.00.

Write the divisor and dividend you would use to estimate the quotient.

1. 63 ÷ 21   2. 89 ÷ 32   3. 78 ÷ 19   4. 47 ÷ 22

5. 58 ÷ 33   6. 67 ÷ 11   7. 81 ÷ 44   8. 92 ÷ 36

9. 594 ÷ 26   10. 905 ÷ 38   11. 825 ÷ 18   12. 652 ÷ 21

13. 452 ÷ 17   14. 395 ÷ 24   15. 6475 ÷ 36   16. 7959 ÷ 43

Estimate the quotient.

17. 95 ÷ 35   **18.** 87 ÷ 43   **19.** 62 ÷ 12   **20.** 59 ÷ 28

21. 49 ÷ 25   **22.** 81 ÷ 21   **23.** 91 ÷ 29   **24.** 67 ÷ 22

25. 644 ÷ 24   **26.** 841 ÷ 19   **27.** 919 ÷ 29   **28.** 592 ÷ 31

29. 799 ÷ 46   **30.** 652 ÷ 38   **31.** 401 ÷ 22   **32.** 423 ÷ 16

33. 8743 ÷ 36   **34.** 7921 ÷ 45   **35.** 5932 ÷ 24   **36.** 6417 ÷ 38

37. $59.75 ÷ 27   **38.** $4.21 ÷ 19   **39.** $91.39 ÷ 34   **40.** $5.56 ÷ 17

PROBLEM SOLVING

41. Last week 896 students came to the Folk Art Museum in buses. About the same number of students traveled on each of 28 buses. About how many students were there on each bus?

42. One class spent $75.05 for lunch in the museum cafeteria. There were 19 students in the class, and each student spent about the same amount. About how much money did each student spend for lunch at the museum?

Skills to Remember

Divide.

43. 7)58   **44.** 5)49   **45.** 3)36   **46.** 6)87

47. 3)745   **48.** 4)936   **49.** 8)277   **50.** 9)545

51. 2)$1.28   **52.** 7)$7.14   **53.** 9)$74.43   **54.** 6)$54.12

Two-Digit Dividends

Terry had 92 flower seeds. He planted 22 seeds in each of his flower baskets. How many flower baskets did Terry have? How many extra seeds were there?

To find how many flower baskets, divide: 92 ÷ 22 = ?

Think: $22\overline{)92}$ 22 > 9 Not enough tens
$22\overline{)92}$ 22 < 92 **Enough ones**

Estimate. 92 ÷ 22
Think: 9 ÷ 2 = ?
Try 4.

Divide the ones.	Multiply.	Subtract and compare.	Check.
4 $22\overline{)92}$	×4 $22\overline{)92}$ 88	4 R 4 $22\overline{)92}$ −88 4 4 < 22	22 × 4 88 + 4 92

Terry had 4 flower baskets.
There were 4 extra seeds.

Study this example.

$.03
$21\overline{)\$.63}$
−63
0

This zero must be written.

When you divide money, remember to write the dollar sign and the decimal point in the quotient.

Copy and complete.

1. $\overset{2}{24\overline{)48}}$
   $\underline{-48}$
   $?$

2. $\overset{3\ \text{R}\ ?}{25\overline{)96}}$
   $\underline{-??}$
   21

3. $\overset{2\ \text{R}\ ?}{44\overline{)89}}$
   $\underline{-88}$
   $?$

4. $\overset{?\ \text{R}\ ?}{21\overline{)94}}$
   $\underline{-84}$
   $??$

Divide and check.

5. $31\overline{)62}$
6. $23\overline{)46}$
7. $42\overline{)84}$
8. $33\overline{)99}$
9. $22\overline{)88}$

10. $32\overline{)64}$
11. $33\overline{)66}$
12. $41\overline{)82}$
13. $23\overline{)92}$
14. $43\overline{)86}$

15. $22\overline{)45}$
16. $42\overline{)88}$
17. $31\overline{)96}$
18. $22\overline{)73}$
19. $34\overline{)69}$

20. $21\overline{)98}$
21. $41\overline{)89}$
22. $32\overline{)99}$
23. $21\overline{)89}$
24. $42\overline{)70}$

25. $51\overline{)65}$
26. $22\overline{)98}$
27. $42\overline{)99}$
28. $21\overline{)96}$
29. $31\overline{)78}$

30. $22\overline{)\$.66}$
31. $45\overline{)\$.90}$
32. $31\overline{)\$.93}$
33. $26\overline{)\$.78}$
34. $33\overline{)\$.66}$

PROBLEM SOLVING

35. Chris set out 96 tomato plants in a vegetable garden. She placed 24 tomato plants in each row. How many rows of tomato plants were there?

36. Mike was putting 95 seed packets in a display. He wanted to put the same number of packets into each of 22 slots. How many packets could he have put into each slot? How many packets would he have had left over?

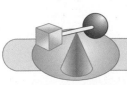

Challenge

Compare. Write $<$, $=$, or $>$. Estimate or find exact answers.

37. $64 \div 32 \ \underline{?} \ 72 \div 24$

38. $84 \div 21 \ \underline{?} \ 96 \div 32$

39. $72 \div 36 \ \underline{?} \ 96 \div 48$

40. $58 \div 29 \ \underline{?} \ 90 \div 45$

Three-Digit Dividends

There are 158 people who want to take a boat ride on the lake. How many trips with 45 passengers can the tour boat make? How many passengers will be on the last trip?

To find how many trips, divide: $158 \div 45 = $?

Think: $45\overline{)158}$ $45 > 1$ Not enough hundreds

$45\overline{)158}$ $45 > 15$ Not enough tens

$45\overline{)158}$ $45 < 158$ **Enough ones**

Estimate. $158 \div 45$
Think: $15 \div 4 = $?
Try 3.

Divide the ones.	Multiply.	Subtract and compare.	Check.
$\begin{array}{r} 3 \\ 4\,5\overline{)1\,5\,8} \end{array}$	$\begin{array}{r} \times\quad 3 \\ 4\,5\overline{)1\,5\,8} \\ 1\,3\,5 \end{array}$	$\begin{array}{r} 3 \text{ R }23 \\ 4\,5\overline{)1\,5\,8} \\ -1\,3\,5 \\ \hline \boxed{23 < 45} \rightarrow 2\,3 \end{array}$	$\begin{array}{r} 4\,5 \\ \times\quad 3 \\ \hline 1\,3\,5 \\ +\quad 2\,3 \\ \hline 1\,5\,8 \end{array}$

The tour boat can make 3 trips with 45 passengers. There will be 23 passengers on the last trip.

Study these examples.

$\begin{array}{r} 6 \\ 6\,3\overline{)3\,7\,8} \\ -\ 3\,7\,8 \\ \hline 0 \end{array}$

$\begin{array}{r} \$\ .0\,5 \\ 7\,2\overline{)\$3.6\,0} \\ -\ 3\,6\,0 \\ \hline 0 \end{array}$

Copy and complete.

$$\begin{array}{r} 6 \\ 51\overline{)306} \\ -??? \\ \hline ? \end{array}$$
1.

$$\begin{array}{r} 9\ R\ ? \\ 46\overline{)419} \\ -??? \\ \hline ? \end{array}$$
2.

$$\begin{array}{r} 4\ R\ ? \\ 83\overline{)392} \\ -??? \\ \hline ?? \end{array}$$
3.

$$\begin{array}{r} ?\ R\ ? \\ 64\overline{)533} \\ -512 \\ \hline ?? \end{array}$$
4.

Divide and check.

5. $22\overline{)176}$ **6.** $32\overline{)160}$ **7.** $43\overline{)258}$ **8.** $57\overline{)285}$ **9.** $74\overline{)222}$

10. $61\overline{)122}$ **11.** $95\overline{)380}$ **12.** $34\overline{)238}$ **13.** $62\overline{)248}$ **14.** $81\overline{)648}$

15. $42\overline{)146}$ **16.** $72\overline{)236}$ **17.** $51\overline{)489}$ **18.** $21\overline{)109}$ **19.** $91\overline{)476}$

20. $63\overline{)456}$ **21.** $54\overline{)237}$ **22.** $83\overline{)229}$ **23.** $75\overline{)474}$ **24.** $32\overline{)266}$

25. $67\overline{)\$1.34}$ **26.** $92\overline{)\$4.60}$ **27.** $71\overline{)\$6.39}$ **28.** $83\overline{)\$3.32}$ **29.** $44\overline{)\$3.08}$

PROBLEM SOLVING

30. Each ticket seller sold 82 tickets to a total of 574 passengers. How many ticket sellers were there?

31. The tickets came in rolls of 150. The ticket sellers sold 35 rolls of tickets. How many tickets did they sell?

32. Each tour bus can carry 64 passengers. What is the least number of buses needed for 595 passengers?

Calculator Activity

Divide. Then use a calculator to check.

$$\begin{array}{r} 4\ R\ 19 \\ 41\overline{)183} \\ -164 \\ \hline 19 \end{array}$$

[4] [×] [4] [1] [+] [1] [9] [=] | 183. |

33. $174 \div 42$ **34.** $218 \div 43$ **35.** $119 \div 23$

36. $213 \div 52$ **37.** $199 \div 62$ **38.** $326 \div 51$

12-6 Trial Quotients

Sometimes the quotient you try is too large. When this happens, you need to change the estimate.

Divide: 172 ÷ 27 = _?_

> Think: 27)‾1̅7̅2̅ 27 > 1; 27 > 17
> 27)‾1̅7̅2̅ 27 < 172 **Enough ones**

Estimate.	172 ÷ 27
	Think: 17 ÷ 2 = _?_ Try 8.

Divide the ones. Multiply.   **Subtract and compare.**   **Check.**

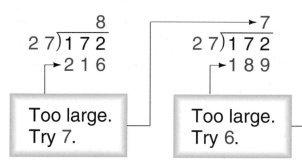   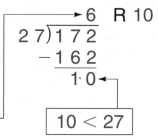

```
      8                    →7                  →6  R 10              2 7
2 7)1 7 2            2 7)1 7 2            2 7)1 7 2            ×     6
  →2 1 6              →1 8 9              −1 6 2                1 6 2
                                            1·0←             +   1 0
                                                                1 7 2
```

| Too large. Try 7. | Too large. Try 6. | 10 < 27 | |

Study this example.

Divide: 281 ÷ 35 = _?_

> Think: 35)‾2̅8̅1̅ 35 > 2
> 35 > 28
> 35)‾2̅8̅1̅ 35 < 281
> **Enough ones**

trial quotient

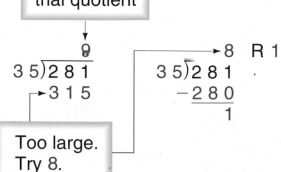

```
         9                      →8  R 1
3 5)2 8 1              3 5)2 8 1
  →3 1 5                −2 8 0
                            1
```

Too large. Try 8.

Estimate: 281 ÷ 35 = _?_
   Think: 28 ÷ 3 = _?_
       Try 9.

Copy and complete.

1.
$$27\overline{)82} \quad \frac{4}{108}$$
Try 3.
→
$$27\overline{)82} \quad \frac{3 \ R \ ?}{-81} \quad \frac{}{?}$$

2.
$$48\overline{)\$2.88} \quad \frac{\$ \ .07}{3 \ 36}$$
Try \$.06.
→
$$48\overline{)\$2.88} \quad \frac{\$ \ .06}{-? \ ??} \quad \frac{}{?}$$

3.
$$64\overline{)310} \quad \frac{5}{???}$$
Try 4.
→
$$64\overline{)310} \quad \frac{? \ R \ ?}{-???} \quad \frac{}{54}$$

4.
$$86\overline{)657} \quad \frac{8}{688}$$
Try ? .
→
$$86\overline{)657} \quad \frac{? \ R \ ?}{-???} \quad \frac{}{??}$$

Divide.

5. $27\overline{)52}$ 6. $36\overline{)91}$ 7. $45\overline{)82}$

8. $35\overline{)124}$ 9. $48\overline{)165}$ 10. $54\overline{)260}$ 11. $79\overline{)221}$

12. $66\overline{)542}$ 13. $94\overline{)638}$ 14. $87\overline{)569}$ 15. $49\overline{)202}$

16. $27\overline{)124}$ 17. $39\overline{)277}$ 18. $76\overline{)571}$ 19. $99\overline{)828}$

20. $28\overline{)\$.84}$ 21. $59\overline{)\$3.54}$ 22. $78\overline{)\$6.24}$ 23. $69\overline{)\$5.52}$

PROBLEM SOLVING

24. Mr. Dean has signed up 180 students for a field trip to the zoo. Each bus can carry 36 students and 4 teachers. How many buses are needed for the field trip?

25. There are 115 monkeys at the zoo. No more than 25 monkeys can be in each environment. What is the least number of environments there could be at the zoo?

26. There are 325 animal spoons in the store. Each display holds 48 spoons. How many displays are there? How many extra animal spoons are there?

Greater Quotients

Divide: 995 ÷ 22 = _?_

Think: 22)995   22 > 9   Not enough hundreds
22)995   22 < 99   **Enough tens**

Estimate.   995 ÷ 22
Think: 9 ÷ 2 = _?_
Try 4.

Divide the tens.	Multiply.	Subtract and compare.	Bring down the ones.

```
      4
22)995
```

```
    ×  4
22)995
    →88
```

```
      4
22)995
   -88
  →  11
```
11 < 22

```
      4
22)995
   -88↓
  1 1 5←
```
partial dividend

Repeat the steps.

Estimate.   115 ÷ 22
Think: 11 ÷ 2 = _?_
Try 5.

Divide the ones.	Multiply.	Subtract and compare.	Check.

```
    4 5
22)995
  -88↓
  115
```

```
×     4 5
22)995
  -88↓
  115
  →110
```

```
      4 5  R 5
22)995
   -88↓
   115
  -110
      5
```
5 < 22

```
    4 5
  ×  2 2
     9 0
  +9 0 0
   9 9 0
  +    5
   9 9 5
```

Copy and complete.

$$\begin{array}{r} 26 \\ 34\overline{)884} \\ -\;?? \\ \hline 204 \\ -\;??? \\ \hline ? \end{array}$$
1.

$$\begin{array}{r} 2?\ \ \text{R}\ ? \\ 42\overline{)890} \\ -\;?? \\ \hline 50 \\ -\;?? \\ \hline ? \end{array}$$
2.

$$\begin{array}{r} ?9 \\ 26\overline{)489} \\ -\;?? \\ \hline 229 \\ \rightarrow 234 \\ \hline \boxed{\text{Try}\ ?\ .} \end{array}$$
3.

$$\begin{array}{r} ??\ \ \text{R}\ ? \\ 26\overline{)489} \\ -\;?? \\ \hline 229 \\ -\;208 \\ \hline ?? \end{array}$$

Divide and check.

4. $40\overline{)840}$ **5.** $70\overline{)920}$ **6.** $60\overline{)954}$ **7.** $30\overline{)629}$

8. $42\overline{)882}$ **9.** $23\overline{)552}$ **10.** $31\overline{)899}$ **11.** $45\overline{)630}$

12. $35\overline{)721}$ **13.** $45\overline{)678}$ **14.** $59\overline{)620}$ **15.** $51\overline{)801}$

16. $61\overline{)827}$ **17.** $82\overline{)963}$ **18.** $35\overline{)745}$ **19.** $52\overline{)856}$

20. $938 \div 21$ **21.** $875 \div 33$ **22.** $882 \div 41$

23. $764 \div 65$ **24.** $770 \div 42$ **25.** $900 \div 64$

26. $\$7.04 \div 32$ **27.** $\$5.52 \div 24$ **28.** $\$6.82 \div 22$

PROBLEM SOLVING

29. Carey picked 865 pears. He put 42 pears into each box. How many boxes did Carey fill? How many pears were left over?

30. Wendell sold one peach to each of 33 customers for a total of $8.25. What was the cost of each peach?

31. Jo picked 535 apples. She put the same number of apples into each of 24 baskets. What is the greatest number of apples Jo could have put into each basket? How many apples would be left over?

12-8 Teens as Divisors

You may have to change your estimate
more than once when the divisor
is a number from 11 through 19.

Divide: 926 ÷ 15 = _?_

Think: $15\overline{)926}$ 15 > 9 Not enough hundreds
$15\overline{)926}$ 15 < 92 **Enough tens**

| Estimate. | 926 ÷ 15 =
Think: 9 ÷ 1 = _?_
Try 9. |

Divide the tens. Multiply.			**Subtract and compare.**

$$\begin{array}{r} 9 \\ 15\overline{)926} \\ 135 \end{array} \longrightarrow \begin{array}{r} 8 \\ 15\overline{)926} \\ 120 \end{array} \longrightarrow \begin{array}{r} 7 \\ 15\overline{)926} \\ 105 \end{array} \longrightarrow \begin{array}{r} 6 \\ 15\overline{)926} \\ -90 \\ \hline 2 \end{array}$$

| Too large. Try 8. | Too large. Try 7. | Too large. Try 6. | 2 < 15 |

Bring down the ones.	**Divide the ones. Multiply.**	**Subtract and compare.**	**Check.**

$$\begin{array}{r} 6 \\ 15\overline{)926} \\ -90\downarrow \\ \hline 26 \end{array} \qquad \begin{array}{r} 61 \\ 15\overline{)926} \\ -90\downarrow \\ \hline 26 \\ 15 \end{array} \qquad \begin{array}{r} 61 \ \ R\,11 \\ 15\overline{)926} \\ -90\downarrow \\ \hline 26 \\ -15 \\ \hline 11 \end{array} \qquad \begin{array}{r} 61 \\ \times\ 15 \\ \hline 305 \\ +610 \\ \hline 915 \\ +\ \ 11 \\ \hline 926 \end{array}$$

| | | 11 < 15 | |

Copy and complete.

$$\begin{array}{r} 8 \\ 12\overline{)96} \\ -?? \\ \hline ? \end{array}$$
1.

$$\begin{array}{r} 9 \text{ R } \underline{?} \\ 14\overline{)127} \\ -??? \\ \hline ? \end{array}$$
2.

$$\begin{array}{r} 2? \\ 17\overline{)357} \\ -?? \\ \hline ?? \\ -?? \\ \hline ? \end{array}$$
3.

$$\begin{array}{r} ?8 \text{ R } \underline{?} \\ 19\overline{)536} \\ -?? \\ \hline ??? \\ -??? \\ \hline ? \end{array}$$
4.

Divide and check.

5. $15\overline{)95}$ **6.** $12\overline{)92}$ **7.** $18\overline{)79}$ **8.** $16\overline{)68}$ **9.** $11\overline{)74}$

10. $14\overline{)112}$ **11.** $16\overline{)128}$ **12.** $17\overline{)139}$ **13.** $14\overline{)127}$ **14.** $15\overline{)146}$

15. $13\overline{)641}$ **16.** $19\overline{)309}$ **17.** $14\overline{)456}$ **18.** $17\overline{)195}$ **19.** $12\overline{)252}$

20. $11\overline{)316}$ **21.** $18\overline{)723}$ **22.** $16\overline{)522}$ **23.** $15\overline{)187}$ **24.** $19\overline{)799}$

25. $17\overline{)435}$ **26.** $12\overline{)651}$ **27.** $18\overline{)563}$ **28.** $13\overline{)286}$ **29.** $16\overline{)344}$

30. $11\overline{)\$1.21}$ **31.** $15\overline{)\$1.80}$ **32.** $19\overline{)\$3.99}$ **33.** $14\overline{)\$4.90}$ **34.** $12\overline{)\$5.04}$

PROBLEM SOLVING

35. The divisor is 13.
The quotient is 64.
The remainder is 10.
What is the dividend?

36. The quotient is 55.
The divisor is 17.
The remainder is 7.
What is the dividend?

37. The divisor is 16.
The quotient is 36.
What is the dividend?

38. The divisor is 11.
The dividend is 852
How many digits will the
quotient contain?

39. The dividend is 747. The divisor is 16.
What is the first digit of the quotient?
the second digit? List the steps you took
to find the answers in your Math Journal.

Math
Journal

12-9 Four-Digit Dividends

Divide: 5481 ÷ 64 = __?__

Think: 64)‾5481‾ 64 > 5   Not enough thousands
        64)‾5481‾ 64 > 54   Not enough hundreds
        64)‾5481‾ 64 < 548 **Enough tens**

Estimate.   5481 ÷ 64
               Think: 54 ÷ 6 = __?__
               Try 9.

Divide the tens. Multiply.		Subtract and compare.	Bring down the ones.

```
      9                 8              8               8
64)5 4 8 1        64)5 4 8 1      64)5 4 8 1       64)5 4 8 1
  5 7 6              5 1 2          -5 1 2           -5 1 2
                                      3 6              3 6 1
```

Too large. Try 8.

36 < 64

Repeat the steps.

Divide the ones. Multiply.		Subtract and compare.	Check.

```
     8 6             8 5             8 5 R 41          8 5
64)5 4 8 1       64)5 4 8 1      64)5 4 8 1         ×   6 4
 -5 1 2            5 1 2           -5 1 2              3 4 0
   3 6 1             3 6 1           3 6 1          + 5 1 0 0
   3 8 4             3 2 0          -3 2 0            5 4 4 0
                                        4 1          +    4 1
                                                      5 4 8 1
```

Too large. Try 5.

41 < 64

398

Copy and complete.

$$
\begin{array}{r}
24 \text{ R } \underline{\ ?\ } \\
51\overline{)1273} \\
-\ 102 \\
\hline
253 \\
-\ ??? \\
\hline
?
\end{array}
\qquad
\begin{array}{r}
46 \text{ R } \underline{\ ?\ } \\
24\overline{)1107} \\
-\ 96 \\
\hline
147 \\
-\ ??? \\
\hline
?
\end{array}
\qquad
\begin{array}{r}
?? \text{ R } \underline{\ ?\ } \\
33\overline{)2179} \\
-\ 198 \\
\hline
199 \\
-\ ??? \\
\hline
?
\end{array}
$$

1. (first) 2. (second) 3. (third)

Divide and check.

4. $40\overline{)2459}$ 5. $80\overline{)5346}$ 6. $70\overline{)6842}$ 7. $50\overline{)4779}$

8. $34\overline{)1180}$ 9. $52\overline{)3115}$ 10. $44\overline{)2106}$ 11. $63\overline{)4914}$

12. $72\overline{)4594}$ 13. $96\overline{)7128}$ 14. $22\overline{)1550}$ 15. $84\overline{)5285}$

16. $64\overline{)5084}$ 17. $48\overline{)4128}$ 18. $38\overline{)2242}$ 19. $55\overline{)3226}$

20. $22\overline{)1810}$ 21. $73\overline{)5808}$ 22. $14\overline{)1248}$ 23. $18\overline{)1200}$

24. $92\overline{)\$21.16}$ 25. $51\overline{)\$16.83}$ 26. $88\overline{)\$22.00}$ 27. $67\overline{)\$56.95}$

PROBLEM SOLVING

28. Each of 45 students bought a copy of *The Great Dinosaurs.* They paid a total of $42.75. How much did one copy of *The Great Dinosaurs* cost?

29. There are 1565 books at the Elmford book fair. If each table can hold 55 books, what is the least number of tables needed for the fair?

30. Each homeroom in Elmford School can seat 36 students. There are 1256 students in the school. What is the least number of homerooms needed for all the students?

31. Students bought 32 copies of *Amazing Science.* Each copy cost $2.98. What was the total amount the students spent to buy *Amazing Science?*

Zero in the Quotient

Divide: 2865 ÷ 14 = ?

Think: $14\overline{)2865}$ 14 > 2 Not enough thousands
$14\overline{)2865}$ 14 < 28 **Enough hundreds**

Estimate. 2865 ÷ 14
Think: 2 ÷ 1 = ?
Try 2.

Divide the hundreds.	Divide the tens.		Divide the ones.

Divide the hundreds.

```
      2
14)2 8 6 5
  -2 8↓
    0 6
```

Divide the tens.

```
      2 0←
14)2 8 6 5
  -2 8↓
      6 |←
    -0↓
      6 5
```

14 > 6 **Not enough tens** Write 0 in the tens place. Bring down the ones.

Divide the ones.

```
      2 0 4  R 9
14)2 8 6 5
  -2 8↓
      6 |
    -0↓
      6 5
    -5 6
        9
```

Check. 14 × 204 = 2856 ⟶ 2856 + 9 = 2865

Study these examples.

```
      3 0 0  R 5
21)6 3 0 5
  -6 3↓
      0 |
    -0 |
        5
      -0
        5
```

```
      $3.0 6
26)$7 9.5 6
   7 8↓
     1 5
   -  0↓
     1 5 6
   -1 5 6
         0
```

Copy and complete.

        60
1. 35)2100
    −210
       ?

       102
2. 57)5814
    −57
     11
    − ?
    1?4
   −???

    $ 2.0?
3. 18)$36.90
    −36
     9
   − ?
    ?0
   −??

    3?? R ?
4. 24)7231
    −72
     3
   − ?
   ?1
  −??
    ?

Divide and check.

5. 45)3600

6. 32)1600

7. 24)2166

8. 56)3930

9. 17)6800

10. 25)5000

11. 41)8214

12. 33)9927

13. 21)2247

14. 19)5852

15. 32)9856

16. 46)9246

17. 15)9097

18. 51)5576

19. 28)8538

20. 34)7068

21. 43)8735

22. 13)9175

23. 62)6736

24. 74)7904

25. 18)$37.44

26. 23)$70.61

27. 85)$92.65

28. 56)$60.48

PROBLEM SOLVING

29. Damon bought a 12-yard length of cloth for $48.72. What was the cost per yard?

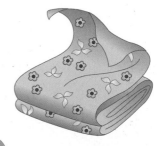

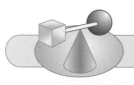

Challenge

Find the quotient and any remainder.

30. 22)22,154

31. 32)64,128

32. 17)61,085

33. 42)84,378

34. 51)51,408

35. 24)96,088

TECHNOLOGY

Calculating with Fractions

You can use a fraction calculator to express an improper fraction as a mixed number in simplest form.

▶ To express $\frac{18}{4}$ as a mixed number in simplest form:

Changes an improper fraction to a mixed number

This key combination simplifies a fraction

Press these keys ▸ 1 8 / 4 Ab/c Simp =

Display ▸ 4⌐2/4 4⌐1/2

So $\frac{18}{4} = 4\frac{2}{4} = 4\frac{1}{2}$.

Simplest form

▶ Find the sum in simplest form.
Add: $\frac{1}{3} + \frac{1}{6} = \underline{?}$

Press these keys ▸ 1 / 3 + 1 / 6 = Simp =

Display ▸ 3/6 1/2

So $\frac{1}{3} + \frac{1}{6} = \frac{3}{6} = \frac{1}{2}$.

Lowest terms

▶ Find the difference in simplest form.
Subtract: $4\frac{3}{5} - 3\frac{1}{5} = \underline{?}$

Press these keys ▸ 4 Unit 3 / 5 − 3 Unit 1 / 5 =

Used to enter the whole number

Display ▸ 1⌐2/5

So $4\frac{3}{5} - 3\frac{1}{5} = 1\frac{2}{5}$.

402

Write as a mixed number in simplest form. Use a calculator.

1. $\frac{37}{3}$ 2. $\frac{99}{7}$ 3. $\frac{82}{6}$ 4. $\frac{46}{4}$ 5. $\frac{57}{8}$ 6. $\frac{71}{3}$

7. $\frac{51}{9}$ 8. $\frac{28}{6}$ 9. $\frac{56}{18}$ 10. $\frac{33}{9}$ 11. $\frac{99}{27}$ 12. $\frac{20}{12}$

Use the Simp **=** **keys to tell if the fractions are equivalent. Write *Yes* or *No*.**

13. $\frac{6}{8}, \frac{3}{4}$ 14. $\frac{9}{12}, \frac{3}{4}$ 15. $\frac{18}{30}, \frac{6}{15}$ 16. $\frac{32}{64}, \frac{4}{6}$

17. $\frac{3}{8}, \frac{27}{72}$ 18. $\frac{16}{28}, \frac{2}{3}$ 19. $\frac{3}{8}, \frac{45}{120}$ 20. $\frac{12}{15}, \frac{144}{180}$

Add or subtract. Write your answer in simplest form.

21. $\frac{4}{7} + \frac{2}{7}$ 22. $\frac{5}{9} + \frac{3}{9}$ 23. $\frac{7}{10} - \frac{2}{10}$ 24. $\frac{9}{12} - \frac{3}{12}$ 25. $\frac{1}{4} + \frac{3}{8}$

26. $\frac{7}{10} - \frac{1}{2}$ 27. $1\frac{1}{4} + 4\frac{1}{4}$ 28. $9\frac{7}{12} - 5\frac{1}{12}$ 29. $28\frac{3}{8} + 52\frac{3}{8}$ 30. $41\frac{1}{3} - 9\frac{1}{3}$

Use a calculator.

$\frac{3}{4}$ of 32 = **?**

| Press these keys | → | 3 2 ÷ 4 = × 3 = |

| Display | → | 24 |

So $\frac{3}{4}$ of 32 = 24.

31. $\frac{1}{8}$ of 72 32. $\frac{2}{5}$ of 25 33. $\frac{3}{7}$ of 49 34. $\frac{4}{6}$ of 54

12-12 Problem Solving: Hidden Information

Problem: Andre needs 14 ft of rope to make a plant hanger. He buys a spool that has 175 in. of rope. Does he have enough rope?

1 IMAGINE Put yourself in the problem.

2 NAME *Facts:* Andre needs 14 ft of rope.
The spool has 175 in.

Question: Does he have enough rope?

3 THINK To solve the problem, you need information that is not stated.

Remember: 12 in. = 1 ft

To find whether he has enough rope:
First divide. 175 in. ÷ 12 in. = ? ft
Then compare. 14 ft ? ? ft

4 COMPUTE

```
      1 4   R 7
1 2)1 7 5
   −1 2
     5 5
    −4 8
       7
```

Think: 14 R 7 means
14 ft 7 in.

Compare. 14 ft < 14 ft 7 in.
Yes, Andre has enough rope.

5 CHECK Multiply and add to check division.

```
     14      →168
   × 12      +  7
     28       175
  + 14
    168
```

The answer checks.

Look for hidden information to help you solve each problem.

1. Winnie made 136 yogurt sundaes. How many quarts of yogurt were used for these pint-sized sundaes?

IMAGINE   Create a mental picture.

NAME   *Facts:*   136 yogurt sundaes
                        1 pint for each sundae

            Question: How many quarts of
                        yogurt were used?

THINK   Is there hidden information
            in the problem?   Yes

> Remember: 2 pints = 1 quart

Since there were 136 sundaes, 136 pints of yogurt were used.
To find out how many quarts were used, divide: 136 ÷ 2 = _?_

COMPUTE ⟶ **CHECK**

2. A nature video is 148 minutes long. Can Saundra watch the video in $2\frac{1}{2}$ hours?

3. Paulo earns the same amount every week for delivering newspapers. He earns $1196 a year. How much does he earn each week?

4. In one full day a satellite transmits 9600 messages. About how many messages can it transmit in 1 hour?

5. Byron deposits $30 in quarters in the bank. How many coins does he turn in at the bank?

Solve each problem and explain the method you used.

1. A store displays 252 different wrapping papers equally on 14 racks. How many papers are on each rack?

2. There are 180 green stripes on a 20-inch sheet of wrapping paper. How many stripes are there per inch?

3. Wrap It Up shop sells 10 ft of ribbon for $1.60. How much does 1 ft of ribbon cost?

4. A pack of 24 party invitations costs $1.98. About how much does each invitation cost?

5. Wrap It Up's window is decorated with ribbon. The manager cuts as many 18-in. strips as he can from a 100-in. roll of ribbon. How much ribbon is left on the roll?

6. Crepe paper streamers are sold in 28-ft rolls. To decorate a gym for a party, 150 ft of crepe paper is needed. How many rolls should be bought?

7. The store sells ready-made bows in packs of 15. Each pack costs $4.20. How much does each bow cost?

8. There are 2892 purple dots on a 12-ft roll of wrapping paper. About how many dots are on 1 ft of paper?

9. A complete party package costs $75.90. If 22 friends share the cost, how much will each friend spend?

Choose a strategy from the list or use another strategy you know to solve each problem.

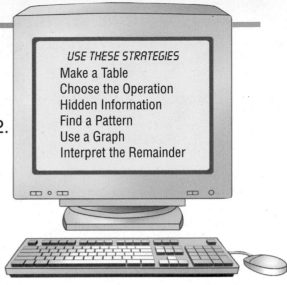

USE THESE STRATEGIES
Make a Table
Choose the Operation
Hidden Information
Find a Pattern
Use a Graph
Interpret the Remainder

10. Birthday candles are sold in packs of 12. How many packs should you buy if you need to put 35 candles on a cake?

11. One wrapping paper shows cats and dogs in 8 rows, following this pattern: 2 cats, 4 dogs, 3 cats, 5 dogs, 4 cats. What is the pattern for the last 3 rows?

12. A 48-in. sheet of safari pattern paper shows 20 lions. About how many lions show on 1 ft of the paper?

13. Wrap It Up's window is filled with 350 balloons. They came in packs of 24. How many packs were used?

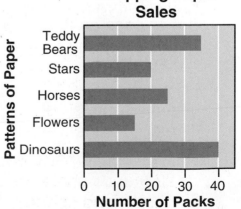

14. Helium balloons are $11.28 a dozen. How much is one helium balloon?

15. On Mondays Wrap It Up gives a discount of 10¢ for every dollar spent. Ted spent $14 on Friday. How much would he have saved if he had shopped on Monday?

16. Al spent $16.28. Julia spent 4 times as much. How much money did she spend?

Use the graph for problems 17 and 18.

17. Of which patterns did the store sell more than 20 but fewer than 40 packs?

18. Of which pattern did the store sell about 15 fewer packs than it did for teddy bears?

Wrapping Paper Sales

407

Divide and check. *(See pp. 382–385.)*

1. 20)66   **2.** 30)75   **3.** 70)351   **4.** 50)299   **5.** 90)319

6. 395 ÷ 70   **7.** 312 ÷ 70   **8.** 139 ÷ 20   **9.** 256 ÷ 20

Estimate the quotient. *(See pp. 386–387.)*

10. 48 ÷ 20   **11.** 82 ÷ 39   **12.** 99 ÷ 47   **13.** 597 ÷ 19

14. 4011 ÷ 38   **15.** $69.03 ÷ 9   **16.** 7482 ÷ 47   **17.** $5.79 ÷ 19

Divide and check. *(See pp. 388–401.)*

18. 22)66   **19.** 45)90   **20.** 31)$.93   **21.** 13)$.65   **22.** 17)85

23. 56)280   **24.** 37)222   **25.** 76)342   **26.** 42)$3.36   **27.** 75)$97.50

28. 17)435   **29.** 14)456   **30.** 13)286   **31.** 15)$1.80   **32.** 12)$5.04

33. 34)1180   **34.** 96)7128   **35.** 14)1248   **36.** 67)5690   **37.** 73)5808

38. 17)6800   **39.** 19)5852   **40.** 13)9175   **41.** 28)8536   **42.** 74)7904

PROBLEM SOLVING *(See pp. 406–407.)*

43. There are 257 sheets of lined paper for 24 students to share equally. How many sheets of paper does each student get? How many sheets are left over?

44. The rental of a school bus for a field trip is $56.00. Thirty-two students are going on the trip. How much money does each student have to pay?

45. The dividend is 80.
The quotient is 4.
What is the divisor?

46. The quotient is 8.
The divisor is 31.
What is the dividend?

(See Still More Practice, p. 471.)

LOGIC

You have to read carefully to be sure you do not draw a false conclusion from true statements.

Read these statements and conclusions. All the statements are true.

statements: All dogs have ears.
Sparky is a dog.

conclusion: Sparky has ears.   TRUE

statements: All dogs have ears.
My cat has ears.

conclusion: My cat is a dog.   FALSE

Read the true statements carefully. Then write *true* or *false* for each conclusion.

1. All ducks have feathers.
A chicken has feathers.
A chicken is a duck.

2. All fish can swim.
A salmon is a fish.
A salmon can swim.

3. Ralph is a 4th-grade boy.
All the 4th-grade boys
wore sneakers on Monday.
Ralph wore sneakers
on Monday.

4. All the 4th-grade boys wore
sneakers on Monday.
Maria wore sneakers
on Monday.
Maria is a 4th-grade boy.

5. All triangles are polygons.
A pentagon is a polygon.
A pentagon is a triangle.

6. All squares are parallelograms.
All rectangles are parallelograms.
All rectangles are squares.

7. Triangle *A* has one right
angle. All triangles with
one right angle are right
triangles.
Triangle *A* is a right triangle.

Check Your Mastery

Performance Assessment

Use these dividends and divisors to make division exercises for the following:

Dividends		Divisors	
1440	2820	40	20

1. A quotient between 60 and 100 with no remainder.

2. A quotient less than 100 with a remainder of 20.

3. A quotient greater than 100.

Estimate the quotient.

4. $58 \div 33$ 5. $825 \div 18$ 6. $395 \div 24$ 7. $7959 \div 43$

8. $29\overline{)919}$ 9. $45\overline{)7921}$ 10. $27\overline{)5975}$ 11. $19\overline{)841}$ 12. $38\overline{)6417}$

Divide and check.

13. $33\overline{)66}$ 14. $31\overline{)96}$ 15. $24\overline{)89}$ 16. $41\overline{)\$.82}$ 17. $23\overline{)\$.92}$

18. $42\overline{)146}$ 19. $34\overline{)\$2.38}$ 20. $83\overline{)\$2.49}$ 21. $24\overline{)109}$ 22. $54\overline{)237}$

23. $66\overline{)542}$ 24. $76\overline{)\$5.32}$ 25. $49\overline{)202}$ 26. $31\overline{)\$1.24}$ 27. $99\overline{)828}$

28. $45\overline{)678}$ 29. $35\overline{)745}$ 30. $42\overline{)770}$ 31. $25\overline{)\$9.00}$ 32. $41\overline{)8565}$

33. $18\overline{)5546}$ 34. $15\overline{)\$1.80}$ 35. $15\overline{)187}$ 36. $88\overline{)9504}$ 37. $73\overline{)\$79.57}$

PROBLEM SOLVING *Use a strategy you have learned.*

38. A bus seats 52 passengers. How many buses are needed to carry 795 passengers from the hotel to the state fair?

39. In one week, a hospital's cafeteria serves 5325 meals. About how many meals are served a day?

Math Class

She talks about the decimal point,
The reasons why—
But on the window, buzzing free,
A fly

With two red eyes
Moves slowly up the pane.
She moves the decimal one place left
And then again

The fly moves up
And up, practiced and slow.
What I have learned of decimal points
Flies know.

Myra Cohn Livingston

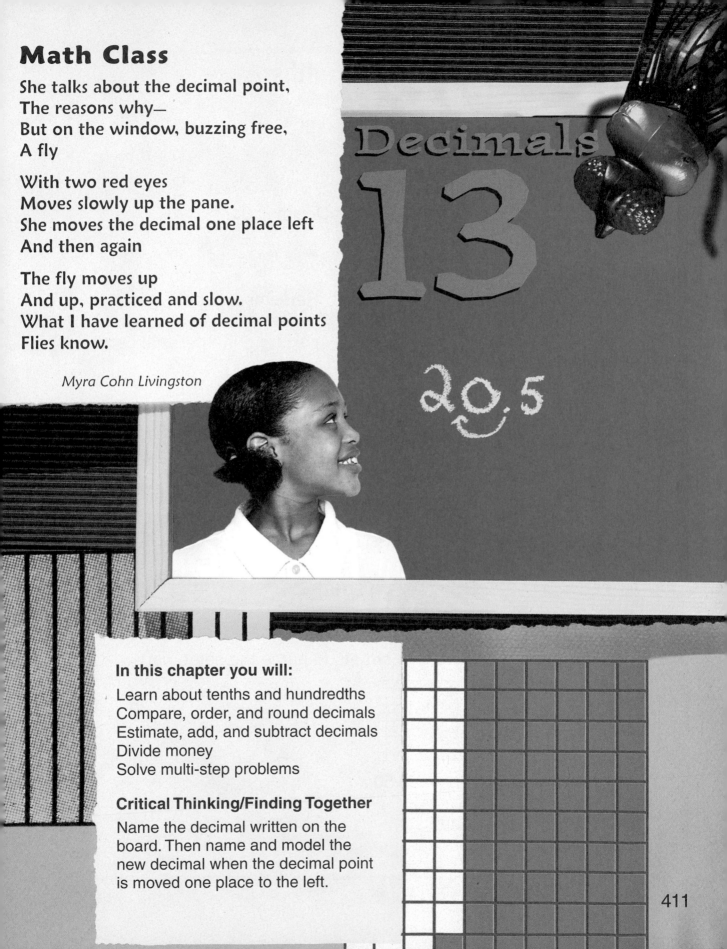

Decimals

13

20.5

In this chapter you will:

Learn about tenths and hundredths
Compare, order, and round decimals
Estimate, add, and subtract decimals
Divide money
Solve multi-step problems

Critical Thinking/Finding Together

Name the decimal written on the
board. Then name and model the
new decimal when the decimal point
is moved one place to the left.

411

13-1 Tenths and Hundredths

▶ You can write tenths as a fraction or as a **decimal**.

Fraction: $\frac{3}{10}$

decimal point

Decimal: 0.3

This 0 means *no ones*.

Read: three tenths

Remember: The *decimal point* separates the whole number part from the decimal part.

▶ You can write hundredths as a fraction or as a decimal.

Fraction: $\frac{45}{100}$

Fraction: $\frac{5}{100}$

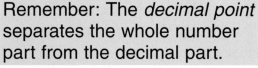
decimal point

Decimal: 0.45

0.05

This 0 means *no tenths*.

Read: forty-five hundredths

five hundredths

▶ You can write **equivalent decimals** to name the same part.

$\frac{5}{10}$ or 0.5

=

$\frac{50}{100}$ or 0.50

Think:
$\frac{5}{10} = \frac{5 \times 10}{10 \times 10} = \frac{50}{100}$

0.5 and 0.50 name the same part. They are equivalent decimals.

Write as a fraction. Then write as a decimal.

1.

2.

3.

4.

5.

6.

Write the word name for each decimal.

7. 0.1 **8.** 0.37 **9.** 0.8 **10.** 0.60 **11.** 0.09

Are the decimals equivalent? Write *yes* or *no*.

12. 0.6; 0.60 **13.** 0.10; 0.1 **14.** 0.7; 0.07 **15.** 0.02; 0.2

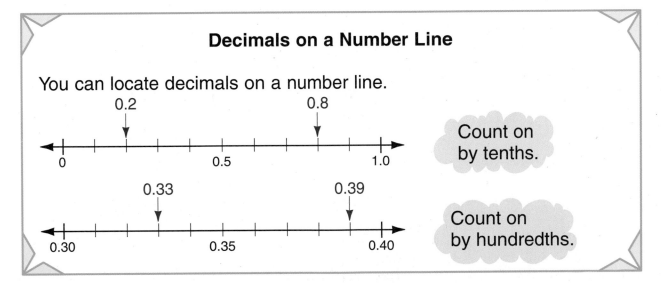

Decimals on a Number Line

You can locate decimals on a number line.

Count on by tenths.

Count on by hundredths.

To what decimal is each arrow pointing?

16.
17.
18.
19.

413

Decimals Greater Than One

You can write a mixed number or a whole number as a decimal.

Mixed Number: $3\frac{25}{100}$

This 3 means *3 ones*.

Decimal: 3.25

decimal point

Read: three and twenty-five hundredths.

Study these examples.

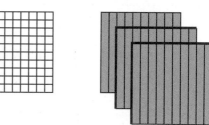

$2\frac{1}{10} = 2.1$

two and one tenth

$1\frac{3}{100} = 1.03$

one and three hundredths

$3 = 3.0$

three

Write as a mixed number. Then write as a decimal.

1.

2.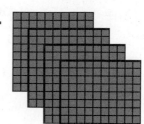

Write each as a decimal.
Then model exercises 3–10 using decimal squares.

3. $5\frac{3}{10}$

4. $8\frac{7}{10}$

5. $4\frac{2}{10}$

6. $7\frac{5}{10}$

7. $9\frac{21}{100}$

8. 10

9. $3\frac{6}{100}$

10. $2\frac{1}{100}$

11. $24\frac{6}{10}$

12. $97\frac{17}{100}$

13. 50

14. $100\frac{9}{100}$

15. three and eight tenths

16. nine and nineteen hundredths

17. twelve and one hundredth

18. one hundred fifty-seven

Write the word name for each decimal.

19. 6.4

20. 4.30

21. 8.08

22. 5.00

23. 60.02

To what decimal is each arrow pointing?

24.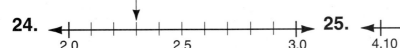

25.

26.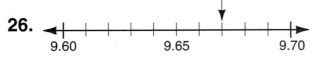

27.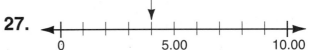

PROBLEM SOLVING

28. Keisha hiked five and two tenths miles. Write this as a decimal.

29. Manny ran in a race that was three and eighty hundredths miles. Write this as a decimal.

30. Seo scored nine and nine hundredths on the balance beam. Write this as a decimal.

31. Ivan jogged 12.3 miles. Write this as a mixed number. Then write it in words.

32. Write a decimal that is:
  a. between 3 and 4.
  c. between 11 and 12.
  b. less than 9 and greater than 8.
  d. greater than 1 and less than 2.

Decimal Place Value

▶ The value of a digit in a decimal depends on its place in the decimal.

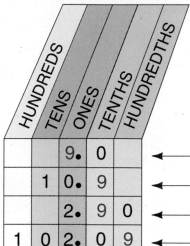

Look at the decimals in the place-value chart.

The value of the digit 9 in each decimal is:

← 9 ones, or 9.

← 9 tenths, or 0.9

← 9 tenths, or 0.9

← 9 hundredths, or 0.09

← 9 hundredths, or 0.09

▶ You can write decimals in standard form or in expanded form.

Standard Form	Expanded Form
24.5	20 + 4 + 0.5
3.60	3 + 0.6
961.04	900 + 60 + 1 + 0.04
87.37	80 + 7 + 0.3 + 0.07

**Write the place of the red digit.
Then write its value.**

1. 2.31   **2.** 0.49   **3.** 62.75   **4.** 11.38   **5.** 129.04

6. 21.59   **7.** 5.04   **8.** 30.03   **9.** 25.15   **10.** 53.96

11. 4.10   **12.** 8.56   **13.** 509.88   **14.** 9.14   **15.** 18.03

Write in expanded form.

16. 0.23   **17.** 4.07   **18.** 9.94   **19.** 1.8   **20.** 205.6

21. 91.05   **22.** 30.8   **23.** 84.73   **24.** 670.01   **25.** 700.60

Write in standard form. Then write each word name.

26. 40 + 2 + 0.9 + 0.07   **27.** 500 + 70 + 5 + 0.2 + 0.06

28. 6 + 0.9 + 0.01   **29.** 8 + 0.2

30. 0.4 + 0.06   **31.** 0.1 + 0.01

32. 800 + 5 + 0.03   **33.** 300 + 20 + 0.8

34. 100 + 0.5 + 0.07   **35.** 50 + 0.3 + 0.04

Write in words.

36. 200 + 30 + 0.6   **37.** 50 + 7 + 0.01

38. 90 + 0.1 + 0.02   **39.** 400 + 9 + 0.09

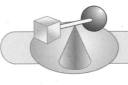

Challenge

The third place to the right of the decimal point is the **thousandths place**.

Ones	Tenths	Hundredths	Thousandths
0.	0	0	2

$$\frac{2}{1000} = 0.002$$

Read: two thousandths

Write as a decimal.

40. $\frac{6}{1000}$ **41.** $\frac{1}{1000}$ **42.** $\frac{9}{1000}$ **43.** $\frac{12}{1000}$ **44.** $\frac{25}{1000}$

45. $\frac{99}{1000}$ **46.** $\frac{702}{1000}$ **47.** $\frac{811}{1000}$ **48.** $\frac{450}{1000}$ **49.** $\frac{940}{1000}$

13-4 Comparing Decimals

Sheena	2.5 km
Noemi	2.58 km

Who rode the greater distance?

To find who rode the greater distance,
compare: 2.5 _?_ 2.58

 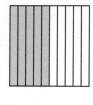   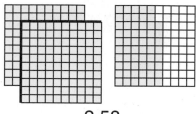

2.5   2.58

To compare decimals:

- Align the digits
  by their place value.

  .2.5
  2.58

- Start at the left. Compare
  the digits in the greatest place.

  2.5   2 = 2
  2.58

- Keep comparing digits until
  you find two digits that
  are *not* the same.

  2.5   5 = 5
  2.58

Think:
0.5 = 0.50

  2.50   0 < 8
  2.58

So 2.5 < 2.58

Noemi rode the greater distance, 2.58 km.

Study these examples.

$.96 _?_ $.92   42.7 ? 4.7

$.96   9 = 9   42.7
$.92   6 > 2   4.7

There are no tens
in 4.7.

So $.96 > $.92   4 > 0

So 42.7 > 4.7

418

Compare. Write <, =, or >. Model exercises 1–8.

1. 0.4 _?_ 0.9

2. 0.22 _?_ 0.18

3. 0.35 _?_ 0.38

4. 0.65 _?_ 0.6

5. 0.7 _?_ 0.70

6. 0.84 _?_ 0.8

7. 2.7 _?_ 1.8

8. 3.5 _?_ 3.9

9. 5.6 _?_ 5.9

10. 5.47 _?_ 5.77

11. 8.03 _?_ 8.30

12. 2.35 _?_ 1.99

13. 23.05 _?_ 8.79

14. 2.17 _?_ 62.1

15. 14.9 _?_ 1.49

16. 100.1 _?_ 100.10

17. 235.0 _?_ 23.5

18. 604.04 _?_ 604.40

19. 839.00 _?_ 839.10

20. 147.5 _?_ 145.7

21. 252.01 _?_ 225.1

22. $5.65 _?_ $3.65

23. $.76 _?_ $.76

24. $20.19 _?_ $2.09

25. $1.04 _?_ $1.40

26. $10.00 _?_ $10.25

27. $3.09 _?_ $3.90

PROBLEM SOLVING

28. Ken's top speed in the bike-a-thon was 32.6 kilometers per hour. Vijay's top speed was 32.65 kilometers per hour. Which boy had the greater top speed?

29. Each week before the bike-a-thon, Misha rode his bike 112.5 km and Luke rode his bike 121.5 km. Who rode his bike the lesser distance each week?

30. Elise had a total of $42.75 in pledges for the bike-a-thon and Andres had a total of $42.05. Who had the greater total pledges?

31. This year the bike-a-thon raised $726.50 for charity. The bike-a-thon last year raised $725.75. Was the greater amount raised this year or last year?

Ordering Decimals

Order the winning times from fastest to slowest.

▶ You can use place value to order decimals from least to greatest.

Winning Times	
Olympic Speed Skating, 500 Meters	
1964 McDermott, U.S.A.	40.1 seconds
1968 Keller, W. Ger.	40.3 seconds
1980 Heiden, U.S.A.	38.03 seconds
1994 Golubev, Russ.	36.33 seconds

Align by place value.	Compare tens. Rearrange.	Compare ones. Rearrange.	Compare tenths. Rearrange if necessary.
40.1	38.03	36.33 ← least	36.33
40.3	36.33	38.03	38.03
38.03	40.1	40.1	40.1
36.33	40.3	40.3	40.3 ← greatest
	30 < 40	6 < 8 0 = 0	0.1 < 0.3

The order from fastest to slowest: 36.33; 38.03; 40.1; 40.3

or

The order from slowest to fastest: 40.3; 40.1; 38.03; 36.33

▶ You can use a number line to order decimals.

Order from least to greatest: 0.6; 0.4; 0.78; 0.65

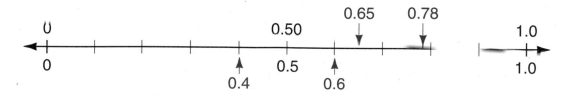

The order from least to greatest: 0.4; 0.6; 0.65; 0.78

or

The order from greatest to least: 0.78; 0.65; 0.6; 0.4

Write in order from least to greatest. You may use a number line.

1. 0.2; 0.9; 0.5

2. 3.5; 3.3; 3.35

3. 1.12; 1.02; 1.2

4. 5; 0.5; 0.05

5. 6.7; 6.77; 6.07; 7.67

6. 2.4; 4.2; 2.44; 4.02

7. 10.03; 1.30; 10.3; 1.33

8. 52.6; 62.5; 6.52; 56.2

9. 83.7; 87.37; 87.3; 83.07

10. 13.3; 33.31; 13.33; 130

Write in order from greatest to least. You may use a number line.

11. 0.1; 0.01; 0.11

12. 2.6; 2.06; 6.26

13. 4.04; 4.40; 4.0

14. 9.99; 9.19; 9.9

15. 1.18; 1.8; 1.81; 1.08

16. 17.6; 16.7; 61.7; 17.76

17. 59.03; 59; 53.9; 53.09

18. 44; 4.04; 40.4; 44.04

19. 90.3; 30.93; 30.09; 39.3

20. 75.01; 75.1; 75.11; 7.51

PROBLEM SOLVING

21. Erhard Keller won the 500-meter speed skating gold medal twice with times of 39.44 and 40.3 seconds. Uwe-Jens Mey also won the gold medal twice with times of 36.45 and 37.14 seconds. Order these winning times from greatest to least.

13-6 | Rounding Decimals

A number line can help you to round decimals.

Round to the nearest tenth: 0.42, 0.45, and 0.48.

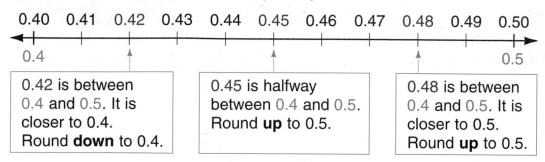

| 0.42 is between 0.4 and 0.5. It is closer to 0.4. Round **down** to 0.4. | 0.45 is halfway between 0.4 and 0.5. Round **up** to 0.5. | 0.48 is between 0.4 and 0.5. It is closer to 0.5. Round **up** to 0.5. |

To round decimals:

- Find the place you are rounding to.
- Look at the digit to its right.

Round to the nearest tenth: 1.35, 5.62, and 2.48.

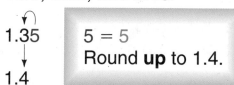
5 = 5
Round **up** to 1.4.

2 < 5
Round **down** to 5.6.

8 > 5
Round **up** to 2.5.

Round to the nearest one: 4.09, 6.7, and 8.54.

0 < 5
Round **down** to 4.

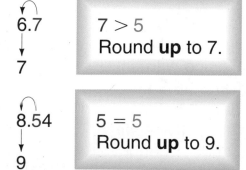
7 > 5
Round **up** to 7.

8.54
5 = 5
Round **up** to 9.

Do not write zeros to the right of the place you are rounding to.

Round to the nearest one.

1. 7.3 **2.** 9.2 **3.** 3.9 **4.** 1.5 **5.** 12.8

6. 16.2 **7.** 28.5 **8.** 62.4 **9.** 30.8 **10.** 19.7

11. 4.64 **12.** 15.35 **13.** 25.78 **14.** 41.23 **15.** 20.91

16. 17.52 **17.** 71.18 **18.** 49.62 **19.** 24.03 **20.** 3.95

Round to the nearest tenth.

21. 6.27 **22.** 4.64 **23.** 9.75 **24.** 2.20 **25.** 1.11

26. 31.37 **27.** 25.65 **28.** 85.06 **29.** 24.75 **30.** 38.33

31. 9.47 **32.** 13.53 **33.** 27.13 **34.** 82.75 **35.** 63.08

36. 52.71 **37.** 30.59 **38.** 81.11 **39.** 55.05 **40.** 44.89

PROBLEM SOLVING

41. What is ten and three tenths rounded to the nearest one?

42. What is six and five tenths rounded to the nearest one?

43. What is seventeen hundredths rounded to the nearest tenth?

44. What is nine and five hundredths rounded to the nearest tenth?

45. Is two and fifteen hundredths rounded to the nearest one 2, 2.2, or 3?

46. Is one and fifty hundredths rounded to the nearest tenth 20, 2.0, or 1.5?

Critical Thinking

Communicate ✓

47. Round 49.92 to the nearest one. Explain how you got your answer.

48. Round 87.99 to the nearest tenth. Explain how you got your answer.

13-7 Adding Decimals

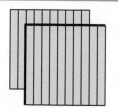

2.1

Shogi walked 2.1 km in the morning
and 1.95 km in the afternoon.
How far did he walk altogether?

1.95

To find how far, add:

Hint: Think of how you add money.

Line up the decimal points.	Add the hundredths.	Add the tenths. Regroup.	Add the ones.
2.1 0 ←	2.1 0	¹ 2.1 0	¹ 2.1 0
+ 1.9 5	+ 1.9 5	+ 1.9 5	+ 1.9 5
	5	0 5	4.0 5

Remember:
2.1 = 2.10

Write the
decimal point
in the sum.

Shogi walked 4.05 km altogether.

Study these examples.

```
  1
  0.8
+ 0.6
  1.4
```

```
  1   1
    5.7 5
    0.9 0
+ 2 8.3 2
  3 4.9 7
```

```
      1
  7 5.0 0
+   6.4 2
  8 1.4 2
```

Add. Use models to help.

1. 0.3
     + 0.6

2. 1.4
     + 6.2

3. 0.9
     + 0.7

4. 5.6
     + 9.8

5. 8.5
     + 10.5

6. 0.12
     + 0.43

7. 3.01
     + 0.57

8. 0.84
     + 0.77

9. $2.9
     + 7.83

10. 4.5
      + 2.86

424

Find the sum.

11.	5.03 + 8.9	12.	83.8 + 47.65	13.	90.41 + 62.7	14.	17.54 + 5.9	15.	45 + 9.24

16.	16.75 4.32 + 10.08	17.	0.7 1.2 + 8.9	18.	92.3 48.05 + 18.39	19.	74.32 10.1 + 0.8	20.	59.11 0.98 + 100.2

Align and add.

21. 0.3 + 8.44   **22.** 12.87 + 34   **23.** 0.95 + 22.6   **24.** 62 + 0.8

25. 32.5 + 575 + 0.49   **26.** 367.92 + 0.09 + 5.1

PROBLEM SOLVING

27. Val ran the first 100 meters of a 200-meter dash in 15.34 seconds. She ran the next 100 meters in 16.9 seconds. What was Val's time in the 200-meter dash?

28. Xavier swam the 100-meter freestyle in 58.95 seconds. If he could keep up that pace for another 100 meters, what would be his time in the 200-meter freestyle?

29. The times for the 4 legs of a relay race were 10.9 seconds, 12.74 seconds, 11.08 seconds, and 10.06 seconds. How long did it take to run the race?

Choose a Computation Method

Communicate ✓

Add. Use mental math, paper and pencil, or a calculator. Explain why you chose your methods.

30.	15 + 5.83	31.	6.75 + 4.99	32.	13.7 + 0.05	33.	71.74 + 86.9	34.	94.7 + 68.5

35.	8.39 + 9.92	36.	10.9 + 10	37.	4.66 + 0.7	38.	133.04 + 0.8 + 3.47

13-8 | Subtracting Decimals

How much farther is it from the Village to Black Rock than from Old Farm to Sam's Beach?

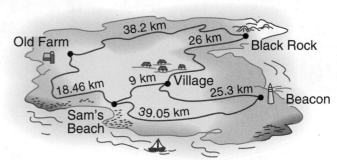

To find how much farther, subtract: $26 - 18.46 = $?

Hint: Think of how you subtract money.

Line up the decimal points. Regroup.	Subtract the hundredths.	Subtract the tenths.	Regroup. Subtract the ones.

$$\begin{array}{r} \overset{9}{} \\ 5\ \overset{}{\cancel{10}}\ 10 \\ 2\,\cancel{6}.\cancel{0}\,\cancel{0} \\ -\ 1\,8.4\,6 \\ \hline \end{array}$$

Remember:
26 = 26.00

$$\begin{array}{r} \overset{9}{} \\ 5\ \cancel{10}\ 10 \\ 2\,\cancel{6}.0\,0 \\ -\ 1\,8.4\,6 \\ \hline 4 \end{array}$$

$$\begin{array}{r} \overset{9}{} \\ 5\ \cancel{10}\ 10 \\ 2\,\cancel{6}.0\,0 \\ -\ 1\,8.4\,6 \\ \hline 5\ 4 \end{array}$$

$$\begin{array}{r} 15\ \ 9 \\ 1\ \cancel{5}\ \cancel{10}\ 10 \\ 2\,\cancel{6}.0\,0 \\ -\ 1\,8.4\,6 \\ \hline 7.5\ 4 \end{array}$$

Write the decimal point in the difference.

It is 7.54 km farther.

Study these examples.

1.2 = 12 tenths

$$\begin{array}{r} 0\ \ 12 \\ \cancel{1}.\cancel{2} \\ -\ 0.8 \\ \hline 0.4 \end{array}$$

$$\begin{array}{r} 9 \\ \cancel{10}\ 10 \\ \cancel{1}\,0.\cancel{0}\,5 \\ -\ \ \ 8.2\,0 \\ \hline 1.8\,5 \end{array}$$

$$\begin{array}{r} 7\ \ 10 \\ 9.\cancel{8}\,\cancel{0} \\ 3.1\,2 \\ \hline 6.3\,8 \end{array}$$

Subtract. Use models to help.

1. 0.9
    − 0.6

2. 0.35
    − 0.02

3. 8.7
    − 1.4

4. 1.48
    − 1.03

5. 5.65
    − 2.43

Find the difference. Use mental math, paper and pencil, or a calculator.

6. 18.7
  − 13.9

7. 24.2
  − 16.7

8. 3.43
  − 2.84

9. 62.19
  − 48.75

10. 75.11
  − 27.25

11. 23.16
  − 15.9

12. 82.6
  − 56.75

13. 64.5
  − 56.48

14. 10
  − 9.07

15. 16
  − 15.5

16. 17
  − 7.4

17. 92.1
  − 0.77

18. 76
  − 8.32

19. 58
  − 9.09

20. 31.2
  − 0.99

Align and subtract.

21. 90.17 − 9.07

22. 40.6 − 2.04

23. 8.34 − 0.5

24. 100 − 55.5

25. 99 − 0.09

26. 76.1 − 75.06

PROBLEM SOLVING Use the map on page 426.

27. How much closer to the Village is the Beacon than Black Rock?

28. How much farther from Old Farm is Black Rock than Sam's Beach?

29. Is the route from Sam's Beach to the Beacon longer or shorter than the distance from Black Rock to Old Farm? How much longer or shorter?

30. How many kilometers would you travel if you went from Old Farm to the Beacon by way of Sam's Beach and the Village?

 **Calculator Activity**

Subtract. Then check by adding.

31. 506.2
  − 175.35

32. 239.07
  − 86.6

33. 400.02
  − 0.8

34. 604
  − 64.91

Estimating with Decimals

Rounding is one way to estimate decimal sums and differences.

To estimate sums or differences with decimals:
- Round the decimals to the greatest *nonzero* place of the smaller number.
- Then add or subtract.

Estimate: 123.6 + 8.43

$$123.6 \longrightarrow 124$$
$$+ 8.43 \longrightarrow + 8$$
$$\text{about} \quad 132$$

Estimate: 78.61 − 0.45

$$78.61 \longrightarrow 78.6$$
$$- 0.45 \longrightarrow - 0.5$$
$$\text{about} \quad 78.1$$

Study these examples.

$$0.92 \longrightarrow 0.9$$
$$+ 0.37 \longrightarrow + 0.4$$
$$\text{about} \quad 1.3$$

$$4.7 \longrightarrow 4.7$$
$$- 0.18 \longrightarrow - 0.2$$
$$\text{about} \quad 4.5$$

$$8.8 \longrightarrow 9$$
$$+ 5.1 \longrightarrow + 5$$
$$\text{about} \quad 14$$

Estimate the sum or the difference. Watch the signs.

1. 5.9
    + 3.2

2. 9.7
    − 4.6

3. 8.75
    − 1.17

4. 9.38
    + 6.04

5. 4.91
    + 6.73

6. 42.3
    − 6.7

7. 38.5
    + 5.8

8. 56.2
    − 4.84

9. 27.8
    + 6.65

10. 85.43
    − 1.7

11. 0.85
    + 0.63

12. 10.3
    − 0.81

13. 62.77
    + 9.84

14. 48.5
    − 0.69

15. 20.21
    + 0.59

16. 74.36
    + 18

17. 62
    − 7.8

18. 49.95
    − 5.2

19. 405.5
    − 5.76

20. 380.4
    + 2.35

21. 4.5 + 39.03

22. 17.03 − 1.5

23. 47 − 6.62

Using Front-End Estimation

Front-end estimation is another way to estimate decimal sums and differences.

To make a front-end estimate with decimals:

- Add or subtract the nonzero front digits.
- Write zeros for the other digits.

$$\begin{array}{r} 83.41 \\ +71.3 \\ \hline \text{about} \quad 150.00 \end{array}$$

$$\begin{array}{r} 9.3 \\ -4.76 \\ \hline \text{about} \quad 5.00 \end{array}$$

$$\begin{array}{r} 0.65 \\ +0.5 \\ \hline \text{about} \quad 1.10 \end{array}$$

Estimate the sum or the difference. Use front-end estimation.

24.	25.	26.	27.	28.
30.98 + 56.44	8.6 + 9.2	43.21 − 12.04	7.4 − 2.9	58.4 − 21.62

29.	30.	31.	32.	33.
0.94 − 0.55	0.26 + 0.77	23.2 + 96.09	48.4 − 18.36	74.6 − 21.09

34.	35.	36.	37.	38.
8.09 + 8.9	6.74 − 1.53	81.2 − 27.35	50.09 + 97.79	59.5 − 24.07

PROBLEM SOLVING Use front-end estimation.

39. Maria jogged 97.5 miles. Audrey jogged 79.37 miles. About how many more miles did Maria jog than Audrey?

Skills to Remember

Find the quotient.

40. $2\overline{)\$37.32}$ **41.** $4\overline{)\$10.40}$ **42.** $9\overline{)\$2.79}$ **43.** $3\overline{)\$9.36}$ **44.** $8\overline{)\$64.16}$

45. $24\overline{)\$87.60}$ **46.** $53\overline{)\$57.24}$ **47.** $39\overline{)\$82.29}$ **48.** $42\overline{)\$41.16}$

Dividing with Money

Martin designs greeting cards. They cost
$.50 each if you buy them separately,
or you can buy a box of 25 cards
for $12. Which is the better buy?

To find which is the better buy, find the cost
of one boxed card. Then compare the cost to $.50.

To find the cost of one boxed card,
divide: $12 ÷ 25 = __?__

Before dividing, write a decimal point and two zeros in the dividend.	Divide as usual. Write the dollar sign and decimal point in the quotient.	Check.

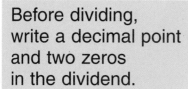

$$25)\overline{\$12.00}$$

$$
\begin{array}{r}
\$\ \ .48 \\
25)\overline{\$12.00} \\
-10\ 0\downarrow \\
\hline
2\ 00 \\
-2\ 00 \\
\hline
\end{array}
$$

$$
\begin{array}{r}
\$.48 \\
\times\ \ \ 25 \\
\hline
2\ 40 \\
+\ \ 96 \\
\hline
\$12.00
\end{array}
$$

$.48 < $.50
So the better buy is a box of 25 cards for $12.

Study this example.

$$8)\overline{\$18.00} \longrightarrow
\begin{array}{r}
\$\ 2.25 \\
8)\overline{\$18.00} \\
-16\downarrow \\
\hline
2\ 0\downarrow \\
-1\ 6\downarrow \\
\hline
4\ 0 \\
-4\ 0 \\
\hline
\end{array}
$$

Find the quotient.

1. $\$27 \div 6$ **2.** $\$41 \div 5$ **3.** $\$54 \div 8$ **4.** $\$38 \div 4$

5. $\$19 \div 2$ **6.** $\$90 \div 8$ **7.** $\$45 \div 6$ **8.** $\$78 \div 8$

9. $\$6 \div 24$ **10.** $\$48 \div 32$ **11.** $\$60 \div 16$

12. $\$8 \div 10$ **13.** $\$21 \div 14$ **14.** $\$32 \div 20$

Divide. Then check.

15. $4\overline{)\$17}$ **16.** $5\overline{)\$2}$ **17.** $8\overline{)\$60}$

18. $2\overline{)\$9}$ **19.** $8\overline{)\$10}$ **20.** $4\overline{)\$5}$

21. $52\overline{)\$65}$ **22.** $25\overline{)\$8}$ **23.** $48\overline{)\$12}$

24. $66\overline{)\$33}$ **25.** $72\overline{)\$54}$ **26.** $84\overline{)\$21}$

PROBLEM SOLVING Tell which is the better buy.

27. 8 erasers for $2.80
or
10 erasers for $3

28. 5 notebooks for $10
or
9 notebooks for $18.45

29. 6 bottles of shampoo for $21
or
8 bottles of shampoo for $22

30. 12 pencils for $3
or
10 pencils for $2

31. 20 plums for $14
or
16 plums for $12

32. 10 melons for $12
or
4 melons for $6

33. 8 juice cartons for $18
or
12 juice cartons for $33

34. 6 boxes of detergent for $27
or
4 boxes of detergent for $17

13-11 | Problem Solving: Multi-Step Problem

Problem: Hector bought 3 jumbo magnets and 1 magnifying glass at the science sale. How much change did he get from $10?

Science Sale	
mini-magnet	$ .45 each
jumbo magnet	$1.19 each
magnifying glass	$5.78 for 2

1 IMAGINE Put yourself in the problem.

2 NAME *Facts:*   3 jumbo magnets—$1.19 each
            1 magnifying glass—2 for $5.78
            paid $10

Question: How much change did Hector get?

3 THINK Plan the steps to follow.

Step 1: *Multiply* to find the cost of
       3 jumbo magnets. $3 \times \$1.19 = \underline{\ ?\ }$

Step 2: *Divide* to find the cost of
       1 magnifying glass. $\$5.78 \div 2 = \underline{\ ?\ }$

Step 3: *Add* to find the total cost.

Step 4: *Subtract* to find Hector's change from $10.

4 COMPUTE First estimate: cost of magnets $3 \times \$1 = \$3$
                  cost of magnifying glass $\$6 \div 2 = \$3$
                  $\$10 - \$6 = \$4$ change

Then compute:

Step 1	Step 2	Step 3	Step 4

$$
\begin{array}{c}
\overset{2}{} \\
\$1.1\,9 \\
\times \quad 3 \\
\hline
\$3.5\,7
\end{array}
\qquad
\begin{array}{r}
\$2.8\,9 \\
2)\overline{\$5\,7\,8} \\
-4 \\
\hline
1\,7 \\
-1\,6 \\
\hline
1\,8 \\
-1\,8 \\
\end{array}
\qquad
\begin{array}{r}
1\ 1 \\
\$3.5\,7 \\
+2.8\,9 \\
\hline
\$6.4\,6
\end{array}
\qquad
\begin{array}{r}
9\ 9 \\
0\ \cancel{10}\ \cancel{10}\ 10 \\
\$\cancel{1}\,0.0\,0 \\
-\quad 6.4\,6 \\
\hline
\$\quad 3.5\,4
\end{array}
$$

3 magnets

1 magnifying glass

total cost

Hector's change

5 CHECK The answer $3.54 is close to the estimate of $4.
The answer is reasonable.

Use the Multi-Step Problem strategy to solve each problem.

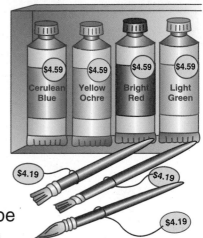

1. Mary wants 4 tubes of oil paint at $4.59 each and 3 brushes at $4.19 each. If she has saved $30.75, how much more money does she need?

IMAGINE   Create a mental picture.

NAME   *Facts:*   4 paint tubes at $4.59 a tube
                         3 brushes at $4.19 a brush
                         Mary has saved $30.75.

           Question: How much more money does Mary
                         need to buy the items?

THINK   Plan the steps to follow.
           Step 1: Multiply to find the cost of 4 paint tubes.
           Step 2: Multiply to find the cost of 3 brushes.
           Step 3: Add to find the total cost.
           Step 4: Subtract $30.75 from the total cost
                   to find how much more money Mary needs.

           COMPUTE ———→ CHECK

2. Mr. Ortiz collects 7.5 lb of honey in one bucket and 5.5 lb in another. He gives 1.2 lb of honey to a neighbor and 2.1 lb each to two workers. How much honey is left?

3. A shelf is 104.5 cm long. A set of encyclopedias uses 64.6 cm of space, and two books use 2.5 cm each. Is there more than 30 cm of space left? how much more or less?

4. It takes Lyn 58.34 s to swim a lap doing the backstroke and 42.15 s to swim a lap doing the crawl. She does 2 laps using the backstroke and 1 using the crawl. How much less than 3 minutes does she swim?

Problem-Solving Applications

Solve each problem and explain the method you used.

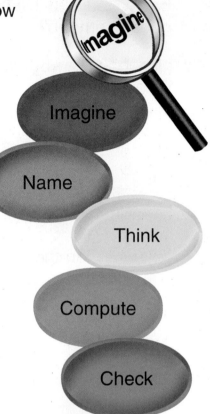

1. On Monday, 2.4 cm of rain fell in the morning and another 1.8 cm fell in the afternoon. How much rain fell on Monday?

2. The time between a bolt of lightning and the sound of thunder was 4.72 s. What is this time rounded to the nearest second?

3. A thunderstorm lasted 78.2 minutes. How much longer than an hour was the storm?

4. A meteorologist found that the diameter of a hail pellet measured 2.28 cm. What is this measurement to the nearest tenth?

5. The meteorologist found hail pellets with these diameters: 2.28 mm, 1.09 mm, 1.9 mm, 0.98 mm, and 1.42 mm. Order the pellets from smallest to largest.

6. The temperature during a hailstorm started at 11.4°C and then dropped by 0.5 degree. What was the temperature then?

7. Ms. Dell's car received 5 dents during the storm. She paid $85.50 to repair the damage. Each dent cost the same amount to fix. How much did it cost to repair each dent?

8. During a snowstorm, 12.3 dm of snow fell. There were already 45.9 dm of snow on the ground. How much snow was on the ground after the storm?

Choose a strategy from the list or use another strategy you know to solve each problem.

USE THESE STRATEGIES:
Multi-Step Problem
Use a Drawing or Model
Working Backwards
Logical Reasoning
Find a Pattern
Use a Graph
Extra Information

9. At 6:00 A.M. the snow was 1.4 cm deep. It snowed 1.4 cm more every half hour. What time was it when the snow was 11.2 cm deep?

10. A gopher dug a tunnel in the snow. The tunnel began at ground level, rose 2.2 ft, fell 0.7 ft, and then rose another 2.8 ft. How high above ground level did the tunnel end?

11. A winter storm warning lasted 4.5 h. It began at 2:30 P.M. The storm brought 4.3 in. of snow. When did the warning end?

12. Lina broke off 1.2 dm from a long icicle. It melted and lost another 0.8 dm. It was 3.5 dm long at the end of the day. How long was the original icicle?

13. Hugh built a snow sculpture with three large snowballs. They weighed 45.2 lb, 32.7 lb, and 20.1 lb. Luke's snow sculpture used three 28.5 lb snowballs. Whose snow sculpture was heavier? by how much?

Use the line graph for problems 14 and 15.

14. Between which two months did the amount of snowfall change the most on Mt. Sloper?

15. Joan did not ski in March. She did ski during a month that received less than 7 in. of snow. During which month did Joan ski?

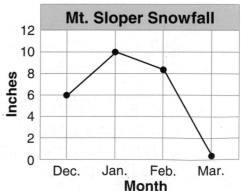

Mt. Sloper Snowfall

435

Write the value of the underlined digit.   *(See pp. 416–417.)*
Then write its place.

1. 3.<u>1</u>   **2.** 2.4<u>2</u>   **3.** 0.<u>9</u>6   **4.** <u>1</u>.92

5. 59.<u>6</u>   **6.** <u>8</u>.5   **7.** 2.2<u>3</u>   **8.** <u>1</u>5.49

Write as a decimal.   *(See pp. 412–415.)*

9. five tenths   **10.** thirty-two hundredths

11. three and four tenths   **12.** eight hundredths

Compare. Write <, =, or >.   *(See pp. 418–419.)*

13. 0.03 <u>?</u> 0.7   **14.** 9.45 <u>?</u> 12.8   **15.** 0.64 <u>?</u> 0.05

16. 12.8 <u>?</u> 12.80   **17.** 7.02 <u>?</u> 7   **18.** 5.06 <u>?</u> 5.6

Add or subtract.   *(See pp. 424–427.)*

19.   0.6   **20.**   4.9   **21.**   23.5   **22.**   44
        $+\,0.2$ $-\,2.73$ $+\,13.95$ $-\,\ \ 6.8$

Round each to the nearest one.   *(See pp. 422–423.)*
Then round each to the nearest tenth.

23. 12.17   **24.** 32.74   **25.** 0.88

Compute.   *(See pp. 430–431.)*

26. $36 ÷ 15   **27.** 8)$\overline{\$2}$ **28.** $5 ÷ 25

PROBLEM SOLVING   *(See pp. 424–429, 434–435.)*

29. The weight of one bag of onions is 2.47 lb.
      The weight of another is 0.73 lb.
      Estimate the weight of the two bags of onions.

(See Still More Practice, p. 472.)

MAGIC SQUARES

In a **magic square** each row, column, and diagonal has the same sum, called the **magic sum**.

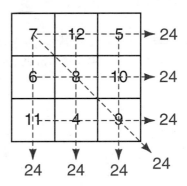

$7 + 12 + 5 = 24$ $7 + 6 + 11 = 24$
$6 + 8 + 10 = 24$ $12 + 8 + 4 = 24$
$11 + 4 + 9 = 24$ $5 + 10 + 9 = 24$

$7 + 8 + 9 = 24$ $5 + 8 + 11 = 24$

Copy and complete each magic square.

1.

6	7	?
?	5	9
?	3	?

2.

9	?	7
4	6	?
5	?	?

3.

?	3	?
7	10	?
?	17	5

4.

2.7	3.8	?
5.2	3.6	?
?	3.4	4.5

5.

3.5	7.5	8.5
11.5	?	?
4.5	5.5	?

6.

8.6	7	6.6
5.4	?	?
8.2	7.8	?

7.

?	63	68
?	67	?
?	71	64

8.

2.42	8	5.96
9	5.46	?
4.96	?	?

9.

?	?	24
?	15	?
6	?	12

(*Hint:* Use multiples of 3.)

437

Check Your Mastery

Performance Assessment

1–4. Draw a number line to show the numbers in the box.

9.08	9.6
9.40	9.89

Write the value of the underlined digit. Then write its place.

5. 4.<u>6</u>9　　　　**6.** <u>4</u>7.33　　　　**7.** <u>2</u>.26

8. 0.1<u>3</u>　　　　**9.** 6.<u>6</u>1　　　　**10.** 55.7<u>4</u>

Write as a decimal.

11. nine tenths　　　　　**12.** nine hundredths

13. four and six tenths　　**14.** seven and seven hundredths

Compare. Write <, =, or >.

15. 0.8 _?_ 0.4　　　　**16.** 0.7 _?_ 0.70

17. 0.46 _?_ 0.64　　　**18.** 2.43 _?_ 2.39

Write in order from least to greatest.

19. 13.4, 6.5, 13.3, 6.05　　**20.** 2.15, 2.51, 2.05, 2.5

Round to the nearest tenth.

21. 3.94　　　　**22.** 17.25　　　　**23.** 12.53

PROBLEM SOLVING　*Use a strategy you have learned.*

24. Last year Michele measured 153.8 cm. During the past year she grew 6.8 cm. How tall is she now?

25. Which is the better buy: 25 stickers for $3 or 20 stickers for $2?

26. What is the difference between 2.5 and 6?

Cumulative Test II

Choose the best answer.

1. Round 6749 to the nearest thousand.
   - **a.** 6000
   - **b.** 6700
   - **c.** 6750
   - **d.** 7000

2. Estimate: 3236 + 5873 + 1884
   - **a.** 8000
   - **b.** 9000
   - **c.** 11,000
   - **d.** 15,000

3. 8000 − 592
   - **a.** 7408
   - **b.** 7518
   - **c.** 7592
   - **d.** not given

4. 85 × 409
   - **a.** 5317
   - **b.** 34,725
   - **c.** 34,765
   - **d.** not given

5. Estimate.

   $42\overline{)7846}$
   - **a.** 200
   - **b.** 300
   - **c.** 2000
   - **d.** 3000

6. 8 ft 4 in.
   + 7 ft 10 in.
   - **a.** 15 ft 4 in.
   - **b.** 15 ft 6 in.
   - **c.** 16 ft 2 in.
   - **d.** 16 ft 6 in.

7. Which type of graph would you use to show changes in data over time?
   - **a.** bar graph
   - **b.** pictograph
   - **c.** circle graph
   - **d.** line graph

8. Is it more likely, less likely, or equally likely that the spinner will land on yellow than on green?
   - **a.** more likely
   - **b.** less likely
   - **c.** equally likely
   - **d.** cannot tell

9. Is the fraction two sevenths closer to 0, closer to $\frac{1}{2}$, or closer to 1?
   - **a.** 0
   - **b.** $\frac{1}{2}$
   - **c.** 1
   - **d.** cannot tell

10. What is the least common multiple (LCM) of 4 and 6?
    - **a.** 2
    - **b.** 24
    - **c.** 36
    - **d.** none of these

11. Which figure has half-turn symmetry?

    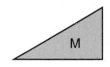 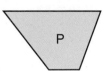
    - **a.** M
    - **b.** N
    - **c.** P
    - **d.** none of these

12. What is the volume of a rectangular prism that is 12 m long, 9 m wide, and 7 m high?
    - **a.** 28 cubic meters
    - **b.** 126 cubic meters
    - **c.** 189 cubic meters
    - **d.** 756 cubic meters

439

Add or subtract. Write the answers in lowest terms.

13. $\frac{3}{5}$
$+\frac{1}{5}$

14. $\frac{7}{8}$
$-\frac{2}{8}$

15. $\frac{1}{2}$
$+\frac{3}{4}$

16. $7\frac{3}{7}$
$+2\frac{2}{7}$

17. $\frac{5}{6}$
$-\frac{2}{12}$

Compute.

18. $41\overline{)88}$

19. $34\overline{)272}$

20. $16\overline{)139}$

21. $25\overline{)\$48.75}$

22. 9.9
-3.6

23. 0.9
$+0.8$

24. 42
-5.9

25. $7.2 + 18.29 + 0.52$

Find the perimeter and the area. Use formulas.

26.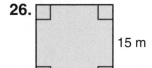
15 m

27.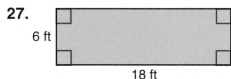
6 ft
18 ft

28.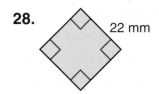
22 mm

PROBLEM SOLVING

29. A wooden box is 8 cm long, 5 cm wide, and 3 cm high. How many cubic centimeters does it hold?

30. Mrs. Gallon packed 432 prizes for the fair. She packed 12 in each bag. How many bags did she use?

31. Of the 27 rosebushes planted, $\frac{1}{3}$ are yellow. How many are yellow? How many are *not* yellow?

32. Which is the better buy: 4 cassettes for $25 or 6 cassettes for $39? Explain your answer.

For Rubric Scoring

Listen for information on how your work will be scored.

33. **a.** Ed wants to carpet a room. (See the diagram.) Carpet costs $12 a square yard. How much will it cost to carpet the room?

b. How can you find the number of square yards in another way? (*Hint:* How can you use the formula A = ℓ × w?)

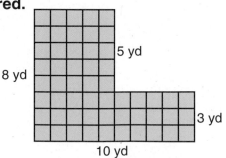
5 yd
8 yd
3 yd
10 yd

Arithmetic

If you take a number and double it and double it again and then double it a few more times, the number gets bigger and bigger and goes higher and higher and only arithmetic can tell you what the number is when you decide to quit doubling.

From "Arithmetic" by Carl Sandburg.

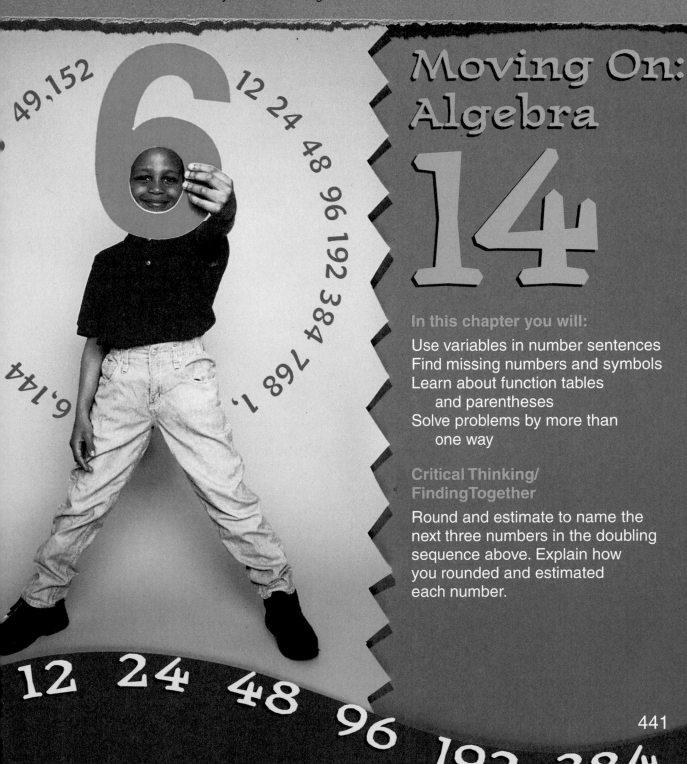

Moving On: Algebra

14

In this chapter you will:

Use variables in number sentences
Find missing numbers and symbols
Learn about function tables
 and parentheses
Solve problems by more than
 one way

**Critical Thinking/
FindingTogether**

Round and estimate to name the next three numbers in the doubling sequence above. Explain how you rounded and estimated each number.

441

Moving-On: Algebra

14-1 Number Sentences

A scout troop is planning a trip to a cave.
They rent a minibus for $17 per hour.
The trip will take 5 hours. How much will
the bus cost?

What do you know?	What do you need to know?	Which operation will you use?
• bus costs $17 per hour • trip takes 5 hours	• how much the bus will cost for 5 hours	• multiplication

Write a number sentence to help you
solve the problem.

- Let *n* stand for the product.

- Write the number sentence. $5 \times \$17 = n$

- Solve for *n*. $\$85 = n$

The bus will cost $85.

Choose the correct number sentence for each
problem. Then solve each problem.

1. The first cave chamber was 18 feet high. The
   second chamber was only 4 feet high. How
   much higher was the first chamber?

   a. $4 + 18 = n$ **b.** $4 \times 18 = n$ **c.** $18 - 4 = n$

2. The scouts discovered 225 bats in the first chamber
   and 172 in the second. How many bats did they
   discover in the two chambers?

   a. $225 + 172 = n$ **b.** $225 - 172 = n$ **c.** $172 \times 225 = n$

442

Problem Solving
Write a number sentence to solve each problem.

3. One chamber was 195 ft below sea level. Another chamber was 119 ft deeper. How many feet below sea level was the second chamber?

4. Each of 24 scouts brought 15 ft of rope. If they laid their ropes end to end to form a long strand, how many feet long would it be?

5. Zack found an arrowhead that was about 1500 years old. Chang found one that was twice as old. About how old was Chang's arrowhead?

6. Lucy's Lunches prepared 24 box lunches for the scouts. The total cost of the lunches was $94.80. What was the cost of each box lunch?

7. Each guide was to lead a team of 5 scouts. There were 24 scouts in all. How many teams of 5 were there? How many guides were needed for all the scouts?

8. The lengths of five passages in the cave are 17.2 mi, 24.5 mi, 18.3 mi, 16.4 mi, and 23.6 mi. What is the total length of the five passages in the cave?

9. Carlsbad Caverns covers 46,755 acres. The Wind Cave covers 28,292 acres. How many more acres does Carlsbad Caverns cover?

10. Mammoth Cave has 144 miles of underground passages. Exploring 3 miles each day, how many days would it take to explore all the passages?

Skills to Remember

Find the missing number.

11. $7 + \underline{\ ?\ } = 15$

12. $\underline{\ ?\ } - 5 = 8$

13. $6 \times \underline{\ ?\ } = 36$

14. $40 \div \underline{\ ?\ } = 8$

15. $12 - \underline{\ ?\ } = 5$

16. $\underline{\ ?\ } \div 7 = 9$

17. $\underline{\ ?\ } \times 4 = 28$

18. $\underline{\ ?\ } \div 8 = 1$

19. $\underline{\ ?\ } + 5 = 5$

Moving-On: Algebra

14-2 Finding Missing Numbers

▶ What number does a stand for?
$8 + a = 7 + 6$

> A letter can be used for an unknown number.

To solve:

- Compute where possible.

$8 + a = 7 + 6$

$8 + a = 13$

> Think:
> missing addend

- Solve for the missing addend.

$a = 13 - 8$
$a = 5$

- Check.

$8 + 5 = 7 + 6$

$13 = 13$

What number does x stand for?
$3 \times 5 = x \div 2$

- Compute.

$3 \times 5 = x \div 2$

$15 = x \div 2$

> Think:
> missing dividend

- Solve.

$15 \times 2 = x$
$15 \times 2 = 30$
$30 = x$

- Check.

$3 \times 5 = 30 \div 2$

$15 = 15$

Write the number that n stands for in each number sentence.

1. $12 - 5 = n - 7$ **2.** $4 \times n = 8 \times 3$ **3.** $40 \div 8 = 30 \div n$

Write the number that y stands for
in each number sentence.

4. $y \div 3 = 63 \div 7$ **5.** $9 + 7 = y + 8$ **6.** $2 \times 10 = y \times 5$

7. $9 + y = 3 \times 6$ **8.** $42 \div 7 = 16 - y$ **9.** $10 + 7 = y - 3$

10. $y \times 3 = 18 \div 3$ **11.** $y - 10 = 7 \times 2$ **12.** $25 - 15 = y \div 4$

13. $8 \times y = 26 - 26$ **14.** $9 \times 8 = y \times 72$ **15.** $100 + y = 9 \times 12$

16. $50 \times 3 = 200 - y$ **17.** $y \div 2 = 10 \times 25$ **18.** $12 \times 12 = 130 + y$

19. $y + 99 = 59 + 40$ **20.** $43 \times y = 0 \div 34$ **21.** $125 \times 2 = 400 - y$

22. $64 + y + 22 = 100 + 20 + 8$ **23.** $8 \times 8 \times y = 2 \times 250 + 12$

24. $500 \div 50 \times 95 = y + 2 \times 450$

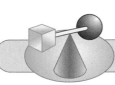

Challenge

Write the number that n stands for
in each number sentence.

25. $n + n = 6$ **26.** $7 - n = 7$ **27.** $n \times n = 25$

> What number
> added to itself
> equals 6?

28. $n \times 4 = n$ **29.** $n \div 5 = n$

30. $n + n = 30$ **31.** $64 \div n = n$

Write the numbers that x and y stand for
in each pair of number sentences.

32. $x + y = 9$
$x + x = 8$

33. $x \times y = 24$
$y \times y = 9$

34. $x \times y = 8$
$y - x = 7$

Moving-On: Algebra

14-3 Functions

▶ The table at the right is called a **function table**.

For each number that you put into the table, there is only one output. You can find the output by following the rule.

The input is 12. What is the output?

Input	Rule	Output

$$12 \quad \times \quad 3 \quad = \quad 36$$

The output is 36.

Rule: × 3	
Input	Output
2	6
4	12
5	15
8	24
12	?

▶ What is the rule for this function table?

Think how each input is related to its output.

$$40 \div 4 = 10 \qquad 32 \div 4 = 8$$
$$28 \div 4 = 7 \qquad 20 \div 4 = 5$$

The rule is ÷ 4.

Rule: ?	
Input	Output
40	10
32	8
28	7
20	5

Copy and complete.

1.

Rule: + 7	
Input	Output
4	11
8	?
25	?
42	?

2.

Rule: − 11	
Input	Output
12	?
20	?
45	?
63	?

3.

Rule: ÷ 2	
Input	Output
250	?
210	?
180	?
100	?

446

Copy and complete each table.

4.

Rule: × 9	
Input	Output
5	?
8	?
10	?
25	?
51	?

5.

Rule: ÷ 20	
Input	Output
500	?
240	?
180	?
120	?
80	?

6.

Rule: × 43	
Input	Output
8	?
15	?
37	?
105	?
232	?

Write the rule for each table.

7.

Rule: ?	
Input	Output
5	40
8	64
12	96
20	160

8.

Rule: ?	
Input	Output
70	55
65	50
58	43
42	27

9.

Rule: ?	
Input	Output
15	40
22	47
36	61
44	69

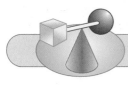

Challenge

Copy and complete.

10.

Rule: × 7	
Input	Output
?	63
?	77
?	98
?	112

11.

Rule: ÷ 9	
Input	Output
?	25
?	22
?	18
?	15

12.

Rule: × 15	
Input	Output
?	75
?	120
?	165
?	225

Moving-On: Algebra

14-4 Missing Symbols

The symbol = means "is equal to."

$$8 = 8$$
$$4 + 5 = 9$$
$$15 = 3 \times 5$$
$$6 + 1 = 5 + 2$$

The symbol ≠ means "is *not* equal to."

$$7 \neq 9$$
$$13 - 4 \neq 12$$
$$6 \neq 20 \div 5$$
$$4 \times 3 \neq 3 \times 5$$

Which symbol completes this number sentence?

$$8 \times 6 \ \underline{?} \ 25 + 25$$

To find the correct symbol:

- Compute both sides of the number sentence.

$$8 \times 6 \ \underline{?} \ 25 + 25$$
$$48 \quad \underline{?} \quad 50$$

- Compare.

$$48 \neq 50$$

So $8 \times 6 \neq 25 + 25$.

Study this example.

$$3 \times 15 \ \underline{?} \ 39 + 6$$
$$45 \quad \underline{?} \quad 45$$
$$45 = 45$$

Write the letter of the correct answer.

1. $6 + 4 \neq \underline{\ ?\ }$

   a. $13 - 3$
   b. $20 \div 2$
   c. 4×2

2. $7 \times 9 = \underline{\ ?\ }$

   a. $87 - 15$
   b. $39 + 24$
   c. $40 + 16$

3. $100 \div 2 \neq \underline{\ ?\ }$

   a. 2×25
   b. $30 + 30$
   c. $62 - 12$

4. $36 \div 6 = \underline{\ ?\ }$

   a. $30 \div 5$
   b. $36 - 6$
   c. 6×6

448

Compare. Write = or ≠ .

5. $10 + 8 \underline{\ ?\ } 9 + 6$ **6.** $13 - 5 \underline{\ ?\ } 11 - 3$ **7.** $5 \times 8 \underline{\ ?\ } 10 \times 4$

8. $54 \div 6 \underline{\ ?\ } 56 \div 8$ **9.** $4 + 5 \underline{\ ?\ } 15 - 6$ **10.** $2 \times 3 \underline{\ ?\ } 30 \div 6$

11. $45 \times 3 \underline{\ ?\ } 125 + 10$ **12.** $225 \div 25 \underline{\ ?\ } 240 \div 30$

13. $7250 + 100 \underline{\ ?\ } 8450 - 200$ **14.** $75 \times 4 \underline{\ ?\ } 900 \div 30$

15. $586 - 139 \underline{\ ?\ } 328 + 160$ **16.** $396 \div 3 \underline{\ ?\ } 12 \times 11$

17. $685 \div 5 \underline{\ ?\ } 5 \times 71$ **18.** $8 \times 525 \underline{\ ?\ } 7 \times 600$

19. $\$4.50 + \$1.15 \underline{\ ?\ } 4 \times \$1.25$ **20.** $6 \times \$5.95 \underline{\ ?\ } 7 \times \$6.95$

Compare. Write <, =, or >.

21. $500 \div 2 \underline{\ ?\ } 200 \div 5$ **22.** $50 \times 600 \underline{\ ?\ } 40 \times 700$

23. $2000 - 1500 \underline{\ ?\ } 50 \times 8$ **24.** $850 - 125 \underline{\ ?\ } 525 + 200$

25. $2 \times 550 \underline{\ ?\ } 5 \times 250$ **26.** $400 \div 5 \underline{\ ?\ } 500 \div 4$

PROBLEM SOLVING

27. Is the product of 8 and 45 equal to the difference of 500 and 140?

28. Is the sum of 534 and 166 equal to the product of 250 and 3?

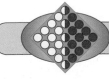

Finding Together

Choose numbers from the box to complete each number sentence. Use each number only once.

29. $3 + 7 > 6 + \underline{\ ?\ }$ **30.** $8 \times \underline{\ ?\ } \neq 4 \times 10$

31. $80 \div 8 > \underline{\ ?\ } + 4$ **32.** $56 \div \underline{\ ?\ } = 5 + \underline{\ ?\ }$

1		5
	3	
7		9

449

Moving-On: Algebra

14-5 Number Line

A number line can help you find sums and differences.

▶ Let $n = 23$
  What is $n + 32$?

To find a sum on a number line, count on from left to right.

Begin at 23.

Think: $32 = 30 + 2$ or 3 tens + 2
So count on 3 tens. Then count on 2.

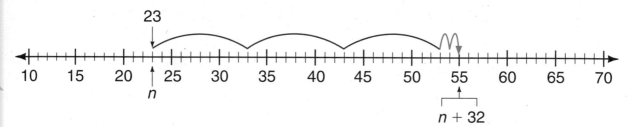

So $n + 32 = 55$.

▶ Let $n = 62$
  What is $n - 14$?

To find a difference on a number line, count back from right to left.

Begin at 62.

Think: $14 = 10 + 4$ or 1 ten + 4
So count back 1 ten. Then count back 4.

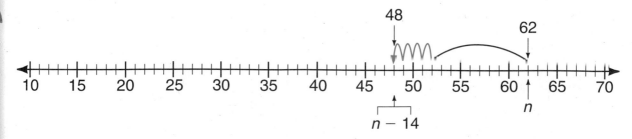

So $n - 14 = 48$.

Copy and complete each table. You may use a number line.

1.

n	n + 25
16	?
22	?
37	?
49	?

2.

n	n − 41
95	?
80	?
66	?
52	?

3.

n	n + 36
28	?
47	?
55	?
69	?

4.

n	n − 28
87	?
73	?
65	?
52	?

5.

n	n + 44
12	?
26	?
37	?
59	?

6.

n	n − 17
73	?
56	?
44	?
38	?

7.

n	n + 65
94	?
123	?
207	?
461	?

8.

n	n − 35
520	?
493	?
452	?
407	?

9.

n	n + 125
25	?
175	?
256	?
595	?

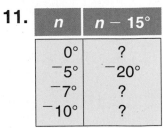

Challenge

Think of the thermometer as a number line.
Write temperatures below zero with a minus sign.

⁻12° "twelve degrees below zero"

Use the thermometer to copy and complete each table.

10.

n	n + 10°
0°	?
⁻5°	5°
⁻8°	?
⁻25°	?

11.

n	n − 15°
0°	?
⁻5°	⁻20°
⁻7°	?
⁻10°	?

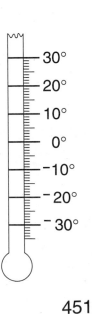

451

Moving-On: Algebra

14-6 Using Parentheses

How would you go about using the order of operations to simplify this problem?

$$40 - 3 \times 5 + (10 \div 2) = \underline{?}$$

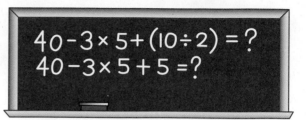

$$40 - 3 \times 5 + (10 \div 2) = ?$$
$$40 - 3 \times 5 + 5 = ?$$

To simplify:

- Always do the operations in parentheses first.

$$40 - 3 \times 5 + \underbrace{(10 \div 2)} = \underline{?}$$

- Multiply or divide.
  Work in order from left to right.

$$40 - \underbrace{3 \times 5} + \quad 5 \quad = \underline{?}$$

- Add or subtract.
  Work in order from left to right.

$$40 - \underbrace{\quad 15 \quad + \quad 5} \quad = \underline{?}$$
$$25 \quad + \quad 5 \quad = 30$$

Study these examples.

$$2 \times \underbrace{(4 + 3)} - 10 + \underbrace{(4 \times 4)} = \underline{?}$$
$$\underbrace{2 \times \quad 7} \quad - 10 + \quad 16 \quad = \underline{?}$$
$$\underbrace{14 \quad - 10 +} \quad 16 \quad = \underline{?}$$
$$4 \quad + \quad 16 \quad = 20$$

$$\underbrace{(4 \times 2)} + \underbrace{(9 \div 3)} - 10 + 1 = \underline{?}$$
$$\underbrace{8 \quad + \quad 3} \quad - 10 + 1 = \underline{?}$$
$$\underbrace{11 \quad \quad - 10} + 1 = \underline{?}$$
$$1 \quad \quad + 1 = 2$$

Simplify.

1. $18 - 6 + 4$

2. $7 + 7 - 5$

3. $6 \times 8 \div 4$

4. $24 \div 6 + 10$

5. $38 - 2 \times 9$

6. $10 \times 10 \div 5$

7. $45 \div (3 \times 3)$

8. $(10 + 10) \times 8$

9. $25 + (5 \times 5) + 50$

10. $(3 + 5) + (10 + 2)$

11. $100 - (100 \div 10)$

12. $(500 \div 2) + 25$

452

Use the order of operations to simplify.

13. $(6 - 2) + (6 \times 2)$

14. $(8 \div 4) \times (9 - 5)$

15. $(56 \div 8) \times (10 + 7)$

16. $(4 \times 12) - (20 - 15)$

17. $9.7 + (6.1 - 5.1)$

18. $20 - (10 - 5.5)$

19. $(8.1 - 8.1) \times (5 + 4)$

20. $(3.2 + 4.6) - (2 \times 2)$

21. $(6 \times 2) + (9.3 - 7.5)$

22. $(45 \div 5) + (10.75 - 2.25)$

23. $\frac{2}{5} + \left(\frac{4}{5} - \frac{2}{5} \right)$

24. $\left(\frac{7}{10} - \frac{4}{10} \right) + \frac{6}{10}$

25. $\frac{1}{2} + \left(\frac{1}{2} - \frac{1}{4} \right)$

26. $\frac{3}{4} + \left(\frac{1}{2} + \frac{1}{4} \right)$

27. $\left(\frac{1}{4} + \frac{1}{4} \right) + \left(\frac{1}{8} + \frac{1}{8} \right)$

28. $\left(\frac{2}{6} + \frac{3}{6} \right) - \left(\frac{1}{3} + \frac{1}{3} \right)$

29. $\left(\frac{7}{8} - \frac{1}{4} \right) - \left(\frac{2}{8} + \frac{1}{8} \right)$

30. $\left(\frac{5}{6} + \frac{1}{6} \right) \times \left(\frac{1}{2} + \frac{1}{2} \right)$

31. $\left(\frac{6}{8} - \frac{3}{4} \right) + \left(\frac{2}{4} + \frac{2}{4} \right)$

32. $\left(\frac{4}{9} + \frac{4}{9} \right) \times \left(\frac{2}{5} + \frac{3}{5} \right)$

Calculator Activity

Use a calculator to simplify.

33. $6.5 + 7.2 \times 3.1$

34. $4.4 - 4.5 \div 1.5$

35. $6.2 \times (8.2 - 5.1)$

36. $10.5 \div 1.05 \times 5$

37. $5 + 5.5 + 5.5 - (9.2 - 3)$

38. $(8.4 \times 10) - 10 + 1$

39. $(7.25 \times 4) - (7.25 \div 5)$

40. $(5.2 \times 4) + (4.8 \div 2)$

453

14-7 Problem Solving: More Than One Way

Problem: Kim is making a rectangular sign that is 3 ft wide. She uses 14 ft of edging to go around the sign. How long is the sign?

1 IMAGINE   Create a mental picture.

2 NAME

Facts: width   =   3 ft
             perimeter = 14 ft

Question: How long is the sign?

3 THINK   There is more than one way to find a solution. Here are 2 ways.

Method 1	*Method 2*
Draw a picture.	Use a formula.
	$P = 2 \times \ell + 2 \times w$
Guess and test to find the length.	Guess and test to find the length.

4 COMPUTE

Method 1	Method 2
First Guess ⟶ 3 ft	Let ℓ = length
3 ft + 3 ft + 3 ft + 3 ft = 12 ft not large enough ⟶	$P = 2 \times \ell + 2 \times w$
	$14 = 2 \times \ell + (2 \times 3 \text{ ft})$
	$14 = 2 \times \ell + 6 \text{ ft}$
Second Guess ⟶ 4 ft	$14 = 2 \times \underline{\ ?\ } \text{ ft} + 6 \text{ ft}$
3 ft + 3 ft + 4 ft + 4 ft = 14 ft correct sum ⟶	$14 = 2 \times 4 \text{ ft} + 6 \text{ ft}$
The sign is 4 ft long.	The sign is 4 ft long.

5 CHECK

Use the formula to check your answer.	Draw a picture to check your answer.

Solve each problem and explain the method you used.

1. The temperature at Beal Beach was
   32.4°C at dawn. It rose 4.7°C by
   noon, and then fell 6.1°C by dusk.
   What was the temperature at dusk?

IMAGINE   Put yourself in the problem.

NAME   *Facts:*   dawn—32.4°C
                         noon— 4.7°C higher
                         dusk— 6.1°C lower

Question: What was the temperature at dusk?

THINK   What method will you use?

Method 1   *Method 2*

Draw and label a   Write a number sentence.
number line.   32.4 + 4.7 − 6.1 = _?_

COMPUTE ⟶ **CHECK**

2. Karl has 25 wheels for wagons and scooters.
   How many of each toy can he make if
   the wagons have 4 wheels and scooters
   have 3 wheels?

3. The digits of a two-digit number have a
   sum of 7 and a difference of 5. The
   number is less than 70 and greater than 20.
   What is the number?

4. The Hoopsters scored 35 points in the first half
   of the game and 18 more than that in the second
   half. The other team scored 90 points in the
   game. Did the Hoopsters win?

Moving-On: Algebra

14-8 Problem-Solving Applications

Solve each problem and explain the method you used.

1. I am a number. If you add me to 28, the sum is 100. What am I?

2. I am a decimal. If I am added together 5 times, the answer equals 43. What am I?

3. I am a decimal equal to the sum of 2.8, 3.2 and 7.4. What am I?

4. What number should you add to complete this sentence? $8\frac{1}{4} + 2\frac{1}{4} + n = 11\frac{1}{2}$

5. Use $=$ or $\neq$ to complete this number sentence. $3 \times 4 - 2$ _?_ $18 \div 2 + 4$

6. Which number sentences equals 25?
   $5 + 10 \times 4 \div 2 = n$ or
   $35 - 10 \div 5 + 5 = n$

7. What is the number halfway between 40 and 70?

8. What is the greatest number less than 65 that is divisible by 3?

9. In a contest, players scored 4 points for each correct answer. How many correct answers did the winner give? the player in 3rd place? (Use the bar graph to the right.)

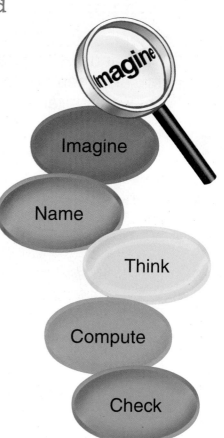

Imagine

Name

Think

Compute

Check

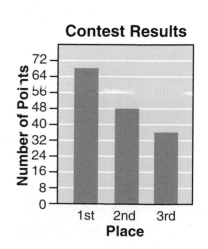

Contest Results

Number of Points: 72, 64, 56, 48, 40, 32, 24, 16, 8, 0

1st   2nd   3rd

Place

Choose a strategy from the list or use another strategy you know to solve each problem.

USE THESE STRATEGIES:
Multi-Step Problem
Hidden Information
Guess and Test
Write a Number Sentence
Logical Reasoning
Working Backwards

10. The winner of a contest may choose from 2 prizes: a dime a day for a year or a dollar a day during March. Which amount is greater?

11. There were three players in the contest. Mel did not win, but he scored more points than Rob. Did Carmen win?

12. A program began at 6:30 P.M. and ended at 7:00 P.M. There were two 4.5-minute commercial breaks. How long was the program itself?

13. The winner received a T-shirt that said, "I'm Number $5 - 2 \times 2$." What does this mean? Create a number sentence for a shirt for the second-place winner.

14. Arrange the numbers in the box so their sum, product, difference, and quotient are equal.

$$\underline{\ ?\ } + \underline{\ ?\ } = \underline{\ ?\ } - \underline{\ ?\ } = \underline{\ ?\ } \div \underline{\ ?\ } = \underline{\ ?\ } \times \underline{\ ?\ }$$

40 19 3 6 16
4   72   5

15. *Math Facts* auditioned students. In the first round, $\frac{1}{2}$ were eliminated. In the second round, 30 more were eliminated. There were 10 students left for the third round. How many students came to the audition?

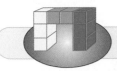

Make Up Your Own

16. Write a problem modeled on problem 10. Have a classmate solve it.

Write the number that *n* stands for in each number sentence. *(See pp. 444–445.)*

1. $28 - n = 4 \times 6$

2. $11 \times 12 = 100 + n$

3. $n \div 4 = 12 \times 2$

4. $32 + 20 + n = 52 \times 4$

Copy and complete each table. *(See pp. 446–447, 450–451.)*

5.

Rule: ×3	
Input	Output
5	?
9	?
33	?
46	?

6.

Rule: ?	
Input	Output
24	4
36	6
48	8
60	10

7.

n	*n* + 39
15	?
67	?
85	?
92	?
98	?

Compare. Write = or ≠. *(See pp. 448–449.)*

8. 36×3 __?__ $24 + 24$

9. $25 - 5$ __?__ $60 - 40$

10. $76 + 2$ __?__ $92 - 31$

Use the order of operations to solve. *(See pp. 452–453.)*

11. $15 + 8 - 2 \times 9$

12. $9 \times 10 \div 5 + 6$

13. $30 + 4 \times 4 + 20$

14. $49 - 3 \times 7$

15. $(54 \div 6) \times (2 + 10)$

16. $\frac{1}{5} + \frac{3}{5} - \frac{2}{5}$

PROBLEM SOLVING
Write a number sentence to solve each problem.

17. Mrs. Lam bought 720 yards of material to make curtains. If 8 yards of material are needed for each pair of curtains, how many windows can she decorate?

18. The school auditorium has 25 rows of seats. Each row has 15 seats. How many seats are in the auditorium?

(See *Still More Practice*, p. 472.)

NEGATIVE NUMBERS

Numbers that are written with a minus sign, such as ⁻6, ⁻25, and ⁻247, are called **negative numbers**.

You already know how to use negative numbers to write temperatures below zero.

⁻15°F   ⁻3°C

You can also use negative numbers to show distances below sea level.

⁻5 ft means "5 feet below sea level."

The scale at the right shows the location of different sites in Crystal Caverns.

Solve. Use the scale of Crystal Caverns.

1. Which site is located at ⁻90 ft?

2. About how many feet below sea level is Stalagmite Garden?

3. Which site is farthest below sea level? About how many feet below sea level is it?

4. How many feet difference is there between Bottomless Pool and Stalactite Chamber?

5. Which site is halfway between sea level and Pirate's Rest? How many feet below sea level is it?

6. Suppose there was a site at ⁻150 ft. How much lower than sea level would it be than Bottomless Pool?

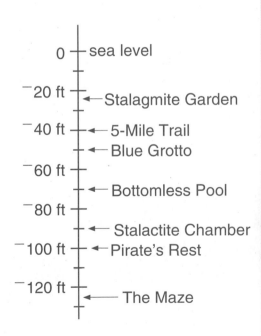

Performance Assessment

1. What is the rule for the table at the right?

2. Make up a table for each of the following:

   a. Rule: $\div 3$ **b.** Rule: $n - 8$

Rule: ?	
Input	Output
5	40
8	64
10	80
12	96

Write the number that *n* stands for in each number sentence.

3. $6 + 8 = n - 5$

4. $27 \times n = 112 - 4$

5. $n \div 20 = 17 - 12$

6. $700 - 7 + 20 = n + 100 - 5$

Compare. Write = or ≠.

7. $65 \div 5$ _?_ $8 + 5$ **8.** 7×8 _?_ $66 - 9$ **9.** 3×36 _?_ 6×18

Use the order of operations to solve.

10. $16 \div 4 + 8$ **11.** $100 - 90 \div 10$ **12.** $9 + 8 - 7 + 5$

13. $250 \div (5 \times 2) + 100$ **14.** $8 \times (25 + 6) - 7$ **15.** $(2 + 6) \times (14 - 8)$

PROBLEM SOLVING   *Use a strategy you have learned.*

Write a number sentence to solve each problem.

16. Brian has 87 rare stamps in his collection. Susan has 127 stamps, and Judy has 95 stamps. How many stamps do they have altogether?

17. Betty has 135 flowers. Each arrangement requires 15 flowers. How many arrangements can she make?

Practice 1-1

Write the number in standard form.

1a. 8 thousands 2 tens
b. twenty-two thousand

2a. four hundred seventy-three million
b. 400,000 + 10,000 + 7000 + 200 + 1

Write the word name for each number.

3a. 1,020,140
b. 80,000 + 4000 + 500

Write each number in expanded form.

4a. 668,850,201
b. 5,884,901

Write the place and value of the underlined digit.

5a. 73,4̲02
b. 608̲,721
c. 2,3̲00,400

Compare. Write <, =, or >.

6a. 3983 ? 3892
b. 120,121 ? 102,101

7. Write the amounts $45.15, $38.06, and $37.05 in order from greatest to least.

8. Write the numbers 5403, 3405, 4340, 3450, and 5430 in order from least to greatest.

9. What number is 100 more than 4,506,722?

10. What number is 1000 less than 439,800?

11. What number is 1000 more than 9,829,432?

12. What is the greatest even four-digit number?

13. The Beekman Library has 23,450 books. The Conrad Library has 24,355 books and the Doral Library has 25,320 books. Put the libraries in order from least books to most books.

Practice 1-2

Write each amount.

1a. 2 dollars, 2 quarters, 1 dime, 3 nickels
b. 5 quarters, 4 dimes, 8 nickels, 3 pennies

Write the fewest coins and bills you would receive as change. Then write the value of the change.

2a. Cost: $4.20 Amount given: $5.00
b. Cost: $8.39 Amount given: $10.00

Round to the nearest hundred or dollar.

3a. 2390
b. 821
c. 56,472

4a. $3.29
b. $12.90
c. $35.85

Round to the nearest thousand.

5a. 54,320
b. 8199
c. 65,328

About what number is each arrow pointing toward?

6.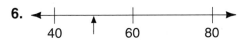

7.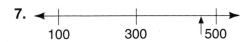

8. What is 4809 rounded to the nearest ten?

9. What is $328.59 rounded to the nearest ten dollars?

10. Suzy has $32.28. Can she buy a fig tree that costs $23.82?

11. Yinka bought a book bag for $15.95. He gave the clerk a twenty-dollar bill. How much change did he receive?

12. What number is halfway between 1000 and 2000?

Practice 2-1

1a. $1 + 0$ **b.** $4 + 4$ **c.** $0 + 7$

2a. $3 + 5 + 4$ **b.** $6 + 1 + 6 + 1$

3a. $3 - 1$ **b.** $8 - 0$ **c.** $7 - 7$

4a. $17¢ - 8¢$ **b.** $11¢ - 6¢$ **c.** $2¢ - 2¢$

Find the missing number.

5a. $9 + \underline{\ ?\ } = 14$ **b.** $9 = 7 + \underline{\ ?\ }$

6a. $7 - \underline{\ ?\ } = 1$ **b.** $5 = \underline{\ ?\ } - 8$

Estimate the sum or difference.

7a. $28 + 22$ **b.** $589 + 612$ **c.** $825 - 592$

8a. $\begin{array}{r} \$1.28 \\ +\ 1.15 \end{array}$ **b.** $\begin{array}{r} \$309 \\ +\ 194 \end{array}$ **c.** $\begin{array}{r} \$8.89 \\ -\ 7.20 \end{array}$

Find the sum or difference.

9a. $38 + 41$ **b.** $211 + 544$ **c.** $\$17 + \$32$

10a. $85 - 40$ **b.** $54 - 43$ **c.** $\$68 - \$55$

11. A quilt has 12 blue squares and 24 green squares. How many squares does it have?

12. Max has 48 comic books. He sells 23 of them. How many does he have left?

13. Jan scored 8 points in a basketball game. Ina scored 19 points. How many more points did Ina score than Jan?

14. There are 18 turtles in a pond. There are 7 adult turtles. How many are not adults?

15. Alma needs $14 to buy a compact disc. She has $11. How much more money does she need?

Practice 2-2

Add. Use the addition properties.

1a. $5 + 0$ **b.** $7 + 6$ **c.** $6 + 7$

2a. $\begin{array}{r} 6 \\ 2 \\ 5 \\ +4 \end{array}$ **b.** $\begin{array}{r} 3 \\ 5 \\ 1 \\ +7 \end{array}$ **c.** $\begin{array}{r} 1 \\ 2 \\ 4 \\ +8 \end{array}$

Add mentally.

3a. $4 + 5$ **b.** $9 + 9$ **c.** $5 + 6$

4a. $\begin{array}{r} 7 \\ 3 \\ +4 \end{array}$ **b.** $\begin{array}{r} 8 \\ 5 \\ 8 \\ +2 \end{array}$ **c.** $\begin{array}{r} 9 \\ 3 \\ +9 \end{array}$

Find the missing number.

5a. $\underline{\ ?\ } + 3 = 11$ **b.** $15 = 7 + \underline{\ ?\ }$

6a. $13 - \underline{\ ?\ } = 7$ **b.** $16 - \underline{\ ?\ } = 8$

Estimate. Then find the sum or difference.

7a. $25 + 73$ **b.** $314 + 622$ **c.** $\$82 - \$47$

8a. $\begin{array}{r} \$.47 \\ +\ .09 \end{array}$ **b.** $\begin{array}{r} \$9.27 \\ -\ 2.93 \end{array}$ **c.** $\begin{array}{r} \$19.10 \\ -\ 4.82 \end{array}$

9. Mark has 14 toy trucks in a carrying case. The case can hold 20 trucks. How many more trucks does Mark need to fill the case?

462

Practice 3-1

1a. 323
    + 679

b. 19
    + 894

c. 695
    + 8126

2a. 94,320
    + 84,002

b. 190,029
    + 870,993

c. $18.26
    + 4.59

3a. 2
    97
    + 73

b. 79
    500
    639
    + 322

c. $ 919
    610
    8120
    + 1293

4a. 25 + 75 + 50   **b.** $45.99 + $68.20

5a. 8550 + 10,203   **b.** 194,344 + 940,277

Make a rough estimate. Then adjust.

6a. 920
    + 735

b. 2402
    + 5111

c. $79.45
    + 60.99

7a. 382
    989
    + 105

b. 277
    184
    + 457

c. $18.95
    27.72
    + 11.08

8. A bulletin board has 19 notes in English and 12 in Spanish. How many notes are on the board?

9. Find the total number of pencils in a box of 24 red, 12 blue, 30 green, and 23 yellow pencils.

10. Mr. Kanin has 1940 postcards from the United States and 2430 from other countries. How many postcards are in his collection?

11. Add 19,200 to the sum of 394 and 377.

12. Mitch uses tiles to cover a floor. He uses 287 black tiles, 78 white tiles, and 118 blue tiles. How many tiles does he use?

13. The sum is 54,000. One addend is 28,250. What is the other addend?

14. A necklace has 26 glass beads, 48 metal beads, and 82 tiny wooden beads. How many beads are in the necklace?

Practice 3-2

1a. 894
    − 190

b. 300
    − 28

c. 738
    − 592

2a. 5493
    − 2500

b. 7000
    − 429

c. 69,504
    − 18,366

3a. $9.29
    − 1.63

b. $43.50
    − 25.70

c. $50.22
    − 8.99

4a. 280 − 223   **b.** 29,302 − 10,233

5a. $35.98 − $7.23   **b.** $600.75 − $240.80

Estimate the difference. Use front-end estimation.

6a. 849
    − 290

b. 8394
    − 2011

c. 73,382
    − 14,006

7a. $51.20
    − 10.75

b. $757
    − 522

c. $98.35
    − 52.20

8. How much greater than 427 is 549?

9. Ms. Brownell has 1327 marbles. There are 272 white marbles; the rest are multicolored. How many multicolored marbles does she have?

10. Ruth is reading a 178-page book. She is on page 67. How many pages does she still have to read?

11. Angie sells seed packs. She starts with a carton of 250 packs. She has 117 packs left. How many has she sold?

12. An adult's T-shirt costs $8.99 and a child's T-shirt costs $5.50. How much more expensive is the adult's T-shirt?

13. Subtract 3405 from the sum of 2847 and 5032.

Reinforcement

Practice 4-1

1a. 3×0 **b.** 1×5 **c.** 0×8

2a. 7×6 **b.** 6×7 **c.** 9×1

3a. 3×21 **b.** 5×18 **c.** 6×94

4a. 7×100 **b.** 4×805 **c.** 2×4500

5a. $8 \times \$1.05$ **b.** $9 \times \$31.59$ **c.** $3 \times \$82.80$

Estimate each product. Use front-end estimation.

6a. 2×148 **b.** 5×822 **c.** 9×704

Find each missing factor.

7a. $2 \times \underline{\ ?\ } = 18$ **b.** $21 = 7 \times \underline{\ ?\ }$

8a. $45 = 5 \times \underline{\ ?\ }$ **b.** $27 \times \underline{\ ?\ } = 27$

9. What is the product of 78 and 7?

10. What is 459 multiplied by 5?

11. The product is 81. One factor is 9. What is the other factor?

12. Which is greater: 7×1 or 0×7?

13. What is the product of 472 and zero?

14. Joel bought 3 boxes of peaches. There were 6 peaches in each box. How many peaches did he buy?

15. There are 8 shelves of books. Each shelf holds 45 books. How many books are there?

16. What is the product of $\$19.95$ and one?

17. Meg bought 6 cassettes. Each cassette cost $\$9.98$. How much did she spend?

Practice 4-2

1a. 10×34 **b.** 10×58 **c.** 10×985

2a. 20×12 **b.** 40×42 **c.** 50×50

3a. $\begin{array}{r} 24 \\ \times 18 \end{array}$ **b.** $\begin{array}{r} 57 \\ \times 21 \end{array}$ **c.** $\begin{array}{r} 61 \\ \times 63 \end{array}$

4a. 96×17 **b.** 27×793 **c.** 63×403

5a. $12 \times \$1.02$ **b.** $41 \times \$3.40$ **c.** $35 \times \$6.50$

Estimate by rounding. Then multiply.

6a. 32×41 **b.** 29×491 **c.** 47×307

7a. $12 \times \$1.25$ **b.** $22 \times \$4.59$ **c.** $84 \times \$8.82$

8. What is the product of 748 and 10?

9. A theater has 24 rows of seats. There are 18 seats in each row. How many seats are there?

10. A compact disc is on sale for $\$7.99$. How much would it cost to buy 11 of the discs?

11. Zenia earns $\$4.15$ an hour. She works 20 hours a week. How much does she earn in one week?

12. Each volume of an encyclopedia has 568 pages. There are 24 volumes. How many pages are in the entire encyclopedia?

13. What is the product of 409 and 89?

14. A pillowcase costs $\$4.25$. How much would cases for 15 pillows cost?

15. A toy store has 52 bags of marbles. There are 35 marbles in each bag. How many marbles does the store have?

16. There are 115 windows on each floor of an office building. The building has 48 floors. How many windows does the building have?

Practice 5-1

1a. 9)0 **b.** 1)8 **c.** 7)7

2a. 5 ÷ 5 **b.** 0 ÷ 4 **c.** 2 ÷ 1

Find the missing number.

3a. 42 ÷ ? = 6 **b.** ? ÷ 9 = 6

Estimate the quotient.

4a. 8)82 **b.** 4)51 **c.** 3)621

5a. 2)6905 **b.** 5)$5.25 **c.** 7)$34.89

Divide.

6a. 7)49 **b.** 5)48 **c.** 3)29

7a. 4)84 **b.** 9)90 **c.** 6)73

8a. 2)868 **b.** 8)969 **c.** 7)865

9. Is 3892 divisible by 2?

10. Is 193 divisible by 5?

11. Is 711 divisible by 3?

12. Is 25,570 divisible by 10?

13. Elena has 98 inches of ribbon. How many 6-inch pieces can she cut? Will there be any ribbon left over? how much?

14. If 3634 is divided by 7, what are the quotient and the remainder?

15. What is the next number in this pattern: 3645, 1215, 405, 135, . . . ?

16. An album has 164 photos. Each full page holds 8 pictures. At most, how many pages are full? How many pages are partly filled?

17. What numbers between 107 and 125 are divisible by 2?

Practice 5-2

1a. 5)325 **b.** 7)421 **c.** 6)598

2a. 9)819 **b.** 4)110 **c.** 8)209

3a. 3)621 **b.** 6)650 **c.** 2)811

4a. 5)515 **b.** 7)745 **c.** 4)839

5a. 8)8968 **b.** 5)1005 **c.** 7)7325

6a. 3)$.21 **b.** 2)$24.40 **c.** 6)$7.20

7a. 4)$31.20 **b.** 9)$9.36 **c.** 8)$7.52

Use the order of operations to solve.

8a. 9 − 2 × 3 **b.** 16 ÷ 2 + 3

9a. 5 × 10 ÷ 2 **b.** 360 ÷ 4 × 2

10a. 15 − 5 × 2 + 1 **b.** 21 ÷ 7 + 9 × 3

11. There are 3727 flyers. What is the greatest number of flyers there could be in each of 8 equal stacks?

12. Michael bought 8 oak saplings for $48.40. How much did each sapling cost?

13. Leila makes 850 muffins for a bake sale. She places them in bags of 8. How many bags can she fill? How many muffins are left over?

14. Zack spent $200.35 during a 5-day vacation. How much did he spend each day if he spent an equal amount daily?

15. What is the average of 104, 205, 47, and 36?

16. In their games this season, the Hoops scored 64, 68, 42, 70, 92, and 54 points. What is their average score per game?

17. A train travels 600 miles in 9 hours. About how many miles per hour does the train travel?

Practice 6-1

Write *in., ft, mi, c, gal,* or *lb* for the unit you would use to measure each.

1a. the length of a finger **b.** the weight of a bowling ball

2a. the capacity of a juice glass **b.** the distance from San Diego to Las Vegas

3a. the height of a door **b.** the capacity of an oil barrel

Add.

4a. 8 ft 5 in.
+ 4 ft 7 in.

b. 6 ft 8 in.
− 3 ft 5 in.

Complete.

5a. 36 in. = ? ft **b.** 4 gal = ? qt

6a. 3 lb = ? oz **b.** 32 c = ? pt

7. Is a shoelace for a pair of sneakers about 3 in., 3 ft, 3 yd, or 3 mi long?

8. Would you need 2 fl oz, 2 c, 2 pt, or 2 gal of water to fill a large bucket?

9. Does a wild rabbit probably weigh 3 oz, 3 lb, or 33 lb?

10. A recipe calls for 3 c of milk. Janet has 1 qt of milk. Does she have enough for the recipe?

11. There are 5 apples in a bag. Each apple weighs 5 oz. Does the bag weigh more than 2 lb?

12. Does a soup ladle probably hold 4 fl oz, 40 fl oz, 4 c, or 4 pt?

13. How many inches are there in 12 ft?

14. Is a 5-lb box heavier than a 90-oz box?

Practice 6-2

Write *cm, m, km, mL, L,* or *g* for the unit you would use to measure each.

1a. the mass of a goldfish **b.** the thickness of a book

2a. the distance from Rome to Madrid **b.** the capacity of a fish tank

3a. the capacity of a teaspoon **b.** the length of a large rug

Compare. Write <, =, or >.

4a. 200 cm ? 20 m **b.** 7 L ? 6000 mL

5a. 6000 g ? 5 kg **b.** 4 km ? 5000 m

Write how much time has passed.

6. from 12:30 A.M. to 4:00 A.M.

7. from 10:20 P.M. to 11:15 P.M.

8. Is a room comfortable when it is 68°F or 68°C?

9. Will ice melt at 2°F or 2°C?

10. What time is it when it is 12 minutes before noon?

11. Does a postcard have a mass of 1 g or 1 kg?

12. Is a pencil about 15 mm, 15 cm, or 15 m long?

Use the map to solve.

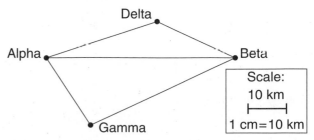

13. How far is it from Alpha to Beta in kilometers?

14. Is Beta closer to Alpha or Delta?

Practice 7-1

Use the list to solve.

Favorite Numbers of Mr. Porter's Class
7, 5, 7, 19, 11, 2, 3, 13, 5, 7, 19, 11, 2, 8, 8, 7, 7, 5

1. Make a tally chart and a table from the data in the list.

2. Which was the most popular number?

3. Which numbers were equally popular?

Use the line graph to solve.

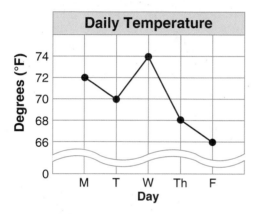

4. Which day was the warmest?

5. On which day was the temperature 70°F?

Use the chart to solve.

Type of Boat	Number
Motor Boat	45
Sail Boat	80
Canoe	60
Row Boat	35

6. Make a pictograph from the data in the chart.

7. What type of boat was second most popular?

Use the bar graph to solve.

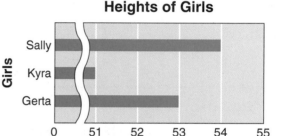

8. Which girl is 2 in. taller than Kyra?

9. How much taller is Sally than Gerta?

Practice 7-2

Use the circle graph to solve.

Cards at Holly's Card Shop

1. Does the shop have more *thank you* or *get well* cards?

2. How many cards in all does the shop have?

3. Molly must wear a blue, red, or white shirt with either black, brown, or blue pants to work. How many combinations of shirt and pants can she wear?

4. A computer picks a random number between 1 and 100. Is it more or less likely to pick a number above 20?

5. Is the computer more likely, less likely, or equally likely to pick an odd number?

6. Andrew tosses a coin. What is the probability that it will land heads up?

7. Irene tosses a nickel. It lands tails up. What is the probability that it will land tails up on her next toss?

Practice 8-1

Write each as a fraction.

1a. two fifths   **b.** three sevenths

Write each fraction in words.

2a. $\frac{3}{4}$ **b.** $\frac{5}{6}$ **c.** $\frac{1}{3}$ **d.** $\frac{2}{8}$

About what fraction of the region is shaded?

3.

Write the equivalent fraction.

4a. $\frac{1}{2} = \frac{?}{12}$ **b.** $\frac{3}{4} = \frac{?}{8}$

5a. $\frac{2}{3} = \frac{?}{9}$ **b.** $\frac{8}{10} = \frac{16}{?}$

6. A carnival wheel is divided into 10 equal parts. Three of the parts are red. Write a fraction to show what part is red.

7. An orange has 9 equal sections. Rose ate 6 sections. Write a fraction to tell what part was eaten.

8. Eight out of 32 students are honor students. What fraction shows how many are honor students?

9. Is $\frac{1}{5}$ closer to 0, $\frac{1}{2}$, or 1?

10. Is $\frac{2}{3}$ closer to 0, $\frac{1}{2}$, or 1?

11. Is $\frac{5}{8}$ closer to 0, $\frac{1}{2}$, or 1?

12. How many sixths are equal to one half?

Practice 8-2

Find the missing factor.

1a. $5 \times \underline{\ ?\ } = 10$ **b.** $16 = \underline{\ ?\ } \times 8$

List all the common factors of each set of numbers. Then circle the GCF.

2a. 8 and 10   **b.** 20 and 30   **c.** 6, 12, and 42

Write each fraction in lowest terms.

3a. $\frac{5}{25}$ **b.** $\frac{3}{9}$ **c.** $\frac{6}{18}$

4a. $\frac{20}{100}$ **b.** $\frac{2}{14}$ **c.** $\frac{8}{12}$

Compare. Write $<$, $=$, or $>$.

5a. $\frac{1}{2} \underline{\ ?\ } \frac{3}{4}$ **b.** $\frac{1}{10} \underline{\ ?\ } \frac{2}{20}$

6a. $\frac{1}{6} \underline{\ ?\ } \frac{1}{12}$ **b.** $\frac{5}{8} \underline{\ ?\ } \frac{1}{8}$

7a. $\frac{4}{5} \underline{\ ?\ } \frac{4}{6}$ **b.** $\frac{7}{8} \underline{\ ?\ } \frac{6}{12}$

Write in order from least to greatest.

8a. $\frac{3}{8}, \frac{5}{8}, \frac{1}{8}$ **b.** $\frac{2}{3}, \frac{7}{12}, \frac{1}{12}$

9. What is the greatest common factor of 8, 12, 20, and 40?

10. A flag shows 15 equal sections, 5 of which are blue. What fraction tells the part of the flag that is blue? Write the fraction in lowest terms.

11. A group of 90 children visit Canada. Nine of the children were born there. What fraction tells how many of the children were born in Canada? Write the fraction in lowest terms.

12. Write nine and two ninths as a mixed number.

13. What whole number is equivalent to $\frac{16}{1}$?

14. What whole number is equivalent to $\frac{22}{22}$?

Practice 9-1

Solve. Write the answer in lowest terms.

1a. $\frac{6}{8} + \frac{1}{8}$ **b.** $\frac{4}{10} - \frac{2}{10}$

2a. $3\frac{3}{5} + 2\frac{1}{5}$ **b.** $9\frac{7}{8} - 4\frac{3}{8}$

3a. $\frac{2}{3} + \frac{4}{6}$ **b.** $\frac{8}{10} + \frac{3}{5}$

4a. $\frac{1}{2} - \frac{1}{4}$ **b.** $\frac{2}{5} + \frac{3}{10}$

List the first six common multiples for each. Circle the least common multiple.

5a. 4, 10   **b.** 2, 6   **c.** 3, 6, and 9

Write as a whole number or mixed number in simplest form.

6a. $\frac{12}{10}$ **b.** $\frac{16}{4}$ **c.** $\frac{22}{4}$

7. Len eats $\frac{1}{8}$ of a pizza and Mia eats $\frac{3}{8}$ of the pizza. What part of the pizza did they eat?

8. A recipe calls for $\frac{3}{4}$ cup of milk. Rachel has $\frac{1}{8}$ cup of milk. How much more does she need?

9. There are 6 red marbles and 3 blue marbles in a bag. Lou picks one without looking. What is the probability that Lou picks a red marble?

10. What is one fourth of 40?

11. What is $\frac{2}{5}$ of 25?

12. Alan makes 20 brownies. He sells $\frac{3}{4}$ of them at a bake sale. How many does he sell?

13. There are 35 horses. One fifth of them are brown. How many of the horses are brown?

Practice 10-1

Name each figure.

1a.   **b.** 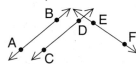   **c.**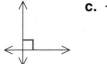

2a.   **b.** 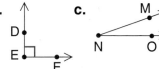   **c.**

3. Which lines are parallel?

4. Which lines are *not* perpendicular?

a.   **b.**   **c.**

5. What shape is formed when two rays share a common endpoint?

6. How many sides does a triangle have? a pentagon? a hexagon?

7. How many vertices does a quadrilateral have? an octagon?

8. Name this figure.

9. Name the diameter and two radii.

10. Is this a simple closed curve?

11. How is a square different from a rectangle?

12. A sign has 4 straight sides and 4 vertices. No 2 sides are the same length. What shape is the sign?

13. Is a circle a simple closed curve? Explain.

Practice 10-2

Write *triangle*, *right triangle*, or *equilateral triangle* to describe each figure.

1a. **b.** **c.**

How is the pattern made? Write *slide* or *flip*.

2a.

2b.

3. Are these figures congruent?

4. Are these figures similar?

5. Which figure is symmetrical?

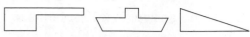

6. Which figure has half-turn symmetry?

Use the grid to answer each question.

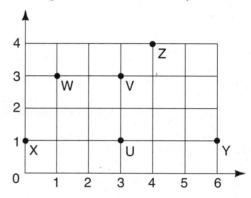

7. What point is located at (1, 3)?

8. What ordered pair gives the location of point *X*?

Chapter 11

Practice 11-1

Find the perimeter of each figure.

1a. 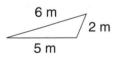 6 m, 2 m, 5 m **b.** 3 in., 2 in., 4 in., 2 in., 5 in.

Find the area of each figure.

2a.  **b.** 22 ft, 11 ft

3. A tabletop is 4 feet long and 5 feet wide. What is the perimeter of the tabletop?

4. A space figure has no faces and a curved surface. What is it?

5. How many faces, edges, and vertices does a cube have?

6. Name the shape of the new flat surface made by the cut.

Find the volume of each figure.

7a. 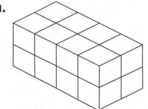 **b.** 2 cm, 2 cm, 4 cm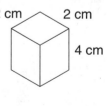

Practice 12-1

1a. 8 ÷ 1   **b.** 80 ÷ 10   **c.** 800 ÷ 10

2a. 420 ÷ 70 **b.** 500 ÷ 50 **c.** 210 ÷ 30

3a. 20)‾4000 **b.** 80)‾640 **c.** 90)‾54,000

Estimate the quotient.

4a. 56 ÷ 11   **b.** 249 ÷ 32   **c.** 109 ÷ 48

5a. 62)‾142   **b.** 74)‾657   **c.** 52)‾$4.80

Divide and check.

6a. 21)‾88   **b.** 31)‾94   **c.** 33)‾$.99

7a. 35)‾73   **b.** 72)‾360   **c.** 91)‾$5.46

8. How many dozens are there in 48?

9. A factory can make 21 toy trains in one hour. How long will it take to make 147 trains?

10. Roger worked 30 hours a week at summer camp. He worked a total of 240 hours. How many weeks did he work?

11. A box can hold 52 cans. How many boxes are needed to hold 260 cans?

12. There are 682 baseball cards and 31 children. If each child takes the same number of cards, what is the greatest number each child will get?

13. Avi buys 11 marbles for $.99. How much does each marble cost?

14. The dividend is 549. The divisor is 61. What is the quotient?

15. Amy earns $44 in 11 hours. How much does she earn in 1 hour?

16. A ship travels 29 miles an hour. How long will it take the ship to travel 87 miles?

Practice 12-2

1a. 28)‾100   **b.** 12)‾90   **c.** 14)‾234

2a. 79)‾229   **b.** 98)‾877   **c.** 38)‾279

3a. 65)‾541   **b.** 72)‾630   **c.** 63)‾371

4a. 86)‾$20.64 **b.** 92)‾5060 **c.** 54)‾2920

5a. 62)‾3000   **b.** 47)‾$9.40   **c.** 24)‾2360

6a. 8)‾832   **b.** 16)‾$32.16   **c.** 25)‾$50.75

7. A carton can hold 24 cans of soup. A diner uses 627 cans in a month. How many full cartons does the diner use?

8. The diner has 576 drinking glasses stored on shelves. Each shelf holds 48 glasses. At most, how many shelves are there?

9. Rita buys 25 postcards for $8.75. How much does each postcard cost?

10. A paper company donates 774 packs of paper to 18 schools. If the packs are shared equally, how many does each school receive? How many are left over?

11. The dividend is 4646. The divisor is 23. What is the quotient?

12. A train travels 68 miles per hour. How long will it take the train to travel 748 miles?

13. Trudy buys a newspaper every day for 14 days. She spends $4.90. How much does each newspaper cost?

14. Glen's dog eats 14 oz of dry food every day. Will a 400-oz bag of dog food last four weeks?

15. Ruth buys 18 yards of ribbon for $18.90. How much does one yard of ribbon cost?

16. What is the remainder when 8244 is divided by 42?

Practice 13-1

Write the value of the underlined digit.

1a. 5.<u>2</u>   **b.** 0.6<u>1</u>   **c.** <u>2</u>5.83

Write as a decimal.

2a. eight hundredths   **b.** $30 + 6 + 0.4 + 0.02$

3a. $\frac{72}{100}$ **b.** $3\frac{5}{10}$

Compare. Write $<$, $=$, or $>$.

4a. 5.54 _?_ 5.45   **b.** 7.12 _?_ 7.1

5a. 21.98 _?_ 22   **b.** 0.80 _?_ 0.8

Compute.

6a. $2.4 + 4.5$ **b.** $3.6 + 5.89 + 4$

7a. $\begin{array}{r} 7.2 \\ -2.7 \\ \hline \end{array}$ **b.** $\begin{array}{r} 5 \\ 4.2 \\ +6.81 \\ \hline \end{array}$ **c.** $\begin{array}{r} 0.57 \\ 0.75 \\ +0.22 \\ \hline \end{array}$

8. Write 25.89 in expanded form.

9. What is 3.28 rounded to the nearest tenth?

10. What is 45.92 rounded to the nearest one?

11. Write 0.1, 1.1, 1.11, and 1 in order from least to greatest.

12. An icicle is 34.8 cm long in the morning. 5.45 cm melt during the day. How long is the icicle at the end of the day?

13. Ben's cat is 28.8 cm tall. Gil's cat is 32 cm tall. How much taller is Gil's cat?

14. Which is a better buy: 18 crayons for $6.12 or 25 crayons for $8?

15. A bean plant is 46.3 cm tall at the end of May. It grows 10.45 cm in June. How tall is it at the end of June?

16. Write 6.5, 65.5, 65.6, and 60.5 in order from greatest to least.

Chapter 14

Practice 14-1

Find each value for n.

1a. $32 + n = 50$ **b.** $100 - n = 19$

2a. $21 \times n = 105$ **b.** $693 \div n = 63$

Copy and complete.

3.

10	8	17	25	64	3	92
60	48	102	?	?	?	?

Rule: Multiply by 6.

Compare. Write $=$ or $\neq$.

4a. 14×5 _?_ $60 + 20$
 b. $210 \div 15$ _?_ $16 - 2$

5a. $2.2 + 1.7$ _?_ 39
 b. 7×7 _?_ $55 - 2 \times 3$

6. There are 8 boxes of books. Each box holds 16 books. Which number sentence will help you find how many books in all: $8 \times 16 = n$ or $16 \div 8 = n$?

7. Which is greater: $100 \div (2 + 3)$ or $100 \div 2 + 3$?

8. Which is equal to zero: $10 - 2 \times 5$ or $(10 - 2) \times 5$?

Use the number line to solve.

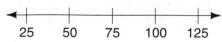

9a. How far is it from 50 to 125?

9b. What number is halfway between 25 and 125?

TEST 1

Compare. Write <, =, or >.
1. 8 + 4 _?_ 18 − 9   2. 16 − 8 _?_ 7 + 6

Compute.
3. (3 × 7) + 1 = _?_   4. (5 × 8) − 7 = _?_
5a. 63 ÷ 7 = _?_   b. 8)48

Give the place and the value of the underlined digits.
6. 5̲2̲8,347,1̲06̲

Give 4 related facts for:
7a. 9, 8, 17   b. 5, 7, 35

Write the number.
8. eighty thousand, forty-nine
9. Stickers cost $.06 each. How much will I pay for 9 stickers?

10. Joan has 356 stickers in her collection. Diane has 365. Which girl has more stickers?
11. Round the sum of 350 + 23 + 126 to the nearest hundred.
12. At $.96 a yard, what is the cost of 8 yards of material?
13. A bookcase has 8 shelves. There are 6 books on each shelf. How many books are in the bookcase?
14. Forty strawberries were divided equally among 5 children. How many did each child receive?
15. How much greater is the product of 6 and 7 than the product of 5 and 8?

TEST 2

Order from least to greatest.
1. 304, 340, 356, 324

Round to the place of the underlined digit.
2a. 9̲2   b. 3̲87

Write in standard form.
3a. one hundred four thousand, three hundred seventy
b. 100,000 + 20,000 + 300 + 4

Compute.
4a. _?_ − 8 = 5   b. 15 = 6 + _?_
5. 23   6. 651   7. 59   8. 738
  + 34   + 728   − 24   − 216
9. $21.50 + $7.25   10. $33.95 − $1.84

11. Helen buys a toothbrush for $.96 and soap for $.45. How much change will she receive from $2.00?
12. What four coins have the same value as one quarter?
13. How many odd numbers are there between 132 and 180? Name them.
14. Write 4,305,060 in expanded form.
15. Jack gave the clerk $1.00 to pay for a $.32 item. The clerk then gave him 2 quarters, 1 dime, 1 nickel, and 2 pennies. Did he receive the correct change? Explain.

TEST 3

Compute.
1. 3 + 6 + 4 + 5 = _?_
2. Double 8 and add 3.

Estimate.
3. 46   4. 371   5. 68   6. 482
  + 22   + 119   − 37   − 245

Compute.
7. 16   8. 572   9. 42   10. 610
  + 25   + 388   − 19   − 436

11. Compute mentally.
75 + 60 + 50 + 40 + 25 = _?_

12. Jan, Sue-ling, and Tanya scored 86, 80, and 100 on the math test. Jan's score was the lowest. Sue-ling had hoped to do better. Give each child's score.
13. Julio bought a sweater for $15.40 and shoes for $22.90. How much change will he receive from $40?
14. Find the total number of days in June, February, December, and July.
15. Mr. Doyle is traveling 682 km from Pensacola to St. Augustine. If he has already traveled 495 km, how much farther must he travel?

Compute.

1.
```
   3475
     63
 + 8468
```

2.
```
  $ 6.95
    15.47
 + 38.56
```

3.
```
  4000
 −  96
```

4.
```
  3060
 − 987
```

5.
```
  4060
 ×    8
```

6.
```
  143
 ×   7
```

7.
```
  809
 × 76
```

8.
```
  $2.56
 ×    10
```

Estimate.

9. 4 × 18

10. 22 × 631

11. Patrick is 18 years old and is 6 ft tall. Bud is 23 years old. How much older is Bud?

12. How long will it take Traci to read a book of 168 pages if she reads 8 pages each day?

13. If a jet travels 300 miles an hour, how far will it go in 13 hours?

14. Each of the 36 students in the graduating class will be inviting 4 guests to the ceremonies. How many guests will be invited in all?

15. If Phillipe earns $4.50 an hour, how much will he earn if he works 20 hours?

Estimate. Then multiply.

1. 403 × 7

2. 3 × 242

Discover the pattern and complete.

3. 6, 8, 10, 7, 9, 11, 8, 10, ? , ? .

Complete.

4. 6 × 7 = 42 is to 42 ÷ 6 = 7 as 4 × 9 = 36 is to ? .

Estimate.

5. 5)‾38‾

6. 7)‾$48.75‾

Compute.

7. 6)‾90‾

8. 4)‾86‾

Which are divisible by 3?

9. 75

10. 82

11. The Kane family drove 1800 miles in five days. How many miles did they average each day?

12. Patsy gave 8 stickers to her sister, and double that amount to each of her 4 friends. She still has 14 stickers left. With how many stickers did Patsy start?

13. Estimate the cost of 7 CDs if each one costs $8.98.

14. A notepad costs $.89 and a pen costs $.59. What is the total cost of six notepads and six pens?

15. If 466 apples are to be put equally into 9 baskets, how many apples will there be in each basket? How many apples will be left over?

Complete.

1. 30 in. = ? ft

2. 6 yd = ? ft

Compute.

3.
```
   2 ft 8 in.
 + 3 ft 9 in.
```

4.
```
   4 yd 2 ft
 − 3 yd 1 ft
```

Compare. Write <, =, or >.

5. 6 qt ? 2 gal

6. 3 pt ? 6 c

Complete.

7. 24 oz = ? lb ? oz

8. 6500 lb = ? T ? lb

Choose the better answer.

9. length of a paper clip: 30 cm or 30 mm?

10. capacity of a swimming pool: 2000 L or 2000 mL?

11. Which is the longer distance: 1800 m or 2 km?

12. There were 936 library books. If an equal number were placed on each of 9 shelves, how many books were on each shelf?

13. Find the average of Rashon's 4 math test scores: 86, 80, 93, 93.

14. Dad needs 95 nails to make a doghouse. If they come packaged 10 nails to a bag, how many bags will Dad need to buy?

15. If the temperature starts at 0°C and drops 4°, what is the temperature? If it then rises 6°, what will the temperature be? If it rises another 3°, what will the temperature be then?

Complete.
1. 3 h = _?_ min   **2.** 96 h = _?_ days

Compute.
3. 365   **4.** 965   **5.** 384   **6.** 3)106
     +279   −298   × 52

Use the graph to answer questions 7 and 8.

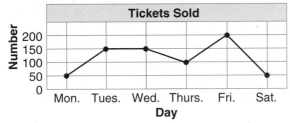

7. How many tickets were sold altogether?
8. How many more tickets were sold on Tues. and Wed. than on Mon. and Thurs.?
9. What is the mass in grams of a 5-kg bag of flour?
10. Find the date of the 40th day after March 8.
11. Two books cost $1.65 and $2.25. What would be the change from $10.00?
12. A point halfway between 20 and 30 on a graph stands for what number?

In a jar with 10 red marbles, 5 each of green and blue, and 1 yellow marble, are you equally, more, or less likely to select:
13. a yellow marble?
14. a green or blue marble?
15. a red marble?

Closer to 0, $\frac{1}{2}$, or 1?
1. $\frac{3}{6}$ **2.** $\frac{7}{8}$ **3.** $\frac{1}{9}$

Give the equivalent fraction.
4. $\frac{2}{3} = \frac{?}{12}$ **5.** $\frac{4}{5} = \frac{?}{15}$ **6.** $\frac{2}{9} = \frac{?}{27}$

Find the GCF.
7. 12 and 36   **8.** 28 and 42

Write each fraction in lowest terms.
9. $\frac{15}{40}$ **10.** $\frac{20}{38}$
11. Latisha has 4 apples. She wants to share them equally with a friend. How many apples does each child get?

12. Make up a question.
Maria baby-sat for 3 hours. She was paid $5.00 an hour. Then she spent $12.00.
13. The box holds $\frac{3}{4}$ cup of raisins. The recipe calls for $\frac{2}{3}$ cup of raisins. Will there be enough raisins for the recipe?
14. Cindy is $4\frac{2}{4}$ ft tall, Desiree is $4\frac{1}{4}$ ft tall, and Emile is $4\frac{3}{4}$ ft tall. Who is the tallest? Who is the shortest?
15. Eight-tenths of the building is above ground level and $\frac{2}{10}$ is below ground level. Write these fractions in lowest terms.

Find the pattern and complete.
1. $\frac{1}{12}, \frac{4}{12}, \frac{2}{12}, \frac{5}{12}, \frac{3}{12}, \frac{?}{?}, \frac{?}{?}$

Compute.
2. $\frac{1}{8} + \frac{1}{8} + \frac{5}{8}$ **3.** $\frac{12}{5} + \frac{8}{5}$
4. $\frac{1}{8} + \frac{3}{4}$ **5.** $\frac{5}{6} - \frac{1}{2}$

Find the LCM.
6. 8 and 10   **7.** 4 and 7
8. To $\frac{1}{3}$ of 36 add 4.
9. Subtract 5 from $\frac{2}{5}$ of 40.
10. A piece of wood measures $4\frac{3}{8}$ ft in length. Another piece is $6\frac{1}{8}$ ft. What is the combined length?

11. On a fair spinner with the numbers 1, 2, and 3, what is the probability of spinning either a 1, a 2, or a 3?
12. Bill had 63 marbles. He gave $\frac{1}{9}$ of them to Chung. How many did Bill have left?
13. At $309 each, what will a dealer pay for 85 television sets?
14. Julie has 3 quarters, 4 dimes, 3 nickels, and 7 pennies in her pocket. Does she have enough to buy a toy that costs $1.29?
15. Daryll did $\frac{3}{8}$ of a project, and Dana did $\frac{1}{4}$ of it. How much of the project is completed? How much still needs to be completed?

Identify each.
1. 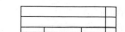 A B   2. •C   3. D E

Draw 3 angles:
4a. a right   **b.** acute   **c.** obtuse
    angle

5. The rungs of a ladder form ___?___ lines.
6. Trace a penny. Then draw a diameter and a radius. Label these line segments.
7. Draw a hexagon. How many angles are there?
8. Draw 2 special quadrilaterals. Label them.
9. Is the figure a flip or a slide? **a.**   **b.**
10. Write *congruent* or *similar figures.* **a.**   **b.**

11. How many rectangles?
12. Find the perimeter of a pentagon whose sides measure: $1\frac{1}{8}$ in., 2 in., $1\frac{5}{8}$ in., 2 in., and $2\frac{1}{8}$ in.
13. The baseball field measures 125 yd long and 75 yd wide. Find the area.
14. Which space figure has 8 edges and 5 faces?
15. A box measures 2 m long, 1 m wide, and 2 m high. Find the volume of the box. Then decide whether you can fit a television that measures 150 cm long, 75 cm wide, and 120 cm high into the box.

TEST 11

1a. $83 + 74 + 36$ **b.** $80 + 24 + 65$
2a. $651 - 289$ **b.** $708 - 498$
3a. $7\overline{)749}$ **b.** $20\overline{)180}$
4a. $30\overline{)241}$ **b.** $23\overline{)74}$
5a. $52\overline{)676}$ **b.** $13\overline{)117}$
6. How many 8s are in: 26; 37; 43; 57?

Estimate the quotient.
7a. $36\overline{)82}$ **b.** $41\overline{)211}$

Write the number.
8. $7 + 0.2 + 0.09$

9. $2.59
     .09
  $+ \ 3.84$

10. $23.50
   $- \ \ 7.65$

11. Vince puts a border around his room, which measures 8 ft by 11 ft. How many feet of border does he need?
12. Mrs. Taylor spent $86.40 to buy 27 identical pairs of scissors. How much did each pair cost?
13. There are 1902 people in line for the roller coaster. Each ride holds 28 people. How many times will the roller coaster need to run so that everyone in line has one ride?
14. Tom's ski run was 61.45 s. Carol's time was 61.39 s. Whose time was faster?
15. How many feet are in a spool of cotton that contains 30 yards?

TEST 12

Round to the nearest one; then to the nearest tenth.
1a. 36.18   **b.** 12.96   **c.** 44.50

Write $+$, $-$, $\times$, or $\div$ to make each sentence true.
2a. $48 \ \underline{\ ?\ } \ 3 = 9 \ \underline{\ ?\ } \ 7$ **b.** $6 \ ? \ 8 = 59 \ ? \ 11$

Solve.
3. $0.7 + 0.6 - (2 \times 0.3) \div (1.9 - 0.9) = \underline{\ ?\ }$

Order from least to greatest.
4. 1.3, 1.36, 0.3, 1.63   **5.** 2.4, 2.43, 2.423

Estimate.
6. $8.6 + 2.9$ **7.** $15.3 - 10.4$

Find the missing number.
8. $\underline{\ ?\ } \times 15 = 25 \times 3$ **9.** $\underline{\ ?\ } \div 4 = 120 \div 6$

Compute. Use a number line.
10. $1 + 7 - 2 + 3 + 4 - \underline{\ ?\ } = 5$
11. Mrs. Riso bought 1 dozen donuts at $.30 each and $\frac{1}{2}$ dozen muffins at $.65 each. How much change will she receive from $10?
12. Complete the pattern.
    0.1, 0.5, 0.7, 0.2, 0.6, 0.8, _?_ , _?_
13. If a ship travels 409 miles in one day, how far will it travel in six days?
14. After Greg paid $40.00 for shoes and $3.50 for socks, he had $20.50 left. How much money did Greg have at first?
15. Maggie had 2 dozen eggs. She used $\frac{2}{3}$ of them for baking. How many eggs were left?

SET 1

1.

3	2	5	3	4	7
+7	+9	+2	+6	+8	+7

2.

7	10	9	16	14	8
−3	−2	−4	−8	−9	−3

3. Give related facts. 8 + 2, 6 + 5,
7 + 4, 5 + 3, 6 + 7, 9 + 7, 1 + 9

4. 2 × 3   2 × 5   3 × 6   3 × 8
2 × 9   3 × 8   2 × 7   3 × 4

5. 4 × 2   4 × 6   5 × 9   5 × 6
4 × 8   5 × 2   4 × 7   5 × 7

6. Don is 9 years old. How old will he be 6 years from now?

7. A farmer had 11 cows. He sold 8 of them. How many cows did he have left?

8. Crackers are 9¢ each. How much will Joey pay for 3 crackers?

9. How many nickels are worth 50 cents?

10. Anna picked 9 flowers. Laura picked 3. How many flowers did they pick in all?

SET 2

1.

9	8	7	5	6	7
+9	+6	+9	+8	+9	+8

2.

16	17	15	18	14	13
−7	−9	−6	−9	−7	−7

3. 7 × 3   6 × 4   7 × 7   6 × 9
7 × 5   7 × 6   6 × 8   6 × 6

4. 8 × 4   8 × 8   9 × 4   9 × 7
8 × 6   9 × 8   8 × 5   9 × 3

5. 8 ÷ 2   10 ÷ 2   12 ÷ 3   18 ÷ 3
15 ÷ 3   4 ÷ 2   21 ÷ 3   14 ÷ 2

6. Josh has 8¢. Therese has twice as much. How much money does she have?

7. At 9¢ each, what will 7 pencils cost?

8. Thirty-five cents is divided equally among 5 students. How much will each student receive?

9. Tom paid 24¢ for 3 balloons. How much did each balloon cost?

10. The dividend is 42. The divisor is 7. What is the quotient?

SET 3

1. Give related facts. 6 ÷ 2, 9 ÷ 3,
10 ÷ 2, 3 ÷ 3, 8 ÷ 4, 6 ÷ 3

2. Subtract 3 from: 21, 18, 15, 12, 9,
6, 3, 24, 27, 30

3. 28 ÷ 7   24 ÷ 4   30 ÷ 5   48 ÷ 6
49 ÷ 7   32 ÷ 4   40 ÷ 5   36 ÷ 6

4. 10 more than: 58, 14, 82, 95, 103,
191

5. Give the value: 5<u>6</u>3; 72<u>1</u>; 3<u>4</u>5;
<u>2</u>97; 6<u>5</u>8; <u>8</u>26; 90<u>8</u>

6. When 67 is divided by 9, what is the quotient? the remainder?

7. What is 78 in words?

8. What is 5000 + 100 + 60 in standard form?

9. What number is ten thousand less than 56,201?

10. There are 3189 adults and 3819 children at the fair. Are there more adults or children?

477

1. Order from least to greatest: 35, 53, 32; 501, 550, 515; 261, 162, 216; 8778, 8887, 7887
2. Round to the nearest ten: 57, 111, 363, 288, 435, 519, 604, 792
3. Divide by 4: 24, 16, 36, 28, 32, 8, 12
4. Multiply by 7, by 8, by 9: 3, 5, 6, 4, 8, 2, 9, 0, 1, 7
5. Round to the nearest hundred: 649, 752, 3150, 2310, 4281, 5399, 5046

6. What is thirteen in standard form?
7. What is 4,000,000 + 500,000 + 30,000 + 2000 + 10 + 8?
8. There are 18 caps. Six are red. How many caps are not red?
9. What number comes between 13,725 and 13,727?
10. One half-dozen cookies costs 72¢. If I give the clerk $1.00, how much change will I get?

1. Round to the nearest dollar: $2.75, $36.10, $42.89, $18.25, $7.60, $1.37
2. Name the period: 74,118; 25,308,433; 8,065,243; 117,589; 608,145
3. Add 7 to: 8, 18, 28, 38, 58, 78, 48, 68
4. Subtract 8 from: 15, 25, 45, 65, 85, 35
5. Count back by 10 from: 200–150, 390–210, 510–380, 220–90, 165–15, 605–505, 412–342, 1110–890

6. How much money: 1 ten-dollar bill, 2 quarters, 3 dimes, 1 nickel?
7. Which is less? by how much? 3575 or 3775
8. What is the value of 3 in 630,241?
9. Tony scored 5 points in the 1st quarter, 6 in the 2nd, and 4 each in the 3rd and 4th quarters. How many points did he score?
10. What must be added to 9 to make a sum of 17?

1. 8 + 0 + 4   7 + 2 + 3   6 + 4 + 1
   5 + 6 + 0   8 + 9 + 2   1 + 9 + 5
2. Double each and add 2: 4, 2, 6, 5, 7, 3, 8, 1, 9
3. $\underline{\ ?\ } + 8 = 11$ $4 + \underline{\ ?\ } = 13$
   $12 - \underline{\ ?\ } = 7$ $9 = \overline{16} - \underline{\ ?\ }$
   $\underline{\ ?\ } - 3 = 9$ $12 = 6 + \underline{\ ?\ }$
4. Add 9 to: 5, 15, 45, 35, 55, 75, 37, 87, 17, 57, 43, 63, 73, 23
5. Estimate. 46 + 21   52 + 38
   12 + 17   29 + 33   13 + 76
   42 − 22   38 − 11   59 − 18
   15 − 11   67 − 45

6. Is the sum reasonable? Check by estimation. 524 + 46 = 984
7. Complete the pattern. 9, 18, 27, _?_, _?_, 54, _?_, 72
8. Nora had 68¢ and spent 35¢. How much money did she have left?
9. Grace is 23 years old. Mary is 11 years older than Grace. How old is Mary?
10. Ned needs $17. He has $8. How much more money does he need?

1. Add 110 to: 34, 134, 244, 354, 424, 564, 634, 714, 844
2. Add 8 to: 7, 17, 57, 37, 47, 27, 67, 77
3. Subtract 9 from: 13, 43, 73, 25, 55, 85, 14, 74, 34, 12, 92, 62, 82, 52
4. Estimate. 123 + 164   185 + 216   351 + 435   694 − 375   716 − 297
5. 4000 + 1200   2300 + 6000   6100 + 3400   5300 + 2400   7500 + 1300

6. Bob's coat cost $67. Ted's coat cost $8 more than Bob's. How much did Ted's coat cost?
7. Gina is 47 in. tall. Don is 5 in. shorter. How tall is Don?
8. Add. 138 + 22 + 19
9. Ramon has $17.30 and Joe has $8.70. How much do the boys have altogether?
10. Rosa had 24 cookies. She gave 7 to Jane. How many cookies did Rosa have left?

SET 8

1. Subtract 5 from: 13, 43, 73, 33, 53
2. $10.00 − $4.00 $12.00 − $6.00 $25.00 − $20.00 $36.00 − $24.00
3. Multiply by 3, then add 4: 4, 8, 0, 9, 1, 5, 3, 6, 2, 7
4. Estimate. 584 − 126   431 − 279   1842 − 1256   3421 − 1538   7186 − 4515
5. Multiply by 7: 2, 4, 5, 7, 9, 1, 0, 3, 6, 8

6. How much greater than 15 is 22?
7. Frank is 7 years old. His sister is 5 years older than Frank. How old is Frank's sister?
8. What is 4 more than the product of 9 times 7?
9. Add 2300 + 3200 + 132.
10. Pedro is 42 in. tall. Dave is 9 in. taller. How tall is Dave?

SET 9

1. 1×6 4×6 7×6 9×6
   6×1 6×4 6×7 6×9
2. 3×0 5×1 4×0 6×0
   1×7 8×0 9×1 2×0
3. $8 \times \underline{?} = 24$ $5 \times \underline{?} = 45$
   $\underline{?} \times 2 = 12$ $\underline{?} \times 6 = 48$
   $7 \times \underline{?} = 35$ $\underline{?} \times 4 = 36$
4. Multiply by 2: 10, 20, 30, 40, 50, 70, 90, 60, 80
5. $3 \times (2 + 5)$ $(1 + 4) \times 4$
   $2 \times (1 + 3)$ $6 \times (2 + 2)$
   $(3 + 2) \times 5$ $(3 + 3) \times 1$

6. Myra pulled out fourteen white socks from the laundry basket. How many pairs of socks can she make?
7. Which is the greater product? 3 times 40   or   4 times 20
8. Paul is 20 years old. Jack is 3 times as old as Paul. How old is Jack?
9. About how much will 5 toys cost if each toy costs $.98?
10. There are 24 stickers on a sheet. How many stickers are on 2 sheets?

1. Multiply by 6, then add 2: 0, 8, 6, 2, 4, 10, 1, 3, 5, 9, 7
2. Multiply by 8, then add 5: 2, 4, 0, 3, 7, 1, 9, 10, 5, 8, 6
3. Estimate. $3 \times \$.48$ $2 \times \$.12$ $4 \times \$.23$ $5 \times \$.36$ $6 \times \$.38$
4. Estimate. 28×21 39×12 13×17 43×36 51×22 14×67
5. 20×100 30×100 20×300 40×200 30×300 20×200

6. Mr. Lass sold 52 tickets on each of the 4 days before the dance. How many tickets did he sell?
7. Tanya bought 2 kites that cost $18 each. How much did she pay for the kites?
8. Velvet costs $8 a yard. How much do 4 yards cost?
9. Complete the pattern. 0, 4, 3, 7, 6, __?__, __?__
10. How much greater is the product of 7 and 6 than the product of 0 and 6?

1. 30×60 90×20 30×31 10×210 10×880 40×31
2. $8)\overline{8}$ $1)\overline{7}$ $6)\overline{0}$ $5)\overline{5}$ $1)\overline{4}$ $3)\overline{0}$ $9)\overline{0}$ $4)\overline{4}$
3. $2)\overline{14}$ $5)\overline{30}$ $7)\overline{28}$ $6)\overline{36}$ $8)\overline{64}$ $9)\overline{72}$ $4)\overline{36}$ $8)\overline{40}$
4. Divide by 4: 25, 17, 37, 29, 33, 9, 13, 21, 26, 18, 38, 30, 34, 10, 22
5. $\underline{\ ?\ } \times 4 = 32$ $6 \times \underline{\ ?\ } = 24$ $\underline{\ ?\ } \times 2 = 18$ $\underline{\ ?\ } \times 7 = 21$ $8 \times \underline{\ ?\ } = 56$ $\underline{\ ?\ } \times 8 = 72$

6. The factors are 23 and 68. Estimate the product.
7. The product is 42. One factor is 6. What is the other factor?
8. What is the remainder when 20 is divided by 9?
9. It took Sam 6 hours to pack 325 cartons. About how many cartons did he pack each hour?
10. Five ties cost $60. Each tie costs the same. How much does 1 tie cost?

1. Divide by 9: 29, 11, 46, 20, 38, 40, 31, 15, 48, 33, 14, 49, 19, 42, 44
2. Which are divisible by 2? by 5? by 10? 12, 25, 42, 90, 63, 75, 110, 68, 130
3. $2)\overline{222}$ $5)\overline{555}$ $3)\overline{363}$ $4)\overline{484}$ $4)\overline{888}$ $2)\overline{462}$ $3)\overline{393}$ $2)\overline{846}$
4. Divide by 8: 73, 74, 78, 79, 69, 71, 65, 12, 19, 21, 30, 31, 35, 38, 37, 59, 61, 57, 63, 49, 52, 55, 53
5. $12 + 4 - 3$ $16 - 9 + 5$ $8 + 7 - 4$ $12 \div 4 \times 5$ $5 \times 6 \div 2$ $40 \div 5 \div 2$

6. Sue has 72¢ to share equally with Meg. How much money will each girl receive?
7. Jim spent $1.40 for 2 feet of wire. How much did each foot cost?
8. A farmer plants 800 corn plants in 4 equal rows. How many plants are in each row?
9. Which is greater? by how much? 2626 or 2662
10. What number comes next after 124,169?

1. Rename as feet: 12 in. 36 in.
   24 in. 48 in. 72 in. 60 in. 84 in.
2. Name the time a half hour later.
   8:15, 10:30, 12:00, 3:45, 5:20,
   6:10, 4:05, 7:15, 9:25
3. Compare. Use $<$, $=$, $>$. 3 c _?_ 2 pt
   3 pt _?_ 6 c 1 gal _?_ 6 qt
   8 pt _?_ 4 qt
4. Name the date 1 week later.
   Jan. 8, Mar. 12, Aug. 23, Oct. 2,
   Nov. 18, Dec. 20
5. Divide by 9: 27, 29, 81, 84, 72, 75,
   9, 13, 45, 49, 53, 63, 64, 69

6. Mr. Jones spends $25 every work week on tolls. He works 5 days a week. How much does he spend each workday on tolls?
7. Which is longer, 1 meter or 98 centimeters?
8. Pete's pet weighs 30 oz. How many more ounces does it need to gain to weigh 2 lb?
9. How much longer is 5 feet than 1 yard?
10. Which distance is longer, 10 kilometers or 1000 meters?

SET 14

1. Multiply by 4: 6, 7, 8, 0, 1, 2, 5, 4, 3
2. Count by 1000: 1400–6400;
   2300–7300; 5900–10,900;
   9700–18,700
3. Divide by 7: 61, 62, 58, 57, 60, 59,
   31, 36, 38, 37, 40, 41, 29, 34
4. Subtract 99 from: 109, 239, 479,
   658, 918, 338, 525, 865, 785
5. Give the fraction for the shaded part of each region.

6. Lori needs 1 L of water. She has 600 mL. How much more does she need?
7. What is 100 + 7 in standard form?
8. Would you go ice skating at 35°C?
9. Can Jan go skiing at 20°F?
10. Randy worked from 11:30 A.M. to 1:00 P.M. on his bike. How long did Randy work?

SET 15

1. What part of a dollar is: 10¢, 50¢,
   25¢, 5¢, 1¢, 75¢, 30¢, 70¢, 20¢
2. $\frac{1}{3} = \frac{?}{6}$ $\frac{1}{4} = \frac{?}{16}$ $\frac{1}{2} = \frac{?}{10}$
   $\frac{2}{5} = \frac{?}{15}$ $\frac{2}{3} = \frac{?}{12}$ $\frac{3}{4} = \frac{?}{20}$
3. $\frac{1 \times 2}{8 \times 2} = \underline{\ ?\ }$ $\frac{3 \times 3}{7 \times 3} = \underline{\ ?\ }$
   $\frac{1 \times 4}{3 \times 4} = \underline{\ ?\ }$ $\frac{4 \times 3}{5 \times 3} = \underline{\ ?\ }$
4. Closer to 0 or to 1? $\frac{1}{8}$, $\frac{6}{7}$, $\frac{2}{6}$, $\frac{1}{9}$, $\frac{8}{10}$,
   $\frac{4}{5}$, $\frac{2}{3}$, $\frac{1}{4}$, $\frac{11}{12}$
5. Name the GCF of: 6 and 12; 3 and 15; 8 and 24; 10 and 12; 9 and 12

6. Would you use centimeters or meters to measure the length of a pencil?
7. One paper clip weighs 1 g. How many paper clips do you need to equal 1 kg?
8. Key: Each ▽ = 10 cones. How many cones does ▽ ▽ ▽ ▿ equal?
9. A _?_ graph is used to show change over a period of time.
10. If 1 mi equals 5280 ft, how many feet are there in 2 mi?

1. Express in lowest terms: $\frac{3}{6}$, $\frac{6}{8}$, $\frac{5}{10}$, $\frac{2}{4}$, $\frac{7}{21}$, $\frac{4}{12}$, $\frac{6}{18}$, $\frac{5}{20}$, $\frac{9}{18}$, $\frac{2}{10}$

2. Fraction or mixed number? $\frac{5}{6}$, $1\frac{2}{3}$, $2\frac{4}{5}$, $\frac{5}{8}$, $3\frac{1}{9}$, $\frac{6}{7}$, $4\frac{7}{8}$, $\frac{9}{10}$

3. Order from least to greatest: $\frac{2}{3}$, $\frac{1}{3}$, $\frac{3}{3}$, $\frac{2}{5}$, $\frac{4}{5}$, $\frac{1}{5}$, $\frac{3}{6}$, $\frac{5}{6}$, $\frac{2}{6}$

4. Multiply by 6, then add 7: 10, 8, 6, 4, 2, 0, 1, 3, 5, 7, 9

5. Subtract a nickel from: 25¢, 18¢, 50¢, $1.35, $2.05, $1.16, $6.96

6. A jar has 10 red beans, 5 green beans, 5 blue beans, and 1 yellow bean. What is the probability of choosing red? yellow? blue? green?

7. In a survey of 100 people, 30 people chose hot dogs. What part of the people chose hot dogs?

8. At 0°C, water ? .

9. Roy rolled a ball 6 yards. Was that more or less than 20 feet?

10. What is 3 more than the product of 9 times 5?

1. $\frac{3}{5} + \frac{1}{5}$ $\frac{2}{7} + \frac{3}{7}$ $\frac{1}{6} + \frac{4}{6}$
   $\frac{1}{3} + \frac{1}{3}$ $\frac{3}{9} + \frac{5}{9}$ $\frac{4}{10} + \frac{5}{10}$

2. $\frac{5}{8} - \frac{2}{8}$ $\frac{2}{3} - \frac{1}{3}$ $\frac{6}{7} - \frac{4}{7}$
   $\frac{3}{5} - \frac{2}{5}$ $\frac{8}{9} - \frac{3}{9}$ $\frac{7}{8} - \frac{3}{8}$

3. Express as a mixed number.
   $\frac{15}{2}$, $\frac{7}{4}$, $\frac{11}{3}$, $\frac{17}{5}$, $\frac{9}{2}$, $\frac{13}{6}$, $\frac{15}{4}$

4. $3 + 1\frac{1}{4}$ $2\frac{1}{2} + 4$ $3 + 5\frac{2}{3}$
   $1 + 3\frac{2}{5}$ $4 + 1\frac{3}{4}$ $2\frac{7}{8} + 6$

5. Multiply by 3, then add 4: 4, 8, 0, 9, 5, 6, 7, 10, 2, 3, 1

6. Tom ate $\frac{1}{3}$ of the pizza. Sal ate $\frac{2}{3}$. Who ate more?

7. Six ninths minus four ninths is ? .

8. In a pet store $\frac{1}{5}$ of the pets are cats and $\frac{2}{5}$ are dogs. What part of the pets are cats and dogs?

9. Sasha bought $5\frac{3}{4}$ lb of chicken. She cooked $3\frac{1}{4}$ lb. How much is left?

10. Estimate the sum of $2\frac{1}{8} + 3\frac{3}{4} + 4\frac{1}{2}$.

1. Subtract 2 from: $3\frac{1}{2}$, $4\frac{1}{5}$, $7\frac{1}{8}$, $8\frac{2}{3}$, $5\frac{2}{5}$, $2\frac{7}{8}$, $2\frac{1}{3}$

2. How many nickels are there in: 25¢, $.20, $.35, $.50, 45¢, 60¢, $.30, $.55, $.40

3. Find $\frac{1}{6}$ of: 6, 18, 42, 54, 24, 36, 12

4. Find half of: 14, 10, 8, 18, 20, 6, 4

5. Add 5 to: 9, 19, 59, 29, 38, 68, 28

6. If 47 is divided by 8, what is the quotient? the remainder?

7. Al did $\frac{1}{6}$ of his homework in school and $\frac{1}{2}$ before dinner. How much of his homework did he do?

8. In a set of 10 pens, 3 are black. What fractional part of the set is black?

9. Of 20 fish, $\frac{3}{4}$ are striped. How many fish are striped?

10. Estimate. $6\frac{5}{10} - 3\frac{1}{3}$

SET 19

1. Double each, then add 3: 10, 20, 30, 40, 50, 60, 70, 80, 90
2. Name the line segments.

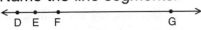

3. Compare to a right angle. <, =, >.

4. Intersecting or parallel?

5. Closed or open?

6. An _?_ is formed by two rays with the same endpoint.
7. Are the sides that meet at each corner of a square frame parallel or perpendicular?
8. Name the radii and the diameter of the circle.
9. How many sides and angles does a pentagon have?
10. Name the parallel sides.

SET 20

1. Congruent or similar? □□   ○○   ◁◁   ▭▭   ○○
2. Flip or slide? →←   →→
3. Divide by 9, then add 2: 18, 9, 36, 81, 27, 54, 72, 63, 36, 45
4. 20 × 200   60 × 20   4 × 800
   40 × 600   50 × 300   30 × 700
5. Multiply by 8: 4, 6, 3, 9, 5, 7

6. Which letter has no line of symmetry, G or M?
7. What number comes next after 9999?
8. Ben's calculator costs $25. Ann's costs $19. What is the difference in cost?
9. Find $\frac{1}{3}$ of 27¢, then add 4¢.
10. What is the value of zero in 30,645?

SET 21

1. Name the space figure.
2. Multiply by 9, then add 3: 2, 3, 7, 5, 0, 1, 10, 6, 4, 9, 8
3. 3)666   4)448   2)684
   1)175   5)505   3)906
4. 3 × 5 × 2   4 × 2 × 6   3 × 3 × 5
   7 × 3 × 2   2 × 6 × 5   8 × 5 × 2
5. Find $\frac{1}{5}$ of: 10, 30, 40, 25, 5, 35, 20, 15, 45, 50

6. What is 50 + 9 in standard form?
7. Find the perimeter of a fenced lot whose sides measure 7 m, 8 m, 5 m, 9 m, and 6 m.
8. What is the perimeter of a square playpen $2\frac{1}{2}$ yd on each side?
9. The bedroom rug measures 10 m by 4 m. What is its area?
10. What space figure has 2 flat surfaces and 1 curved surface?

SET 22

1. How many tens are in: 370, 420, 550, 600, 780, 190, 830, 240, 960
2. $2\overline{)140}$ $3\overline{)210}$ $4\overline{)160}$ $7\overline{)350}$
   $20\overline{)180}$ $30\overline{)150}$ $40\overline{)200}$ $50\overline{)450}$
3. How many 20s are in: 49, 67, 84, 182, 121, 165, 108, 114, 143
4. How many 7s are in: 45, 66
5. Estimate the quotient. $24\overline{)42}$, $31\overline{)89}$, $47\overline{)99}$, $43\overline{)82}$, $20\overline{)85}$, $34\overline{)69}$, $27\overline{)88}$

6. Find the volume of a dollhouse that is 3 ft long, 2 ft wide, and 2 ft high.
7. Dan put 480 soccer cards on the floor. He put them into 20 equal rows. How many cards were in each row?
8. Bus fare to the zoo was $18. About how much did the driver collect from 19 children?
9. Express 6 feet as yards.
10. Each box holds 8 crayons. How many crayons are in 4 boxes?

SET 23

1. How many 9s are in: 56, 19, 12, 39, 46, 68, 76, 29, 84, 65
2. How many 30s are in: 95, 62, 159, 277, 158, 243, 126, 214, 181
3. $16 = \underline{?} \times 8$ $56 = \underline{?} \times 8$
   $40 = \underline{?} \times 8$ $24 = \underline{?} \times 8$
   $64 = \underline{?} \times 8$ $48 = \underline{?} \times 8$
   $32 = \underline{?} \times 8$ $72 = \underline{?} \times 8$
   $80 = \underline{?} \times 8$
4. Estimate the quotient. $32\overline{)124}$, $51\overline{)98}$, $16\overline{)135}$, $23\overline{)144}$, $49\overline{)152}$, $62\overline{)188}$
5. Multiply by 6: 1, 2, 5, 8, 9, 6, 0, 7, 4, 3, 10

6. A box of cupcakes costs $2.40. If there are 24 cupcakes in a box, how much does each cupcake cost?
7. How many dimes are in $4.00?
8. What is the difference in cents between 1 quarter and 3 nickels?
9. Ramon earns $63 a week. He saves $\frac{1}{7}$ of this amount. How much does he save weekly?
10. What is the sum of 9, 17, and 11?

SET 24

1. $4\overline{)200}$ $5\overline{)300}$ $6\overline{)300}$ $7\overline{)700}$
   $8\overline{)400}$ $2\overline{)100}$ $3\overline{)900}$ $9\overline{)900}$
2. Multiply by 7, then add 4: 6, 3, 4, 1, 9, 2, 0, 5, 7, 8
3. How many tens are in: 85, 62, 77, 43, 38, 22, 15, 94, 51
4. Read. 0.4, 0.9, 0.07, 0.5, 0.03, 0.46, 0.72, 0.01, 0.35, 0.11
5. Read. 1.6, 3.7, 8.6, 4.9, 12.5, 5.03, 8.07, 26.3, 6.18, 35.01

6. A parking garage holds a total of 480 cars with an equal number of cars on 4 levels. How many cars does each level hold?
7. What is one fifth of 20 cents?
8. Dan is 36 years old. David is 9 years old. How much older is Dan than David?
9. Name a decimal between 0.1 and 0.3.
10. What is 0.2 more than 1?

1. Give the value. 0.4<u>2</u>, <u>6</u>.23, 14.<u>3</u>, 0.<u>05</u>, <u>36</u>.1, 8.0<u>7</u>, 1.<u>48</u>
2. Compare. $<, =, >$. 0.7 <u>?</u> 0.3
   0.16 <u>?</u> 0.19   2.36 <u>?</u> 2.63
   6.35 <u>?</u> 6.3   1.7 <u>?</u> 1.72
3. Order least to greatest: 0.3, 0.1, 0.6; 0.13, 0.25, 0.20; 3, 0.3, 0.03
4. Complete the pattern. 0.1, 0.4, 0.7, <u>?</u>; 0.05, 0.15, 0.25, <u>?</u>; 1.1, 2.1, 3.1, <u>?</u>; 3.4, 3.6, 3.8 <u>?</u>; 5.9, 5.6, 5.3, <u>?</u>
5. 0.5 + 0.2   0.6 + 0.2   1.3 + 1.4
   2.1 + 1.6   0.8 + 0.1   1.7 + 1.2

6. Name two decimals between 3 and 4.
7. Jesse ran 3.25 m and Thanh ran 3.55 m. Who ran farther? by how many meters?
8. Round $2.87 to the nearest dollar.
9. Missy spent 2.3 min on the first problem and 3.5 min on the next. How long did she spend on both problems?
10. What is the rule for this pattern? 0.3, 0.1, 0.5, 0.3, 0.7, 0.5, 0.9

1. Add 0.2 to: 1.2, 0.3, 2.7, 1.4, 3.9, 2.1, 0.6, 1.5
2. Add 5 cents to the sum of:
   $.04 + $.06 $.25 + $.50
   $.02 + $.03 $.18 + $.02
3. Round to the nearest one: 8.6, 4.9, 6.2, 7.8, 2.3, 3.4, 5.5, 0.7
4. Round to the nearest tenth: 4.18, 5.61, 3.22, 2.73, 7.45, 1.55
5. <u>?</u> + 2 = 7 − 1
   <u>?</u> ÷ 4 = 3 × 3
   6 × <u>?</u> = 9 + 9
   24 ÷ <u>?</u> = 10 − 2

6. The finishing times for the race were 59.1 s for 1st place and 59.6 s for 2nd place. What is the difference in the times?
7. Milk costs $2.89 and bread costs $1.64. About how much money do both items cost in all?
8. Pam bought 2 six-packs of soda. She spent $6.00. What did each can cost?
9. What is 4 + 0.5 in standard form?
10. How much greater than 0.2 is 0.48?

1. Subtract 0.1 from: 9.6, 0.4, 6.3, 1.8, 0.6, 5.4, 3.3, 2.7
2. 3 + 4 − 2 = <u>?</u>
   10 − 7 + 2 = <u>?</u>
   7 + 8 − 9 = <u>?</u>
   9 + 9 − 10 = <u>?</u>
3. Divide by 5, then subtract 2: 35, 20, 45, 15, 25, 40, 30, 10, 50
4. (3 × 3) ÷ 9   10 × (4 − 4)
   (6 − 2) ÷ 1   (4 + 4) × 2
5. 2)$1.80   3)$2.70   4)$4.00
   5)$4.50   6)$4.80   7)$4.20

6. Tom had $1.40. He spent $.50. Then he found $1.00. How much money does he have now?
7. Multiply 6 and 4, add 2, subtract 5.
8. 1 quarter, 2 dimes, 3 pennies = <u>?</u> ¢
9. Three friends share $1.86 equally. How much does each friend receive?
10. There are 25 cookies in each of 3 bags. Tony eats 2 from each bag. How many cookies are left?

A

addend   A number that is added to another number or numbers.

$$3 + 4 = 7$$
addends

A.M.   Letters that indicate time from midnight to noon.

angle   The figure formed by two rays that meet at a common endpoint.

area   The number of square units needed to cover a flat surface.

associative (grouping) property   Changing the grouping of the addends (or factors) does not change the sum (or product).

average   A quotient derived by dividing a sum by the number of its addends.

axis   The horizontal or the vertical number line of a graph.

B

balance   The tool used to measure mass.

bar graph   A graph that uses bars to represent data. The bars may be of different lengths.

benchmark   An object of known measure that can be used to estimate the measure of other objects.

C

capacity   The amount, usually of liquid, that a container can hold.

circle   A simple closed curve; all the points on the circle are the same distance from the center point.

circle ⟶

circle graph   A graph that uses sections of a circle to represent data.

common factor   A number that is a factor of two or more products.

common multiple   A number that is a multiple of two or more numbers.

commutative (order) property   Changing the order of the addends (or factors) does not change the sum (or product).

compatible numbers   Two numbers, one of which divides the other evenly.

composite number   A whole number other than 1 that has more than two factors.

cone   A space or solid figure that has one circular base.

cone ⟶

congruent figures   Figures that have the same size and the same shape; congruent line segments have the same length.

cube   A space or solid figure with six congruent square faces.

cube ⟶

customary system   The measurement system that uses inch, foot, yard, and mile; cup, pint, quart, and gallon; and ounce and pound.

cylinder   A space or solid figure that has two congruent circular bases.

cylinder ⟶

486

D

data Facts or information.

decimal A number in base ten that is written with a decimal point.

2.04 ← decimal
decimal point

degree (°) A unit used to measure angles.

degree Celsius (°C) A unit for measuring temperature. The freezing point of water is 0°C.

degree Fahrenheit (°F) A unit for measuring temperature. The freezing point of water is 32°F.

denominator The numeral below the bar in a fraction; it names the total number of equal parts.

$\frac{1}{2}$ ← denominator

diameter A line segment that passes through the center of a circle and has both endpoints on the circle.

difference The answer in subtraction.

digit Any one of the numerals 0, 1, 2, 3, 4, 5, 6, 7, 8, or 9.

distributive property The product of a number and the sum of two addends is the same as the sum of the two products.

dividend The number to be divided.

24 ÷ 4   4)24
dividend

divisible One number is divisible by another if it can be divided by that number and yield no remainder.

divisor The number by which the dividend is divided.

36 ÷ 9   9)36
divisor

E

edge The line segment where two faces of a space figure meet.

edge

endpoint The point at the end of a line segment or ray.

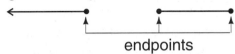
endpoints

equation (See **number sentence.**)

equilateral triangle A triangle whose three sides are congruent.

equilateral triangle →

equivalent decimals Decimals that name the same amount.

0.4 = 0.40

equivalent fractions Different fractions that name the same amount.

$\frac{1}{2} = \frac{2}{4}$

estimate An approximate answer; to find an answer that is close to an exact answer.

even number Any whole number that has 0, 2, 4, 6, or 8 in the ones place.

event A set of one or more outcomes.

expanded form A way to write a number that shows the place value of each of its digits.

400 + 20 + 8 = 428

F

face A flat surface of a space figure surrounded by line segments.

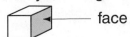
face

factors Two or more numbers that are multiplied to give a product.

$$3 \times 5 = 15$$

factors

family of facts A set of related addition and subtraction facts or multiplication and division facts that use the same digits.

flip The movement of a figure over a line so that the figure faces in the opposite direction.

fluid ounce (fl oz) A customary unit of capacity. 8 fluid ounces = 1 cup

formula A rule that is expressed by using symbols.

fraction A number that names part of a whole or part of a set.

front-end estimation A way of estimating by using the front, or greatest, digits to find an approximate answer.

function A quantity whose value depends on another quantity.

G

graph A pictorial representation of data.

greatest common factor (GCF) The greatest number that is a factor of two or more products.

H

half-turn symmetry A figure that matches its image when it is turned halfway around has half-turn symmetry.

hexagon A polygon with six sides.

hexagon ⟶ ⬡

I

identity property (property of one) The product of one and a number is that number.

improper fraction A fraction whose numerator is greater than or equal to its denominator.

intersecting lines Lines in the same plane that meet or cross.

K

key A symbol that identifies the meaning of each picture in a pictograph.

L

least common denominator (LCD) The least common multiple of two or more denominators.

least common multiple (LCM) The least number that is a multiple of two or more numbers.

line A straight set of points that goes on forever in both directions.

line graph A graph that uses points on a grid connected by line segments to represent data.

line plot A graph of data on a number line.

line segment A part of a line that has two endpoints.

lowest terms A fraction is in lowest terms when its numerator and denominator have no common factor other than 1.

M

mass The measure of the amount of matter an object contains.

metric system The measurement system that uses centimeter, decimeter, meter, and kilometer; milliliter and liter; and gram and kilogram.

millimeter (mm) A metric unit of length. 10 millimeters = 1 centimeter

minuend A number from which another number is subtracted.

missing addend (or factor) An unknown addend (or factor) in addition (or multiplication).

$$\underline{\ ?\ } + 8 = 17$$
$$9 \times \underline{\ ?\ } = 27$$

mixed number A number that is made up of a whole number and a fraction.

$$1\tfrac{1}{2} \longleftarrow \text{mixed number}$$

mode The number that appears most frequently in a set of numbers.

multiple The product of a given number and any whole number.

N

net The shape made by opening a solid figure and laying it flat.

number line A line that is used to show the order of numbers.

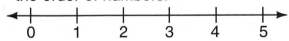

number sentence An equation or inequality.

$$16 = 9 + 7 \qquad 28 < 52$$

numerator The numeral above the bar in a fraction; it names the number of parts being considered.

$$\tfrac{3}{4} \longleftarrow \text{numerator}$$

O

octagon A polygon with eight sides.

octagon $\longrightarrow$

odd number Any whole number that has 1, 3, 5, 7, or 9 in the ones place.

order of operations The order in which operations must be computed when more than one operation is involved.

ordered pair A pair of numbers that is used to locate a point on a grid or coordinate graph.

outcome The result of a probability experiment.

P

parallel lines Lines in the same plane that are always the same distance apart.

parallelogram A quadrilateral whose opposite sides are parallel and congruent.

parallelogram $\longrightarrow$

pentagon A polygon with five sides.

pentagon $\longrightarrow$

percent (%) The ratio or comparison of a number to 100.

perimeter The distance around a figure.

period A group of three digits set off by commas in a whole number.

perpendicular lines Lines in the same plane that intersect at right angles.

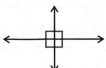

pictograph A graph that uses pictures or symbols to represent data.

place value The value of a digit, depending on its position in a number.

plane A flat surface that never ends.

P.M. Letters that indicate time from noon to midnight.

point An exact location.

polygon A simple closed flat figure made up of three or more line segments.

prime factorization The expression of a composite number as the product of prime numbers.

prime number A whole number other than 1 that has exactly two factors, itself and 1.

probability The chance or likelihood of an event occurring.

product The answer in multiplication.

protractor The tool used to measure angles.

pyramid A space or solid figure that has a polygon for a base and has triangular faces that meet at a point. A square pyramid has a square base.

pyramid ⟶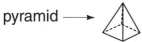

Q

quadrilateral Any four-sided polygon.

quotient The answer in division.

R

radius A line segment with endpoints at the center of a circle and on the circle.

range The difference between the greatest and least numbers in a set of data.

ratio The comparison of two numbers, often expressed as a fraction.

ray A part of a line that has one endpoint and goes on forever in one direction. •⟶

rectangle A parallelogram with four right angles.

rectangle ⟶

rectangular prism A space or solid figure with six rectangular faces.

rectangular prism ⟶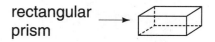

regrouping Trading one from a place for ten from the next lower place, or ten from a place for one from the next higher place.

remainder The number left over after dividing.

$$\begin{array}{r} 7 \quad \text{R}2 \\ 3\overline{)23} \\ -21 \\ \hline 2 \end{array}$$ ← remainder

right angle An angle that measures 90°. It forms a square corner.

right angle ⟶

right triangle A triangle that has one right angle.

right triangle ⟶

Roman numerals Symbols for numbers used by the Romans.

rounding Writing a number to the nearest ten or ten cents, hundred or dollar, and so on.

S

scale The tool used to measure weight.

set A collection or group of numbers or objects.

side A line segment that forms part of a polygon.

similar figures Figures that have the same shape. They may or may not be the same size.

simple closed curve A path that begins and ends at the same point and does not cross itself.

simplest form (See **lowest terms.**)

skip counting Counting by a whole number other than 0 or 1.

slide The movement of a figure along a line.

space (or solid) figure A figure that is not flat but that has volume.

sphere A space or solid figure that is shaped like a ball.

sphere ⟶ ◯

square A parallelogram that has four right angles and four congruent sides.

square ⟶ ▢

square number The product of a number multiplied by itself. It can be represented by a square array.

standard form The usual way of using digits to write a number.

subtrahend A number that is subtracted from another number.

sum The answer in addition.

T

tally A count made by using tally marks.

ton (T) A customary unit of weight. 2000 pounds = 1 ton

triangle A polygon with three sides.

triangle ⟶ △

triangular prism A space or solid figure with two parallel triangular faces.

triangular prism ⟶ ◇

turn The movement of a figure around a point.

V

variable A letter or other symbol that replaces a number in an expression, equation, or inequality.

vertex A common endpoint of two rays or line segments. In a space figure, the point at which three or more edges meet.

volume The number of cubic units needed to fill a space figure.

W

weight The heaviness of an object.

whole number Any of the numbers 0, 1, 2, 3, 4, . . .

Z

zero (identity) property of addition The sum of zero and a number is that number.

zero property of multiplication The product of zero and a number is zero.

=	is equal to	.	decimal point	$\overleftrightarrow{AB}$ line AB
≠	is not equal to	°	degree	$\overline{AB}$ line segment AB
<	is less than	+	plus	$\overrightarrow{AB}$ ray AB
>	is greater than	−	minus	∠ABC angle ABC
$	dollars	×	times	‖ is parallel to
¢	cents	÷	divided by	⊥ is perpendicular to
				(3, 4) ordered pair

Table of Measures

Time

60 seconds (s)	= 1 minute (min)
60 minutes	= 1 hour (h)
24 hours	= 1 day (d)
7 days	= 1 week (wk)
12 months (mo)	= 1 year (yr)
52 weeks	= 1 year
365 days	= 1 year
366 days	= 1 leap year

Money

1 nickel	= 5¢ or $.05
1 dime	= 10¢ or $.10
1 quarter	= 25¢ or $.25
1 half dollar	= 50¢ or $.50
1 dollar	= 100¢ or $1.00
2 nickels	= 1 dime
10 dimes	= 1 dollar
4 quarters	= 1 dollar
2 half dollars	= 1 dollar

Metric Units

Length

10 millimeters (mm)	= 1 centimeter (cm)
100 centimeters	= 1 meter (m)
10 centimeters	= 1 decimeter (dm)
10 decimeters	= 1 meter
1000 meters	= 1 kilometer (km)

Capacity

1000 milliliters (mL) = 1 liter (L)

Mass

1000 grams (g) = 1 kilogram (kg)

Customary Units

Length

12 inches (in.)	= 1 foot (ft)
3 feet	= 1 yard (yd)
36 inches	= 1 yard
5280 feet	= 1 mile (mi)
1760 yards	= 1 mile

Capacity

8 fluid ounces (fl oz)	= 1 cup (c)
2 cups	= 1 pint (pt)
2 pints	= 1 quart (qt)
4 quarts	= 1 gallon (gal)

Weight

16 ounces (oz)	= 1 pound (lb)
2000 pounds	= 1 ton (T)